BIOLOGY IN FOCUS

SKILLS AND ASSESSMENT WORKBOOK

Julie Fraser
Kirsten Prior
Evan Roberts

Biology in Focus: Skills and Assessment Workbook Year 12
1st Edition
Julie Fraser
Kirsten Prior
Evan Roberts
ISBN 9780170449625

Senior publisher: Sarah Craig
Editorial manager: Simon Tomlin
Editor: Kelly Hannah
Proofreader: Jane Fitzpatrick
Original cover design by Chris Starr (MakeWork), Adapted by: Justin Lim
Text design: Ruth Comey (Flint Design)
Project designer: Justin Lim
Permissions researcher: Liz McShane
Production controller: Alice Kane
Typeset by: MPS Limited

Any URLs contained in this publication were checked for currency during the production process. Note, however, that the publisher cannot vouch for the ongoing currency of URLs.

Acknowledgements

Cover image: iStock.com/akesak.

Inquiry questions pages 3, 23, 39, 51, 64, 78, 88, 101, 115, 132, 144, 154, 177, 189, 193, 198 and 203 are from the Biology Stage 6 Syllabus © NSW Education Standards Authority for and on behalf of the Crown in right of the State of New South Wales, 2017.

For product information and technology assistance,
in Australia call **1300 790 853**;
in New Zealand call **0800 449 725**

For permission to use material from this text or product, please email
aust.permissions@cengage.com

ISBN 978 0 17 044962 5

Cengage Learning Australia
Level 7, 80 Dorcas Street
South Melbourne, Victoria Australia 3205

Cengage Learning New Zealand
Unit 4B Rosedale Office Park
331 Rosedale Road, Albany, North Shore 0632, NZ

For learning solutions, visit **cengage.com.au**

Printed in Malaysia by Papercraft.
2 3 4 5 6 7 25 24

CONTENTS

ABOUT THIS BOOK

FEATURES

- Questions are provided to review prior knowledge from Year 11 at the start of each module and check your understanding of key concepts at the end of each module.
- Learning goals are stated at the top of each worksheet to set the intention and help you understand what is required.
- Chapters clearly follow the sequence of the syllabus and are organised by inquiry question.
- Page references to the content-rich student books provide an integrated learning experience.
- Brief content summaries are provided where applicable.
- Hint boxes provide guidance on how to answer questions effectively.
- A complete practice exam is provided.
- Fully worked solutions appear at the back of the book to allow you to work independently and check your progress.

ORGANISATION OF YOUR WORKBOOK

Each chapter begins with the relevant inquiry question and follows the sequence of the syllabus. Worksheets have been designed to complement the student book and provide additional opportunities to apply and revise your learning. Completion of these worksheets will provide you with a solid foundation to undertake assessments and depth studies.

The workbook ends with a practice exam for you to complete independently. Use the solutions and marking criteria in the answers section to self-evaluate and plan for improvement.

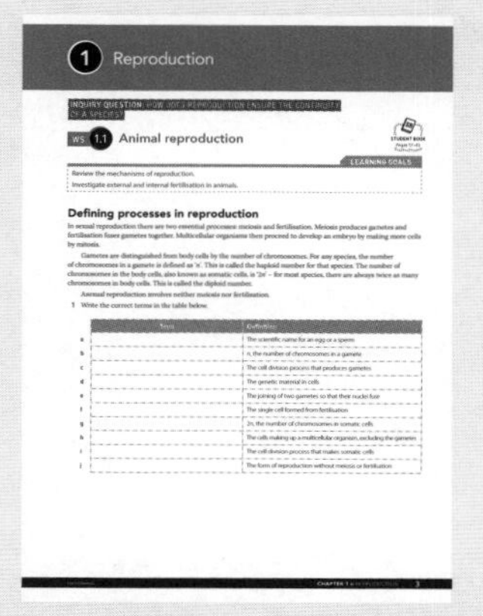

1 Reproduction

WS 1.1 Animal reproduction

Defining processes in reproduction

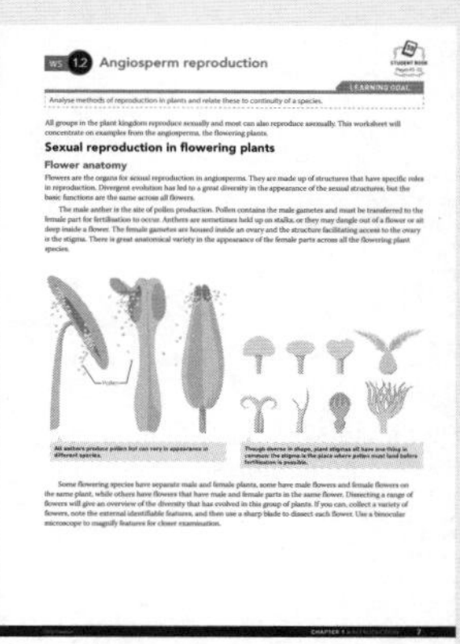

WS 1.2 Angiosperm reproduction

Sexual reproduction in flowering plants

Flower anatomy

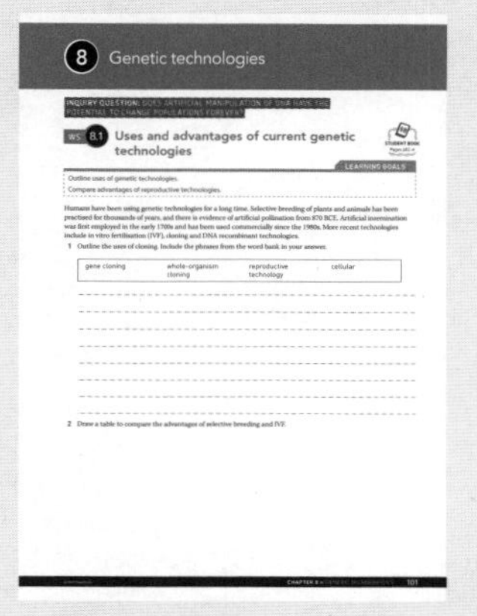

8 Genetic technologies

WS 8.1 Uses and advantages of current genetic technologies

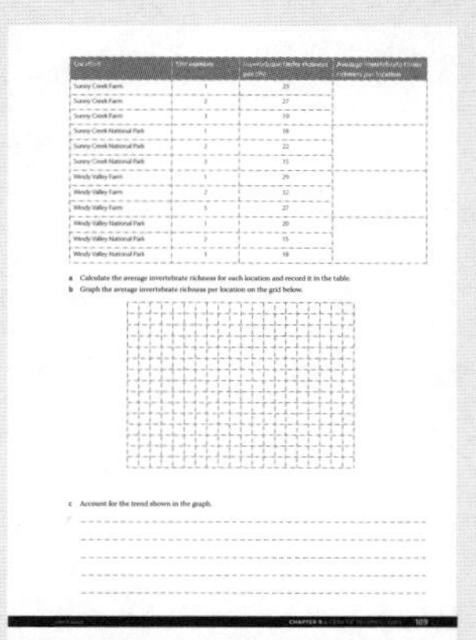

MODULE FIVE »
HEREDITY

Reviewing prior knowledge

1 Distinguish between eukaryotic and prokaryotic cells, giving an example of each.

2 a Complete the table to correctly match structures from the list of words with their descriptions in the table.

Golgi body	Ribosomes	Endoplasmic reticulum	DNA

	Structure	Description	Found in	
			prokaryotic cells	eukaryotic cells
i		A type of nucleic acid that is the genetic material in all cells; can appear in different forms, depending on the type of cell		
ii		Consists of stacks of membrane-bound sacs; receives synthesised products (such as proteins), modifies them and exports them to a variety of destinations		
iii		A membrane structure that branches out through the cytoplasm, enclosing an internal space (lumen); products such as protein can be transported around the cell within the lumen		
iv		Structures that appear as black spheres under a microscope; used for making protein		

b In the table, indicate whether the structure is found in prokaryotic or eukaryotic cells, or both.

3 Proteins and nucleic acids are two groups of biomacromolecules that are constructed by cells from smaller molecules. Outline your knowledge of proteins and nucleic acids in eukaryotic cells from Year 11.

4 Compare the processes of asexual and sexual reproduction.

> **HINT**
> The verb *compare* requires you to look for similarities and differences. Use a table or Venn diagram to compare two concepts.

5 Explain how sexual reproduction increases genetic diversity in a population.

> **HINT**
> The verb *explain* requires you to consider cause and effect.

6 Identify the nucleotides present in DNA.

7 Explain how the sequence of nucleotides relates to the phenotypic traits of an organism.

8 Explain why genetic variation in a population is essential for its long-term survival.

1 Reproduction

INQUIRY QUESTION: HOW DOES REPRODUCTION ENSURE THE CONTINUITY OF A SPECIES?

WS 1.1 Animal reproduction

STUDENT BOOK
Pages 37–43

LEARNING GOALS

Review the mechanisms of reproduction.

Investigate external and internal fertilisation in animals.

Defining processes in reproduction

In sexual reproduction there are two essential processes: meiosis and fertilisation. Meiosis produces gametes and fertilisation fuses gametes together. Multicellular organisms then proceed to develop an embryo by making more cells by mitosis.

Gametes are distinguished from body cells by the number of chromosomes. For any species, the number of chromosomes in a gamete is defined as 'n'. This is called the haploid number for that species. The number of chromosomes in the body cells, also known as somatic cells, is '$2n$' – for most species, there are always twice as many chromosomes in body cells. This is called the diploid number.

Asexual reproduction involves neither meiosis nor fertilisation.

1 Write the correct terms in the table below.

	Term	Definition
a		The scientific name for an egg or a sperm
b		n, the number of chromosomes in a gamete
c		The cell division process that produces gametes
d		The genetic material in cells
e		The joining of two gametes so that their nuclei fuse
f		The single cell formed from fertilisation
g		$2n$, the number of chromosomes in somatic cells
h		The cells making up a multicellular organism, excluding the gametes
i		The cell division process that makes somatic cells
j		The form of reproduction without meiosis or fertilisation

2 Create a flow chart that illustrates the production of an embryo from a male and female parent where $2n = 46$. Include the number of chromosomes involved at each stage and the words *fertilisation, diploid, haploid, zygote, embryo, mitosis* and *meiosis* on your flow chart, where appropriate.

3 Write a paragraph about asexual and sexual reproduction and continuity of the species using detail extracted from the diagram below.

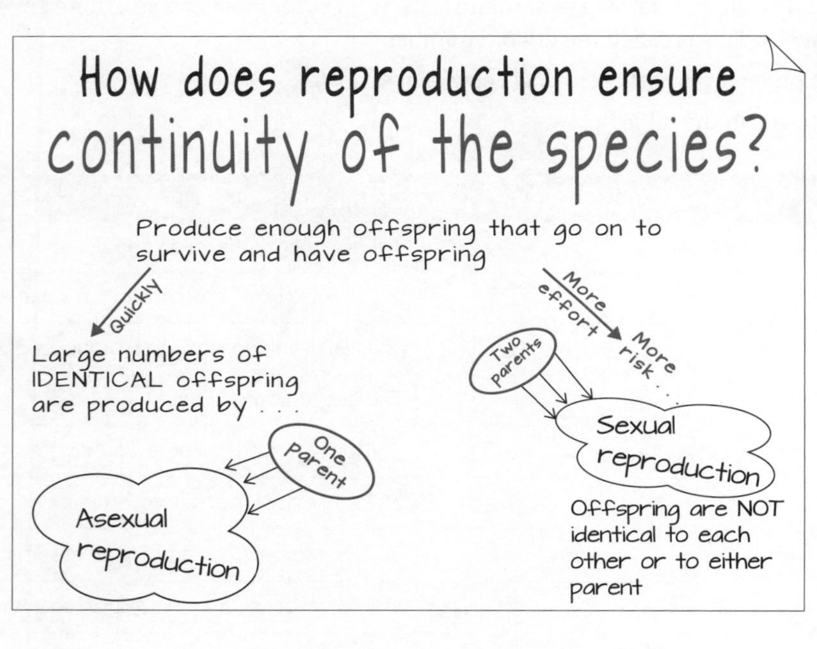

HINT

Think about what you know from Year 11, the theory of evolution by natural selection, and variation in a population.

Internal and external fertilisation

In aquatic environments, males and females can release their gametes directly into the water. This is called spawning. Spawning can be as simple as broadcasting gametes into water. When the gametes meet and fuse in the water, fertilisation occurs externally to the parent bodies. Watery environments are crucial so that the eggs and sperm do not desiccate (dry out). External fertilisation is typical in amphibians, aquatic invertebrates and most fish.

Reproductive strategies to increase the chance of successful external fertilisation include:

- spawning that is timed to seasons, lunar cycles or other environmental events
- courtship rituals followed by copulation.

HINT

In animals, *copulation* can simply refer to a male and female in contact with each other and the simultaneous release of gametes, resulting in external fertilisation; or it can involve male deposition of sperm into the female, resulting in internal fertilisation. The term *intercourse* is usually reserved for human sexual reproduction.

Internal fertilisation involves sperm deposition inside the female's body and, hence, fertilisation occurs internally. Since the internal environment of the female is suitable for cell survival, desiccation of the gametes is not an issue. Land invertebrates, all reptiles, birds and mammals use internal fertilisation, even if they live in aquatic environments.

4 Compare the conditions for external and internal fertilisation.

	External fertilisation	Internal fertilisation
Differences in environment		
Similarities in environment		

5 Explain how some reproductive strategies increase the chances of external fertilisation.

6 Some aquatic organisms such as sponges and corals are sessile (immobile); they are fixed to a substrate. For these organisms, broadcast spawning is the only mechanism for fertilisation and colonisation of new environments. Discuss the strategies used by such organisms to ensure continuity of the species.

HINT

The verb *discuss* means to look at positives and negatives, but you also need to explain the features in your answer.

Shelled eggs

Internal fertilisation and the evolution of shelled eggs in the reptile group allowed the colonisation of the drier parts of Earth. The embryo develops inside the shelled egg, which is waterproof and does not desiccate after being laid in a nest to complete development. These conditions were inherited by the birds and monotreme mammals. Placental and marsupial mammals use internal fertilisation but do not lay eggs; instead, development of the embryo continues in the uterus of the mother. This period is called gestation and is followed by live births. The length of gestation is much shorter for marsupials than for most other mammals.

7 Use secondary sources to investigate examples of frog and reptile reproduction. Explain why a shelled egg is an adaptation to terrestrial life. Justify your answer by contrasting reproduction in frogs and reptiles.

WS 1.2 Angiosperm reproduction

LEARNING GOAL

Analyse methods of reproduction in plants and relate these to continuity of a species.

All groups in the plant kingdom reproduce sexually and most can also reproduce asexually. This worksheet will concentrate on examples from the angiosperms, the flowering plants.

Sexual reproduction in flowering plants

Flower anatomy

Flowers are the organs for sexual reproduction in angiosperms. They are made up of structures that have specific roles in reproduction. Divergent evolution has led to a great diversity in the appearance of the sexual structures, but the basic functions are the same across all flowers.

The male anther is the site of pollen production. Pollen contains the male gametes and must be transferred to the female part for fertilisation to occur. Anthers are sometimes held up on stalks, or they may dangle out of a flower or sit deep inside a flower. The female gametes are housed inside an ovary and the structure facilitating access to the ovary is the stigma. There is great anatomical variety in the appearance of the female parts across all the flowering plant species.

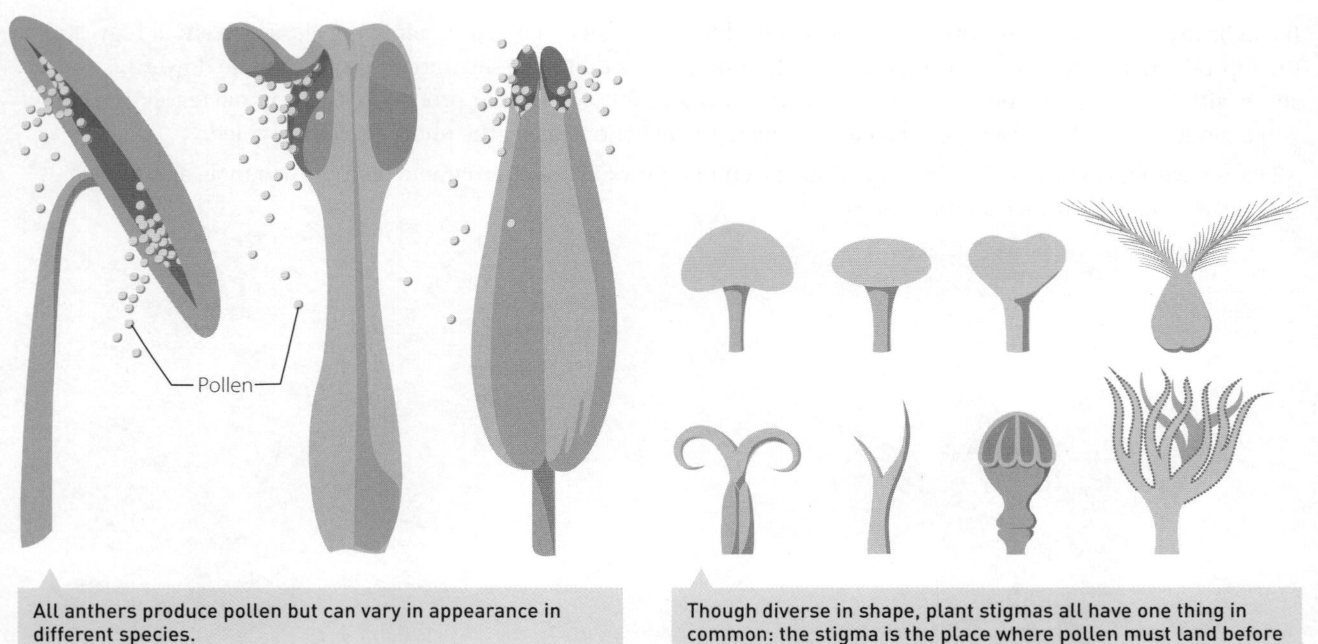

All anthers produce pollen but can vary in appearance in different species.

Though diverse in shape, plant stigmas all have one thing in common: the stigma is the place where pollen must land before fertilisation is possible.

Some flowering species have separate male and female plants, some have male flowers and female flowers on the same plant, while others have flowers that have male and female parts in the same flower. Dissecting a range of flowers will give an overview of the diversity that has evolved in this group of plants. If you can, collect a variety of flowers, note the external identifiable features, and then use a sharp blade to dissect each flower. Use a binocular microscope to magnify features for closer examination.

1 Before any dissection, a risk assessment should be conducted. Construct a table of potential risks and strategies to minimise the risks of a flower dissection.

Dissecting flowers

If you have access to flowers, dissecting tools and a binocular microscope or magnifying glass, you should investigate the reproductive structures for yourself. Try to find specimens that have separate male and female flowers as well as some with both sexes in one flower. The next few questions follow a similar procedure to an actual dissection, but, if you cannot carry out a dissection, you can complete the questions using the photographs provided.

2 Start with an external examination of the zucchini flowers in the photograph, or your own male and female flowers. Label any identifiable features.

Shutterstock.com/ppl

A male zucchini flower

iStock.com/Barcin

A female zucchini flower

9780170449625

3 The next step is to reveal the inside of the flower by removing some of the petals, as shown in the photo below. Label the identifiable features.

Alamy Stock Photo/Custom Life Science Images

Male (left) and female (right) zucchini flowers with petals removed to show the interior of the flowers

4 Repeat the external examination for any available flowers that have male and female parts in the one flower, or use the photographs below.

Shutterstock.com/Tish1

An oriental lily

Shutterstock.com/Huy Thoai

A fuschia

5 A more thorough examination requires cutting the parts of the flower. A binocular microscope was used to take this photo of an ovary. If you are doing an actual dissection, carefully make a longitudinal cut through the ovary and use a binocular microscope or magnifying glass to examine the ovules.

Outline and label the ovary, style and one ovule on the photograph.

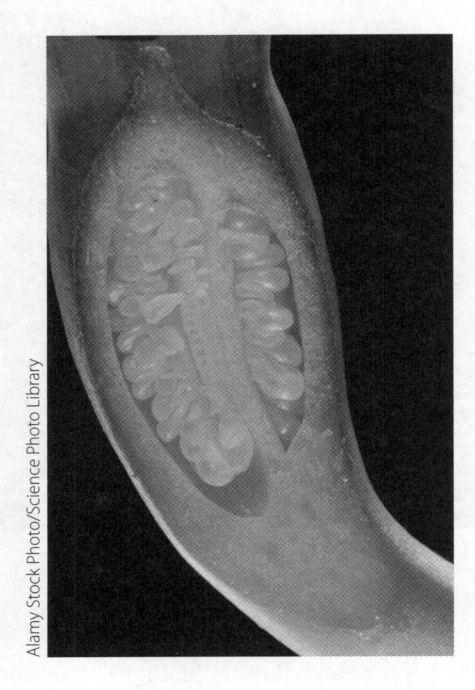
Alamy Stock Photo/Science Photo Library

Pollination and fertilisation

The pollination process occurs when pollen grains from the male part of one flower (anther) are transferred to the female part (stigma) of another flower. Each pollen grain contains sperm nuclei. A pollen tube grows from the pollen grain sitting on the stigma and down into the style. When the end of the pollen tube reaches the ovary, the sperm nucleus can fuse with the nucleus of an ovum inside an ovule. The fusion of these two gametes is called fertilisation. Each fertilised ovule develops into a seed. The ovary, now containing the seeds, develops into a fruit that protects the seeds.

6 Compare pollination and fertilisation in a flowering plant.

7 Use the information in the text above to annotate the diagram below, describing the processes of pollination and fertilisation, including the locations where pollination and fertilisation occur.

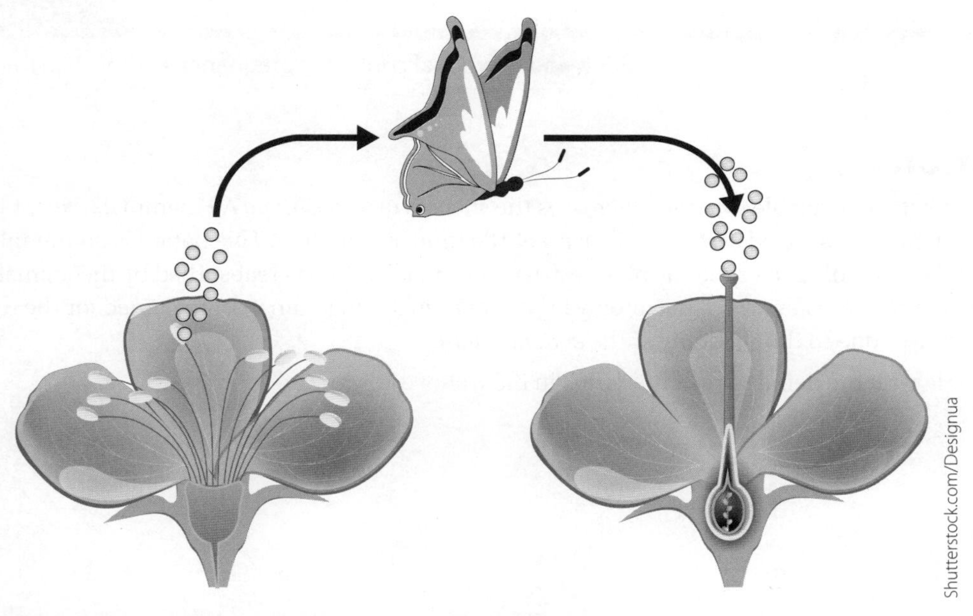

Shutterstock.com/Designua

Asexual reproduction in flowering plants

8 All angiosperm plants have flowers for sexual reproduction. Most angiosperms can also reproduce asexually.

a Use information in the caption and your own knowledge to annotate the diagram below as fully as possible.

Adapted from Springer: Evolutionary *Ecology: The evolutionary ecology (evo-eco) of plant asexual reproduction*, Niklas, K.J., Cobb, E.D., 2017

Schematic representations of some forms of asexual reproduction. These are all classified as clonal growth. **a** Bulbil formation and release of preformed juvenile plants; **b** Adventitious rooting and subsequent separation (not shown); **c** Underground stems or rhizomes; **d** Above-ground stems or runners.

b Explain why the drawings in the figure above are classified as asexual reproduction. Support your answer using two examples from the diagram.

WS 1.3 Fertilisation and pregnancy

STUDENT BOOK
Pages 60–71

LEARNING GOAL

Analyse the features of fertilisation, implantation and hormonal control of pregnancy and birth in mammals.

Fertilisation

Timing sexual activity with female *ovulation* increases the success of *fertilisation.* All mammals except for higher primates display behaviours called *oestrus* exclusively at the time of ovulation. This period is commonly referred to as being 'in heat'. In females of species that display oestrus, the endometrium is reabsorbed by the animal if fertilisation does not occur. *Menstrual cycles,* which occur only in primates, including humans, are named for the regular appearance of menses due to the shedding of the *endometrium.*

1 Design a table with the definitions of the terms in italic above.

Implantation

In mammals, the sperm and ovum meet in one of the fallopian tubes. If fertilisation occurs, the two nuclei fuse to make one nucleus. The outer layer of the ovum hardens as soon as one sperm nucleus enters, so that it becomes impermeable to any other sperm. The resulting zygote travels down the fallopian tube, undergoing several rounds of cell division, becoming a ball of 16 cells, called a morula. In placental mammals the ball of cells, now 32 cells in size, enters the uterus and the cells start to rearrange, becoming a blastocyst. A blastocyst has an inner cell mass that becomes the embryo, a cavity and an outer cell layer called the trophoblast that begins to grow into the endometrium. The blastocyst burrows into the endometrium in a process called implantation.

2 The diagram below represents the ovum's interaction with a sperm cell and the changes as it moves to the uterus. Add labels to the diagram, including the location where events occur.

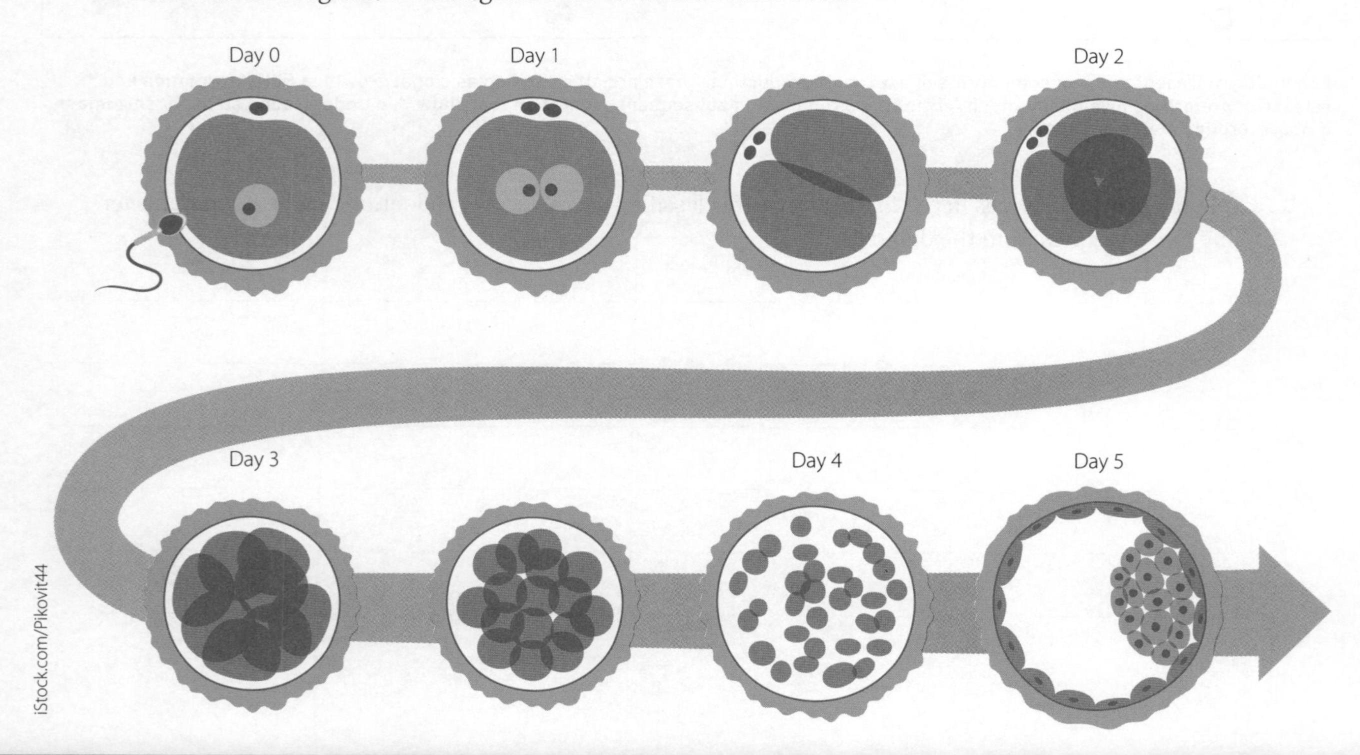

iStock.com/Pikovit44

9780170449625

Hormones in gestation and birth

Human chorionic gonadotropin (HCG)

The most significant hormones in maintaining the pregnancy are human chorionic gonadotropin (HCG), progesterone and oestrogen. Among other roles, these hormones suppress any further egg production and maintain the state of the uterus. HCG is the first hormone of pregnancy; it is secreted by the trophoblast cells in the blastocyst after implantation. These cells go on to form the placenta. HCG is detectable in the urine and blood of the pregnant mother.

HCG levels in urine during human pregnancy

Event	Time (weeks)	Maximum level of HCG (pharmacological units/mL)
Last menstrual period	0	5
Ovulation and fertilisation	2	5
Implantation	3	50
First trimester	4	500
	6	10 000
	7	30 000
	8	50 000
	9	100 000
	10	150 000
	12	200 000
Second trimester	13	180 000
	16	50 000
	18	40 000
	26	50 000
Third trimester	27	52 000
	36	46 000
	40	45 000
Delivery		4000
One week after delivery	41	5

3 a Plot a line graph of the data. Your graph should have a title and all axes labelled; points should be joined, and events should be labelled.

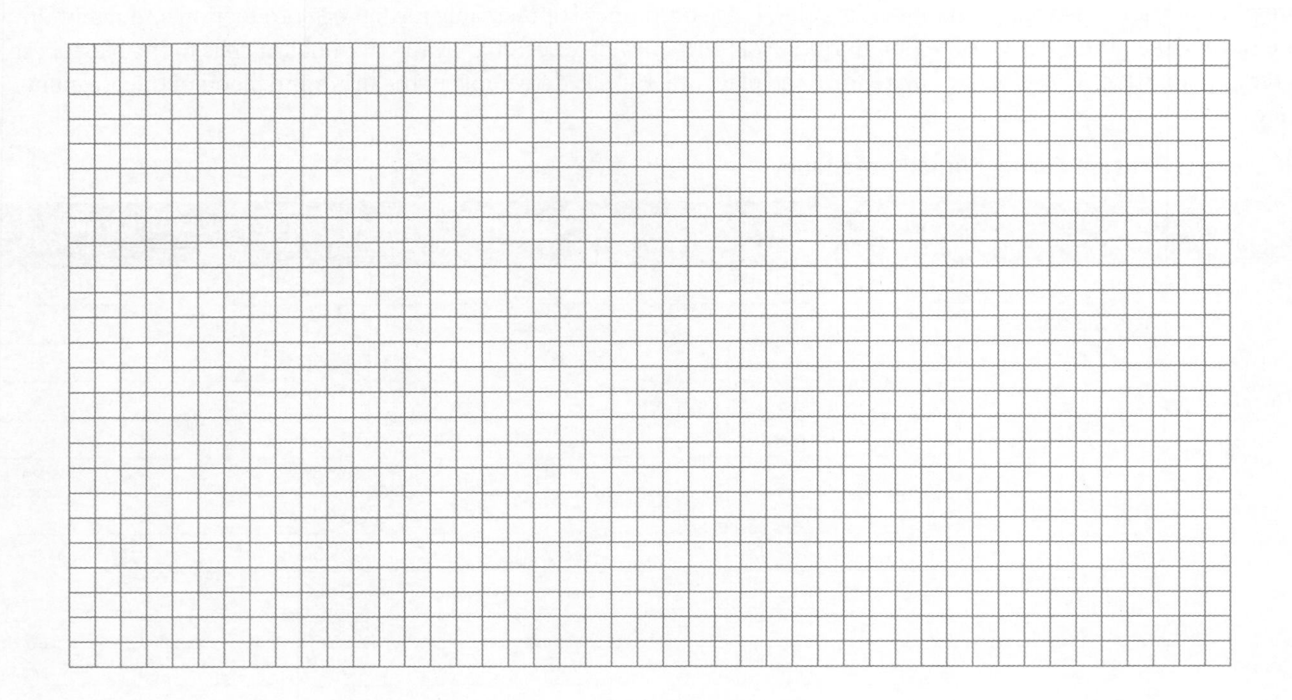

b Describe the trend in HCG levels during a human pregnancy.

c Home pregnancy tests react to levels of HCG in urine. Use information from the data table to explain why people are advised to wait until at least four weeks after the last menstrual period, or at least two weeks after ovulation, before using a home pregnancy test.

Progesterone and oestrogen

Before pregnancy, progesterone and oestrogen are produced by the ovaries, but after implantation the placenta takes over and levels rise throughout pregnancy. High levels of progesterone prevent uterine contractions. Around the seventh month, progesterone levels plateau and then drop. Oestrogen levels continue to rise. The increasing difference between oestrogen and progesterone makes the uterine muscles more sensitive to the hormone that promotes contractions.

The birth process, labour, begins with the release of the hormone oxytocin from the pituitary gland, which is near the brain. This hormone triggers various processes, such as dilation of the cervix and the periodic contractions of the uterine muscles. These contractions get stronger and more frequent until they expel the baby through the cervix and vagina. The placenta separates from the uterine wall and is also delivered through the vagina. All hormones drop to normal levels after delivery.

4 a Based on the information provided, sketch the approximate shape of the curves of HCG, progesterone, oestrogen and oxytocin hormones during pregnancy on the axes below. Include annotations of significant events.

HINT

Sketching the shape does not require you to put a scale on the axes.

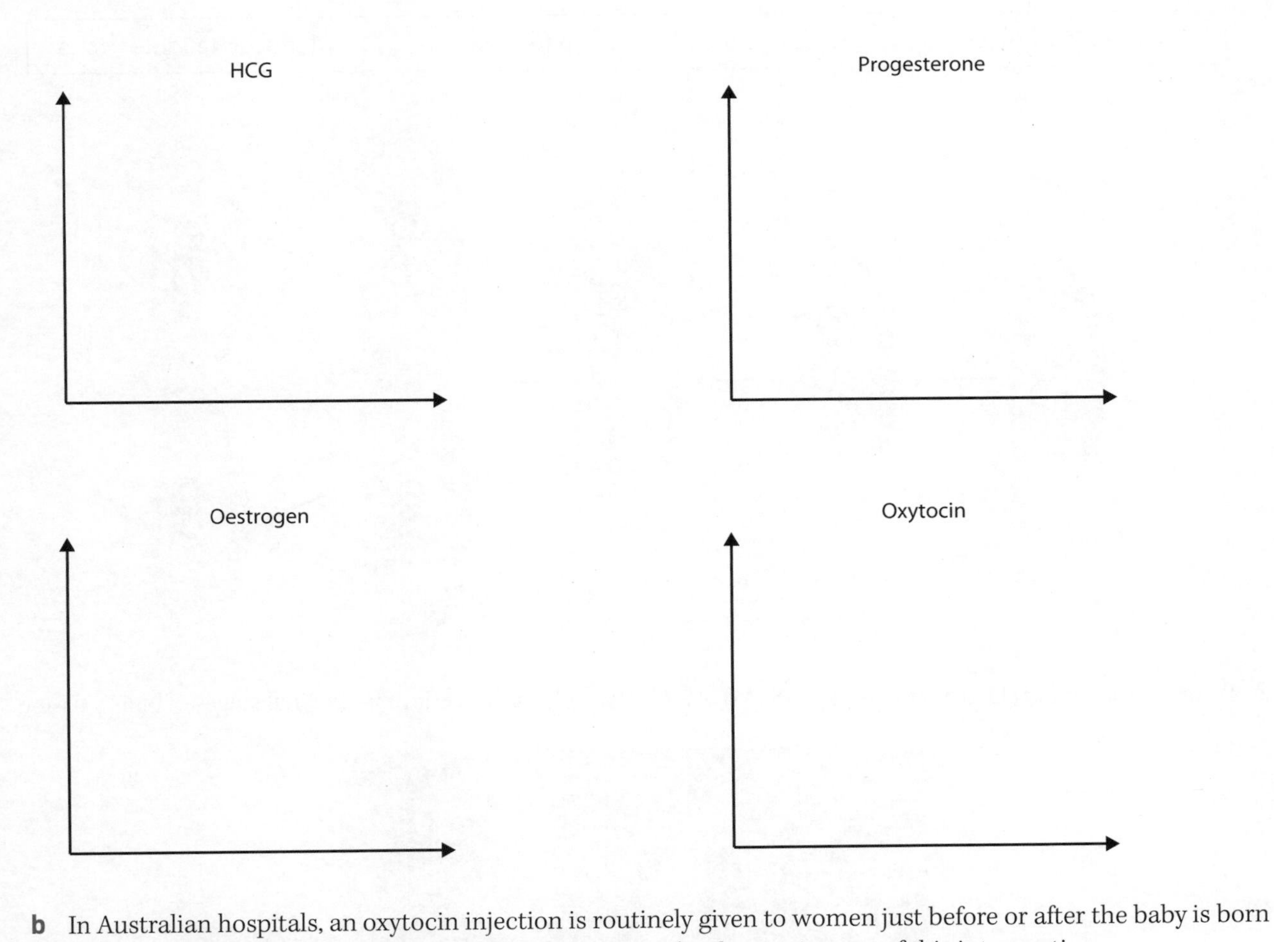

b In Australian hospitals, an oxytocin injection is routinely given to women just before or after the baby is born to hasten the delivery of the placenta. Suggest reasons for the routine use of this intervention.

WS 1.4 Asexual reproduction in simple organisms

STUDENT BOOK
Pages 34–6, 52–60

LEARNING GOAL

Analyse asexual reproduction in bacteria, fungi and protists.

Binary fission

Binary fission simply refers to cells splitting in two. It is a type of asexual reproduction and is the primary method of reproduction in bacteria, but also occurs in some protists. A bacterium must provide each offspring (daughter cell) with a complete copy of its genetic material. In the diagram in question **1**, the original cell has a loop of DNA, the bacterial chromosome, represented as a pink squiggle. A protein helps cut the two cells apart after the cytoplasm has been duplicated and divided between the two daughter cells (cytokinesis).

1 Annotate the diagram below to describe the events in binary fission in bacterial cells. Include the terms in the word bank.

bacterial chromosome	cytokinesis	parent cell	identical daughter cells

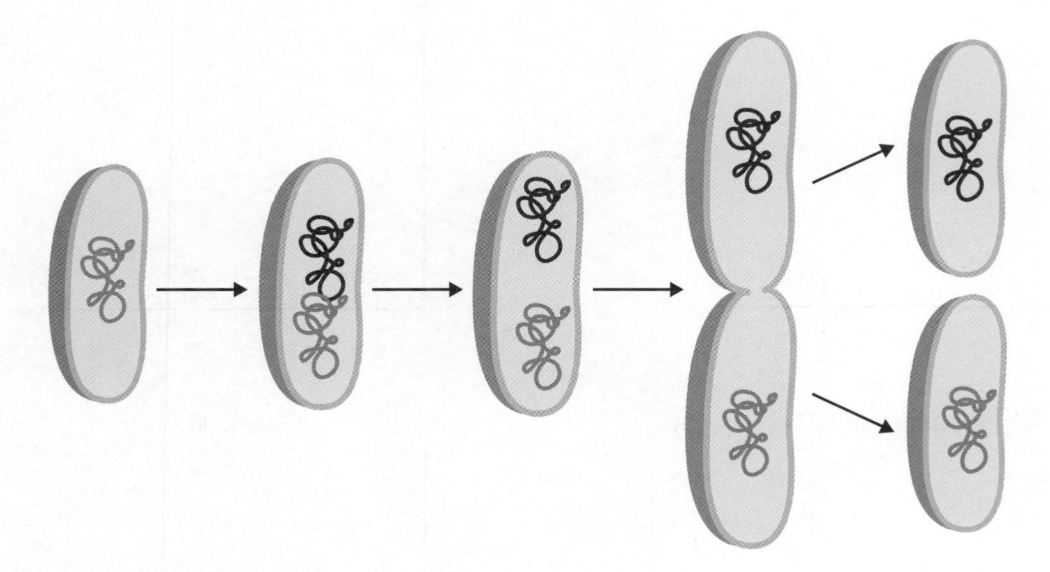

2 Below is a scanning electron micrograph (SEM) of a *Proteus vulgaris* bacterium in the final stages of binary fission.

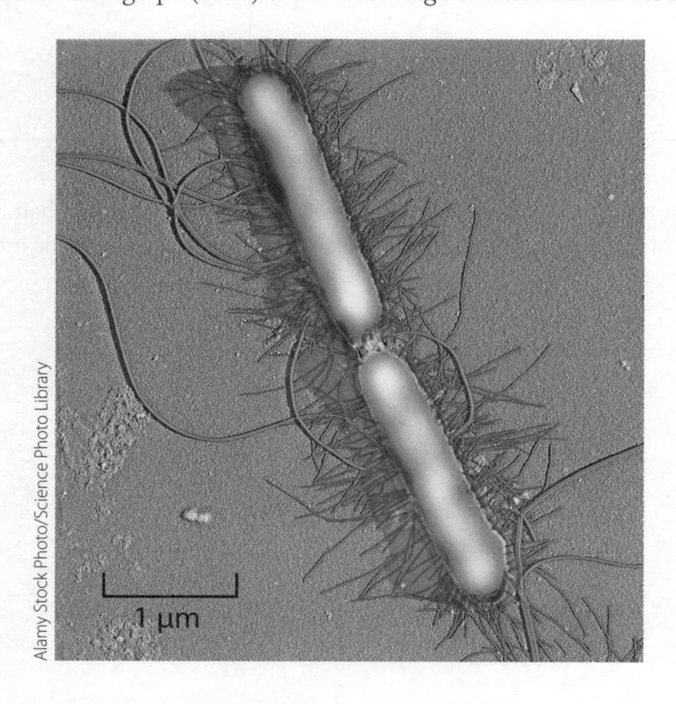

Alamy Stock Photo/Science Photo Library

a Use the scale bar to estimate the length of one *P. vulgaris* cell.

b On the image of the *P. vulgaris* cell, outline the cell wall of one cell. Label one pilus and one flagellum. Draw in a bacterial chromosome.

Budding

Budding is another type of asexual reproduction, in which a new organism develops from an outgrowth on one side. Just as with binary fission, the newly created individual is a clone and is genetically identical to the parent organism. The parent cell and the newly formed cell are similar in size during binary fission but, in budding, the parent cell is always larger than the newly formed bud. Budding is found in single-celled fungi and protists (and occasionally in simple multicellular organisms). The presence of the nucleus in eukaryotic cells complicates the processes of budding and binary fission. The nucleus must divide by a process called mitosis.

3 Starting with a simple hypothetical cell such as the one shown below, create a labelled diagram that distinguishes between binary fission and budding.

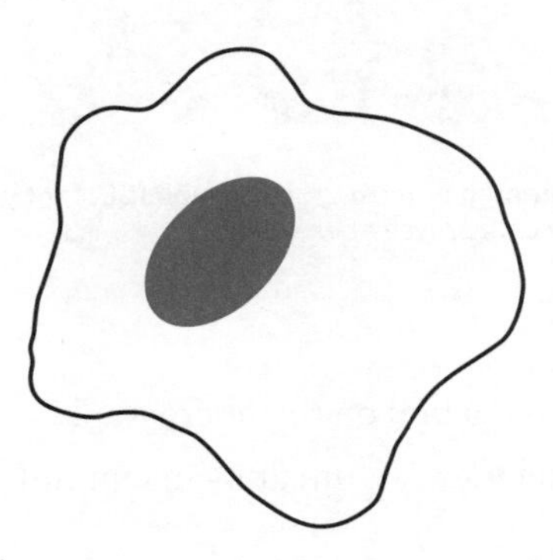

4 Yeast are single-celled fungi that reproduce asexually by budding. While yeast can vary in size, they typically measure 3–8 μm in diameter. *Saccharomyces cerevisiae* is the most commonly used strain in scientific research, in baking and in fermentation.

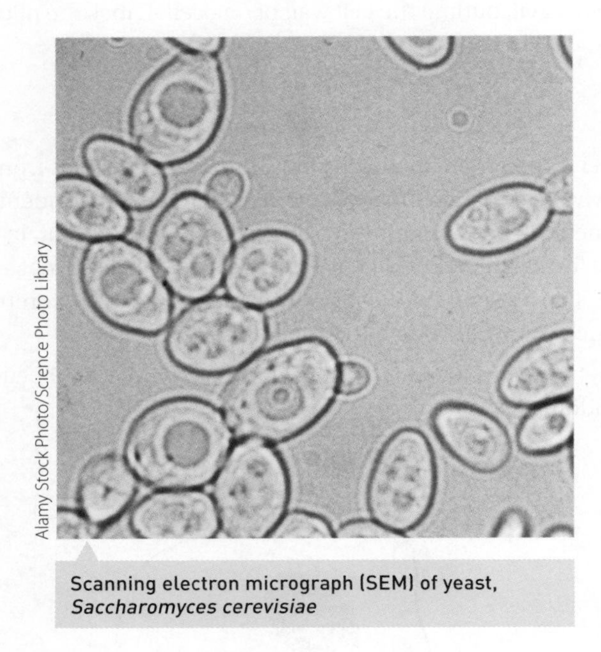

Scanning electron micrograph (SEM) of yeast, *Saccharomyces cerevisiae*

a Outline and label a parent cell and a bud on the micrograph.

b Assume the largest cell in the diagram is 8 μm at its maximum axis. Draw a 1 μm scale bar on the micrograph.

Spores

Spores are minute, single-celled, reproductive units capable of giving rise to a new individual without sexual fusion. They are used by lower plants, fungi and protozoans for asexual reproduction. (These groups can also reproduce sexually.) In unfavourable environmental conditions, a bacterial cell can produce a protective coat that surrounds its DNA. When this layer is present, the bacterial cell is called a spore. Becoming a spore protects the bacterium from desiccation and extreme temperatures and pH. The spore coating dissolves and the cell functions resume when favourable conditions return.

5 Distinguish between the purpose of spore production in bacteria and fungi.

6 *Penicillium* is a genus of multicellular fungi commonly known as blue or green mould. The body of the organism has multi-branched filamentous structures called hyphae that attach to a surface. Periodically, vertical aerial hyphae arise from the horizontal hyphae. Each aerial hypha has multiple branches. Spherical spores form at the end of each branch. While this may sound like a description of a tree, the structure is microscopic. Fungal spores carry genetic material identical to that of the parent. They are produced in enormous numbers and are extremely light, enabling widespread dispersal by air once they detach from the parent. Spores attach to a suitable substrate and germinate, which involves absorption of water and division by mitosis, becoming new hyphae. The aerial structures grow again from hyphae and the cycle continues. This process enables fungi to reproduce rapidly and colonise a wide area.

Use the information from the previous page to annotate the diagram of the *Penicillium* life cycle below.

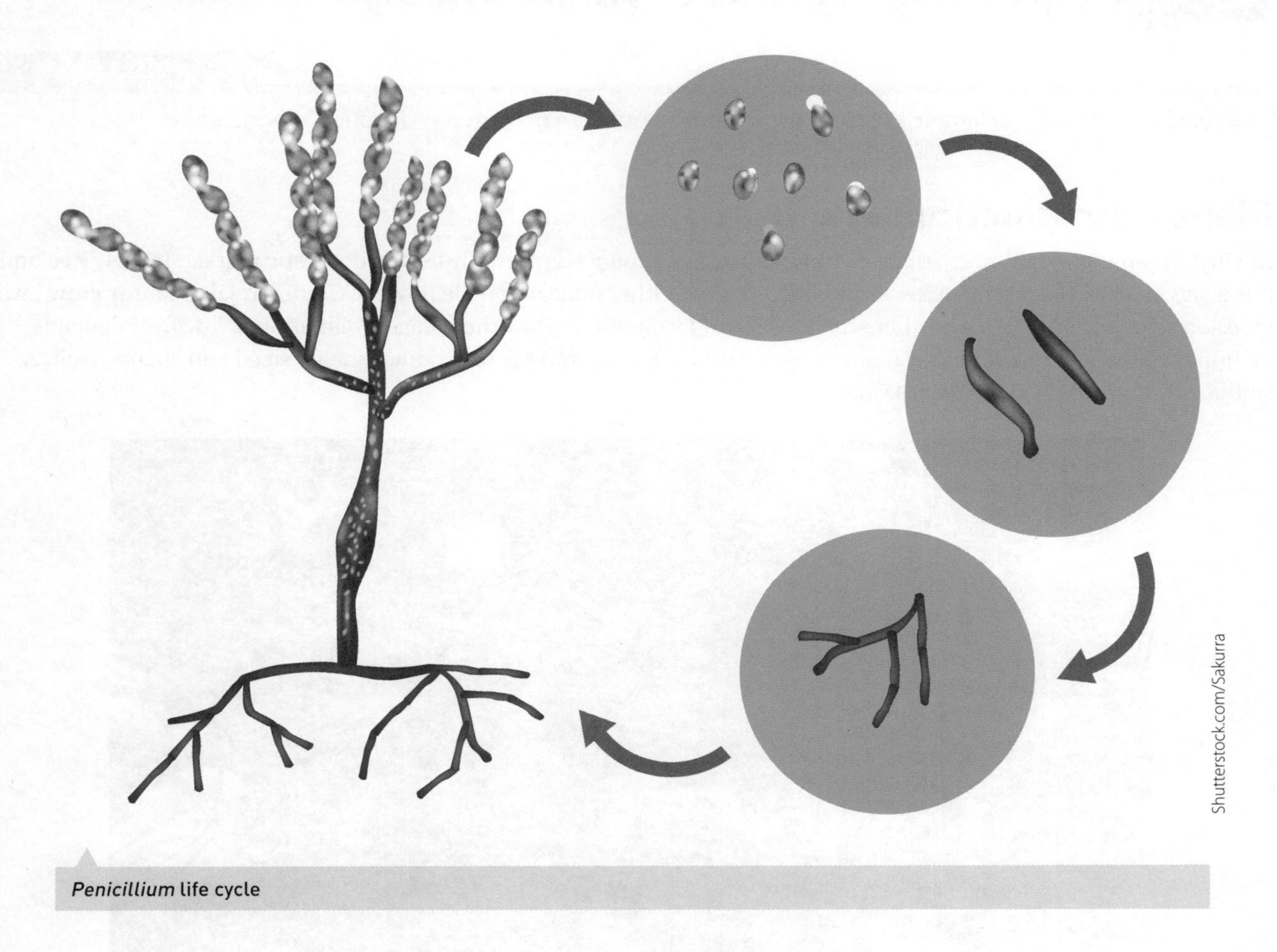

Penicillium life cycle

7 Explain how two types of asexual reproduction ensure the continuity of the species in fungi.

Applying reproductive knowledge

STUDENT BOOK
Pages 72–3

LEARNING GOAL

Evaluate the impact of scientific knowledge on the manipulation of plant reproduction in agriculture.

Plant reproduction: kiwifruit

Kiwifruit is an important horticultural crop in Australia, produced primarily for the domestic market. In New Zealand it is a significant export crop. A kiwifruit vine produces either male or female flowers. Commercial orchards grow kiwi vines on rows of raised trellises where the flowers and fruit hang below the foliage. Pollination of kiwifruit depends on honeybees and wind. Kiwifruit requires high levels of pollen transfer to produce quality sized and shaped fruit. A suitable fruit contains 1000–1400 seeds.

Alamy Stock Photo/Jana Grimova

1 a Explain the relationship between the size of a kiwifruit and pollen transfer.

b A successful fruit production industry needs to understand plant reproduction to optimise yield. Describe the scientific knowledge of flowering plant reproduction that would be relevant to a kiwifruit producer.

Pollination interventions

Distribution of male flowers is an important part of orchard management. Scientific trials have found that, in commercial plantings, 10–12% of the vines must be males and the males should be scattered evenly throughout the block.

Intervening in pollination is an avenue for orchardists to increase yield because there are problems with bee populations. The male flowers are harvested, the pollen is processed, and then it is applied to the female flowers. Spreading the pollen as a dry powder by hand can lead to fertilisation but is impractical for an entire orchard. Scaling up has been necessary. Vehicles are driven under the vines, blowing the collected pollen upward from containers, resulting in fertilisation. Correctly timing the collection and application of pollen for the different varieties of kiwifruit must be precise if fertilisation and fruit formation is to occur.

The Wrangler Ltd/Waverley Klein-Ovink featuring PollenSmart pollinator

Mass artificial pollination of kiwi fruit

A robotic technology, known as RoboBee, is being developed to add to the pollen distributing vehicle that will sense which flowers are female and spray pollen directly on to them. Currently, pollen is sprayed over the whole vine, so a lot of it is wasted when it lands on leaves. Collecting and processing male pollen is expensive, so this technology could save a lot of pollen wastage.

Trials found flowers that were successfully pollinated produced fruit that were comparable in quality to commercially grown kiwifruit. However, the overall fruit set was found to be well below commercial requirements and further work on increasing the overall yield is required.

2 Evaluate the impact of scientific knowledge of reproduction on kiwifruit production.

HINT

Evaluating the *impact* requires you to make a judgement about how important scientific knowledge has been and support your answer with evidence. You need to give details about relevant knowledge and weigh up the benefits and limitations before you make a judgement.

2 Cell replication

INQUIRY QUESTION: HOW IMPORTANT IS IT FOR GENETIC MATERIAL TO BE REPLICATED EXACTLY?

WS 2.1 Cell division basics

STUDENT BOOK
Pages 79–83

LEARNING GOAL

Consolidate the context of cell division in eukaryotic cells.

Cell division and chromosomes

In multicellular organisms, the somatic cells are defined as any cell in the body that is not a gamete. Cell division by mitosis leads to the formation of two new somatic cells that contribute to the growth of the organism. From zygote to embryo, from infant to adult, growth relies on mitosis followed by cell differentiation to support specialised functions. Mitosis occurs in any part of the organism that is growing or being repaired and needs new cells.

If cell division were as simple as splitting in two, the amount of material in the cell would be smaller and smaller at every division. Instead, the amount of DNA in the nucleus and the organelles in the cytoplasm must double before the cell divides. The term ***mitosis*** technically refers to the nucleus dividing, and cytokinesis refers to the cytoplasm dividing and two new cells forming.

Somatic cells in the sexual organs undergo meiosis instead of mitosis to make the gametes. Meiosis is a different type of cell division that results in haploid gametes.

1 **a** Which type of cell division creates gametes?

____

b Where and when does mitosis occur in an organism?

____

____

____

____

c Name the organs of flowering plants and animals where meiosis occurs.

____

____

d Explain the difference between haploid cells and diploid cells in an organism where $n = 8$.

____

____

____

____

e Justify why DNA synthesis is necessary before a cell divides by mitosis.

2 DNA occurs as long strands in the nucleus and cannot be seen through normal laboratory microscopes. Before a cell divides, there is a period when the DNA strands wrap around histone proteins. The histone/DNA combination then coils up on itself. Once the DNA strands are more condensed like this, they become visible in the nucleus. These formations are called chromosomes. Genes are sections of DNA and their location can be described with reference to light and dark bands along chromosomes. Each time a cell divides by mitosis, new daughter cells end up with identical chromosomes, which are identical to those of the original parent cell.

Use the text to annotate the diagrams.

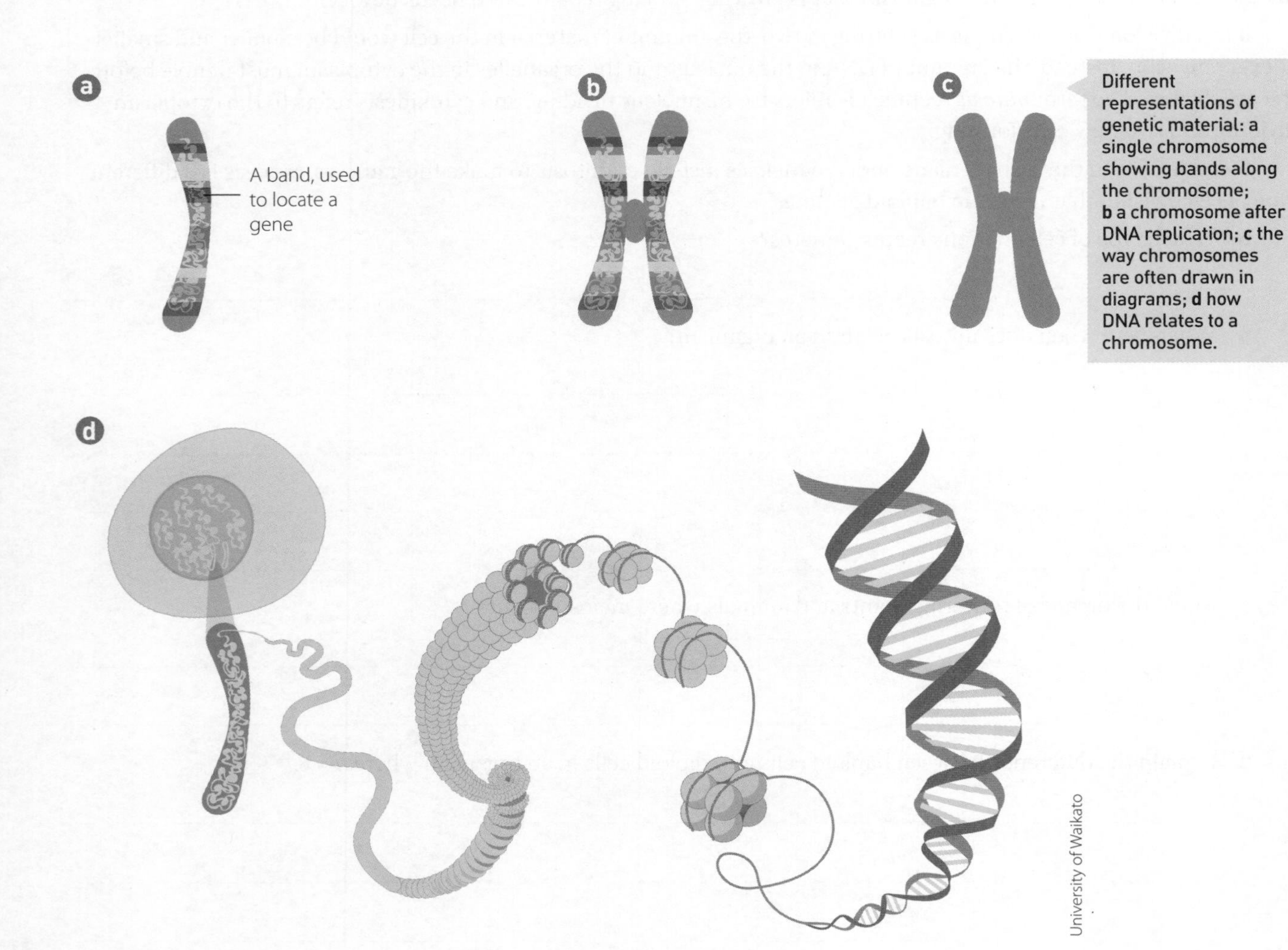

Different representations of genetic material: **a** single chromosome showing bands along the chromosome; **b** a chromosome after DNA replication; **c** the way chromosomes are often drawn in diagrams; **d** how DNA relates to a chromosome.

The cell cycle

Cells have functions to attend to, so they cannot be dividing constantly; there must be time to do their work, such as photosynthesis or respiration. The lower layers of skin cells divide regularly to replace the outer layers that are sloughed off, and cells lining the digestive tract divide rapidly for the same reason. Some adult cells do not divide frequently and others, such as nerve cells, may last a lifetime and never divide by mitosis. The diagram below describes the place of mitosis in the cell cycle.

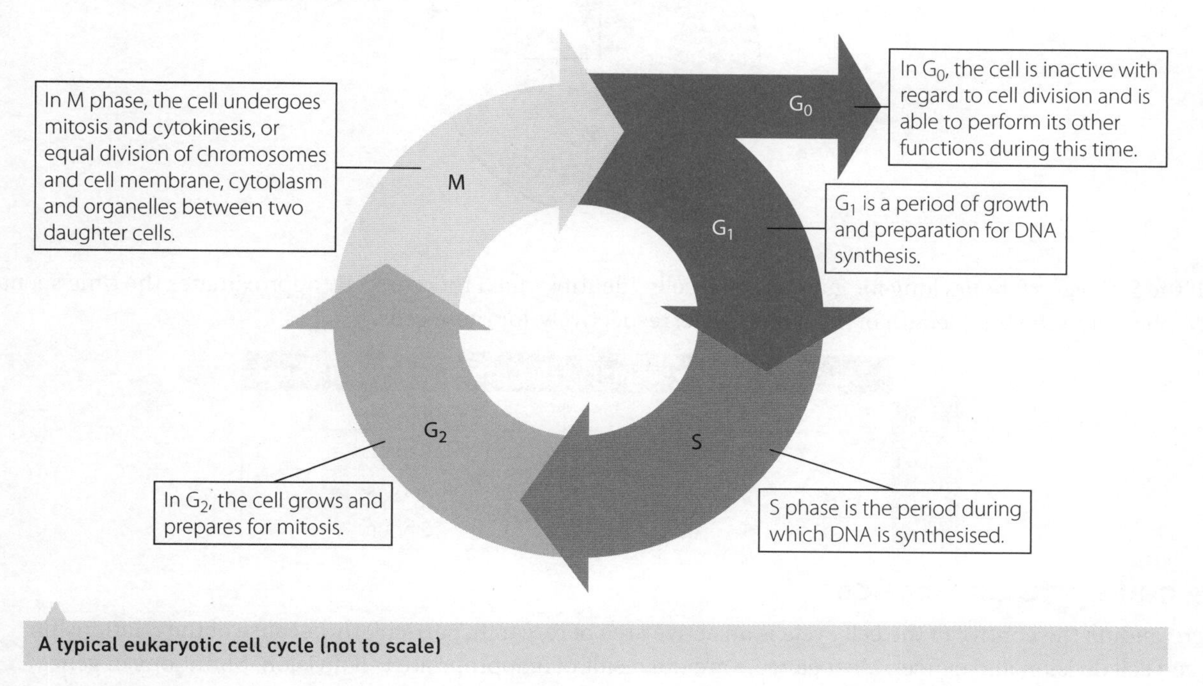

A typical eukaryotic cell cycle (not to scale)

3 Use the annotations on the diagram and the text to answer the following questions.

a Distinguish between phases G_0, G_1 and G_2.

b In which phase does the amount of DNA double?

c Compare the time a nerve cell and a skin cell spend in phase G_0.

d Predict in which phase the mass of the parent cell would significantly increase and decrease. Justify your answer.

4 The diagram in question **3** suggests that the four phases of the cell cycle are equal in length. This is not accurate. Time in each phase depends on the cell type and the species.

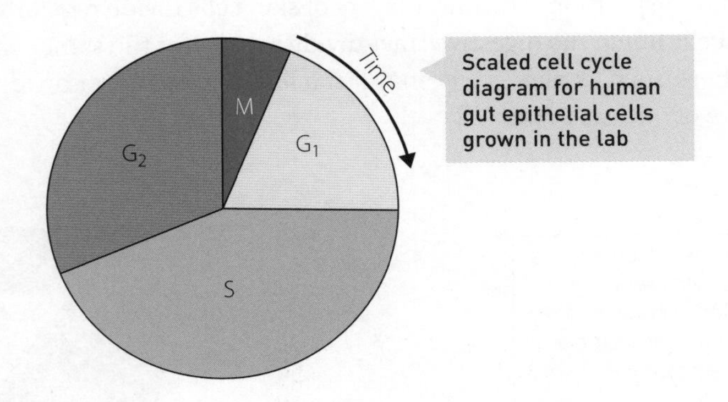

If the S phase is 7 hours long for gut epithelial cells, identify which row correctly approximates the time spent in M phase and the total length of the whole cycle, respectively, for these cells.

	Length of mitosis (hours)	Length of cell cycle (hours)
A	0.5	14 hours
B	2	20 hours
C	0.5	7 hours
D	1	16 hours

The cell cycle and cancer

Understanding the control of the cell cycle is an active area of research, particularly because of the relationship between cell division and cancer. Most cancers are the result of inappropriate cell division. Mutations in genes can cause cancer by accelerating cell division rates or inhibiting the normal controls on the system, such as cell cycle suppression or programmed cell death. Masses of cancer cells are called tumours.

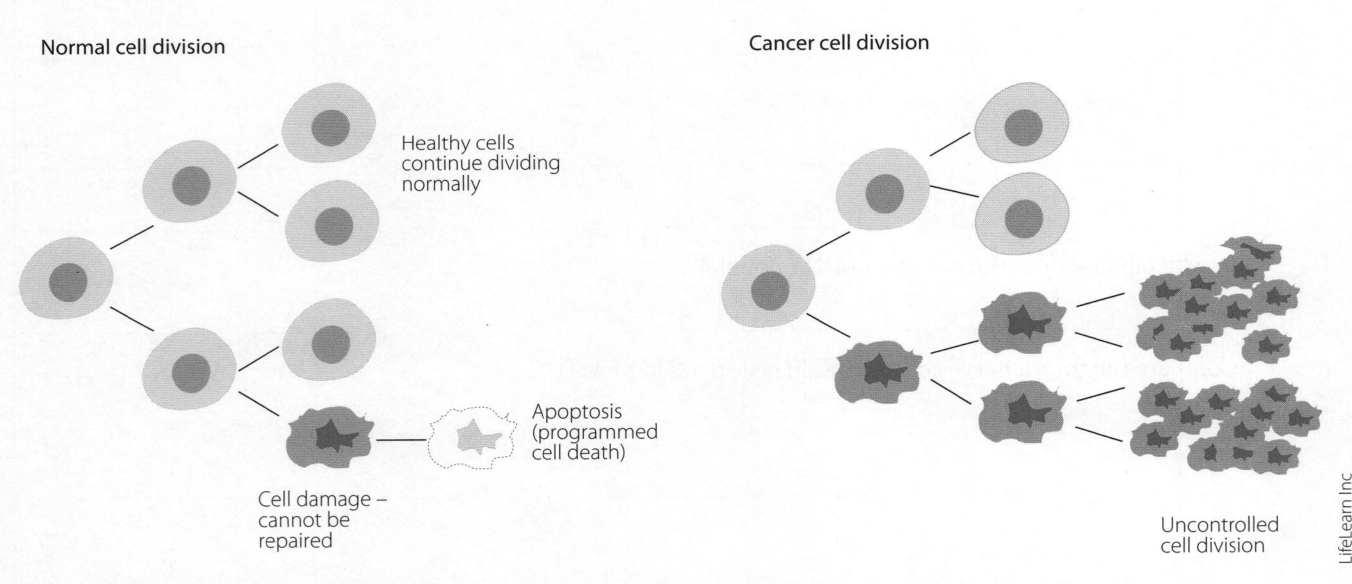

5 Describe how failures in the cell cycle can cause tumours. Include the term *apoptosis* in your answer.

Meiosis is not a cycle

The cells in the anthers, testes and ovaries are diploid, and at specific times for each sex these cells produce the haploid gametes. Some of the processes involved are similar to the cell cycle described earlier, except meiosis is not part of a cycle. The haploid gametes do not go on to divide again and make more gametes. Meiosis is part of a linear progression. It is sometimes called reduction division.

6 Complete a linear flow chart for the production of gametes. Include the terms from the word bank in your answer.

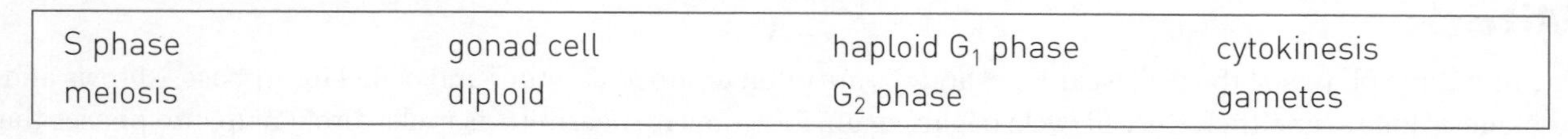

S phase	gonad cell	haploid G_1 phase	cytokinesis
meiosis	diploid	G_2 phase	gametes

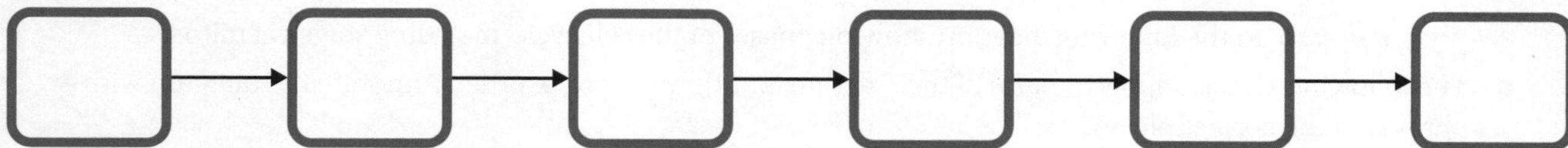

7 During S phase the amount of DNA doubles, just as for mitosis. However, in meiosis there are actually two rounds of divisions, so the gametes end up with half the normal amount of DNA after the second division. Suggest why meiosis is also called reduction division.

8 Explain why gametes should only have half as much DNA as somatic cells.

Mitosis and meiosis

STUDENT BOOK
Pages 79–89, 151–8

LEARNING GOAL

Model the processes involved in mitosis and meiosis.

Mitosis

The first three phases of the cell cycle, G_1, S and G_2, are often grouped together and called interphase. Mitosis and cytokinesis follow, and then the cell cycle begins again. For convenience, mitosis is also broken up into phases that match the behaviour of the chromosomes.

1 a Add *interphase* to the following diagram showing phases of the cell cycle, including stages of mitosis.

b The following phrases apply to parts of the diagram, but they are out of order. Annotate the diagram with phrases in the correct places.

- Spindle fibres attach to chromosomes
- Chromatids split at the centromere and start to move toward opposite ends of the cell
- Two nuclear membranes form around the chromosomes
- Nuclear membrane breaks down
- Single chromosomes are now in two bunches
- Chromosomes become visible in the nucleus
- Growth, DNA synthesis and preparation for mitosis
- Duplicated chromosomes line up across the middle of the cell
- Cytoplasm and organelles duplicate and new cell membranes form around daughter cells
- G_0 phase: some cells enter G_0 and do not divide for some time

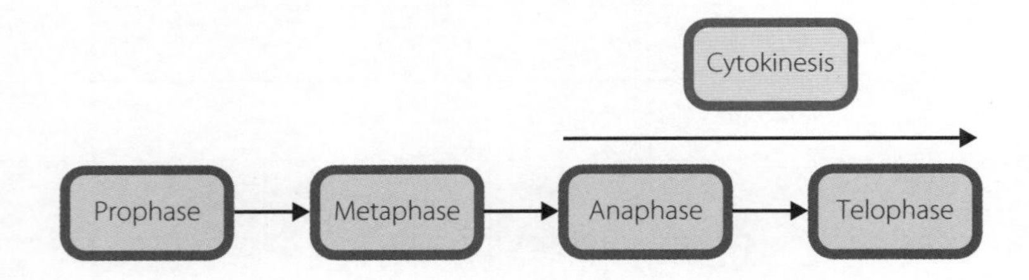

2 Refer to the photomicrographs of root tip cells below.

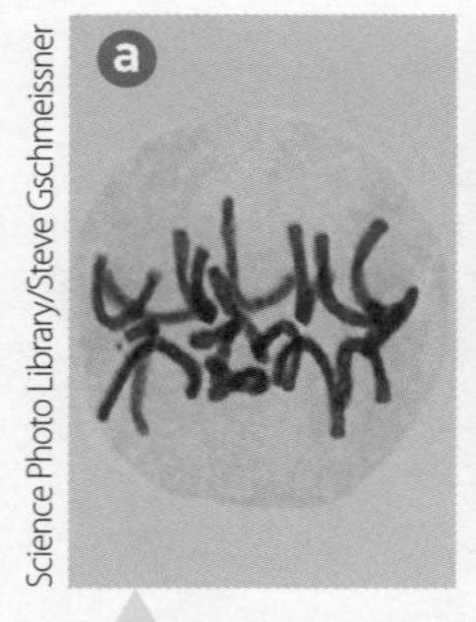

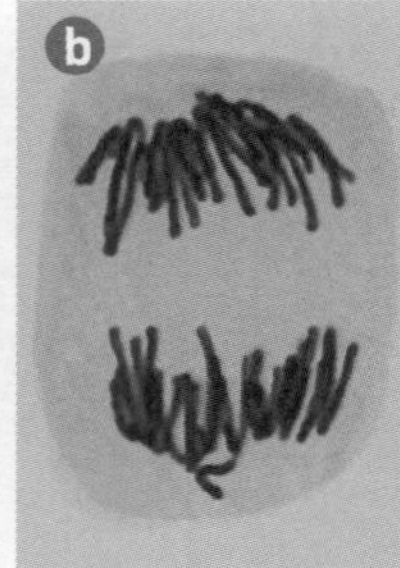

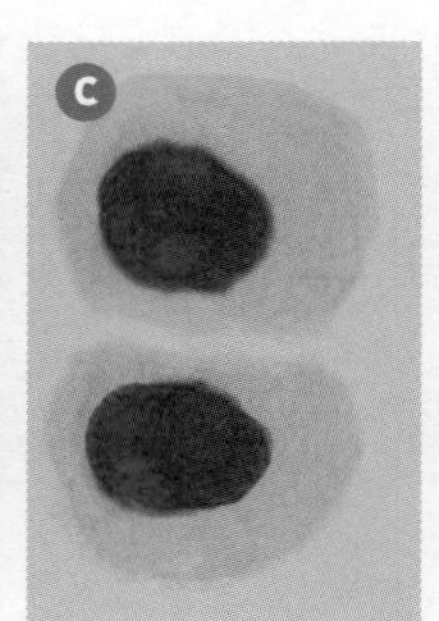

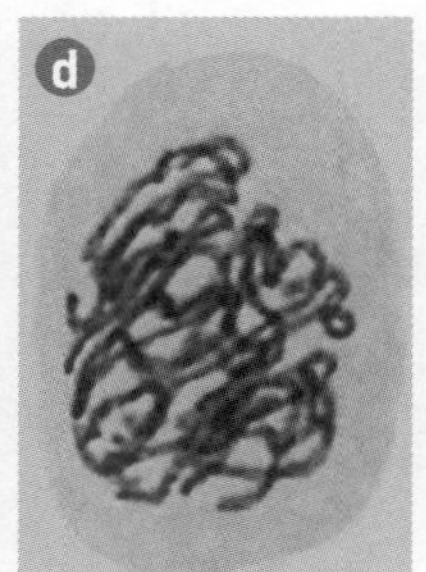

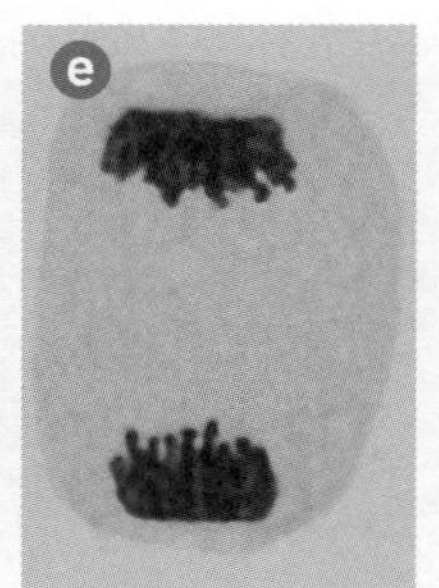

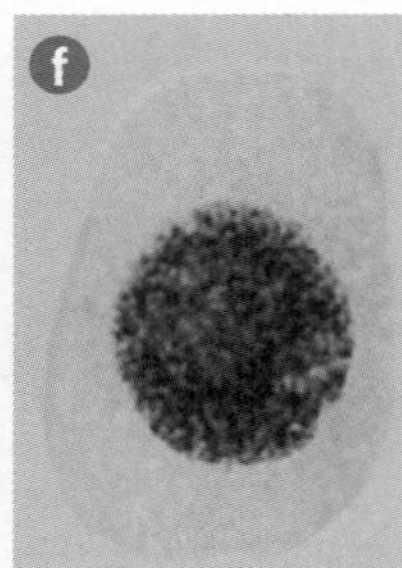

Science Photo Library/Steve Gschmeissner

Photomicrographs of root tip cells. Chromosomes have been stained with toluidine blue (shown here as darker grey).

a Identify which stage of the cell cycle or mitosis is represented by each photograph. You can use 'early' or 'late' in your answer if the photo seems to be between stages.

b Photo **e** was the most difficult to classify. Describe the problem.

Meiosis

The models of meiosis make more sense when you are familiar with the chromosomes in gametes and fertilisation.

In the zygote shown at the right, you can see chromosomes of two different colours and two sizes. The different colours represent the chromosomes from the male parent and the female parent. A pair of homologous chromosomes, each consisting of a single chromatid, has alleles (genetic information) from the father and from the mother.

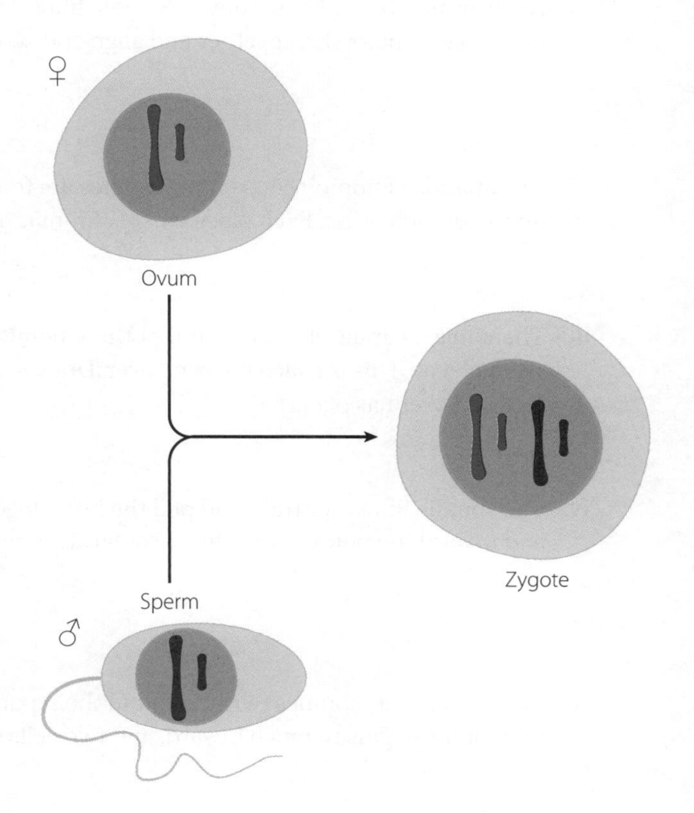

3 a How many chromosomes are in the zygote of this organism?

b What is the n number for the organism in the diagram?

c Create a drawing that arranges the homologous chromosomes in pairs for this organism. Label pair 1 and pair 2. Use two obviously different colours and sizes of chromosomes.

4 A zygote goes through many rounds of mitosis to become a multicellular organism. At some time in life, meiosis will occur in the gonads. Gonad cells are diploid; they have two homologous copies of each chromosome, like every other somatic cell. Just before beginning meiosis, a gonad cell duplicates its DNA. Briefly, the cell contains four DNA copies distributed between a pair of homologous chromosomes. The parts of the chromosome are called sister chromatids and they are attached by a centromere. Two sister chromatids are still called one chromosome because they have alleles inherited from only one parent.

Draw the chromosome pairs as described after duplication, before meiosis. Label the features.

5 **a** Complete the text by filling in the missing words.

The model cell pictured is about to begin meiosis. There are ________________ chromosomes. There are ________________ homologous pairs. Few organisms have such a ________________ number of chromosomes, because models always ________________ a real situation to make it ________________ to understand. The ________________ membrane is intact, and the ________________ are not currently ________________ in any way.

b Use the following descriptions to complete the model of meiosis. Continue to use two obviously different colours and sizes of chromosomes.

Meiosis I (aka 'reduction division')

i Nuclear membrane breaks down. Spindle fibres attach to chromosomes. Members of homologous pairs stay together and align end-to-end across the equator.

ii Chromatids of homologous chromosomes are touching because the pairs are actually on top of each other. Each place where chromatids touch is called a chiasma.

iii There may be multiple chiasmata. At these points some of the parts of the chromatids swap places. This is called crossing over. Draw in the homologous chromosomes after crossing over has occurred.

iv The spindle fibres contract and pull the homologous pairs away from each other and toward the poles. The sister chromatids remain attached at the centromere.

v The nuclear membranes reform around the separated chromosomes. The cell membrane begins to pinch inward, and two cells begin to form.

vi Cytokinesis occurs. Two cells are formed.

Meiosis II

Continue to draw in the features of each cell as meiosis continues.

HINT

Meiosis II is essentially the same as mitosis, except the cells are already haploid instead of diploid. Cells enter meiosis II with haploid cells that contain two identical copies (sister chromatids) of each chromosome.

vii The nuclear membranes break down and the spindle fibres attach to the chromosomes. Chromosomes line up across the centre of the cell.

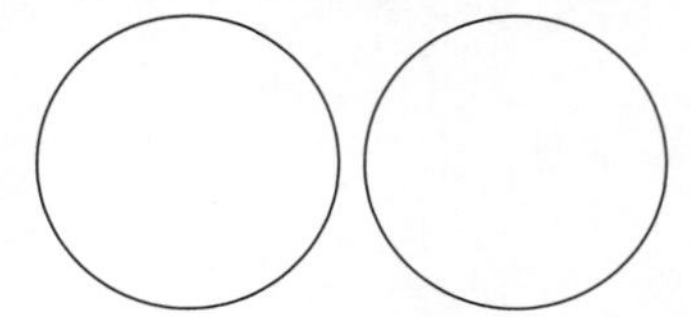

9780170449625

viii Spindle fibres contract, chromatids split at the centromere are pulled toward the poles.

ix The nuclear membranes reform around the separated chromosomes; the cell membranes begins to pinch inward; four cells begin to form.

x Cytokinesis occurs, four cells are formed.

c The four gametes are all different. Describe how crossing over contributed to the diversity of these four cells.

d In the first steps of Meiosis I (part **bi** on page 30), members of homologous pairs stay together and align end-to-end across the equator. Assume you made an arbitrary decision to draw one member of each pair above the other. In the first cell below, copy your drawing from part **bi**. In the other three cells draw the alternative ways these chromosomes could have correctly been drawn.

or or or

e Describe some other points in the model where arbitrary decisions affected the gametes produced in your model.

f Explain how completing the model helps you understand random segregation (aka independent assortment) in meiosis.

WS 2.3 DNA structure and replication

LEARNING GOAL

Work with the Watson and Crick model of DNA structure and replication.

DNA structure

Deoxyribonucleic acid (DNA) is a large polymer made of repeating monomers. The monomers are called nucleotides. There are four nucleotides that are part of the structure of DNA. Each nucleotide consists of a phosphate molecule joined to a sugar molecule called deoxyribose, and a nitrogenous base molecule attached to the sugar.

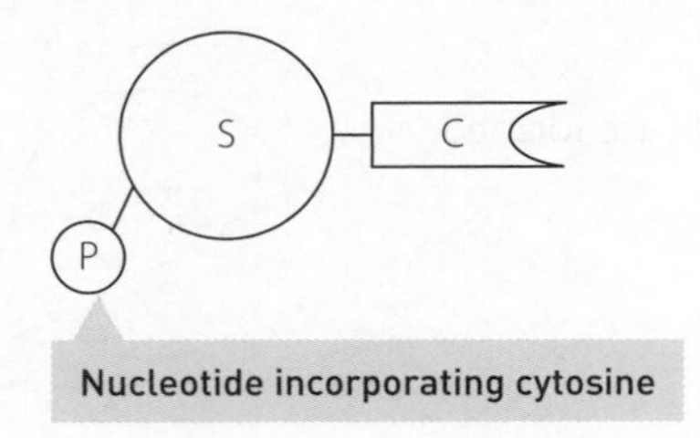

Nucleotide incorporating cytosine

1 Here is a key for drawing DNA bases.

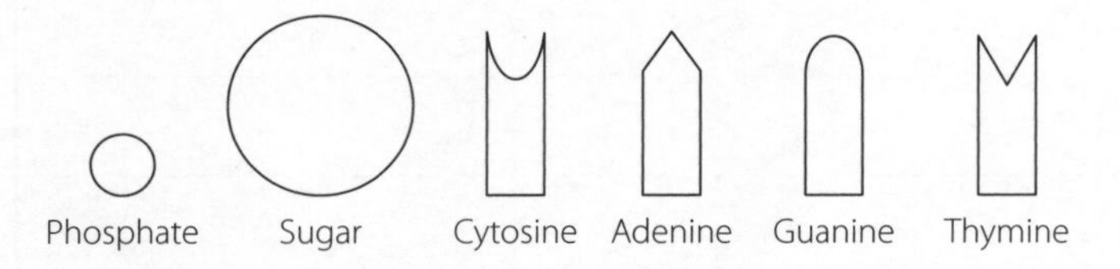

a Using the key, draw three other nucleotides.

b Apply colours or patterns to the key above and then use your key to colour the DNA molecule below.

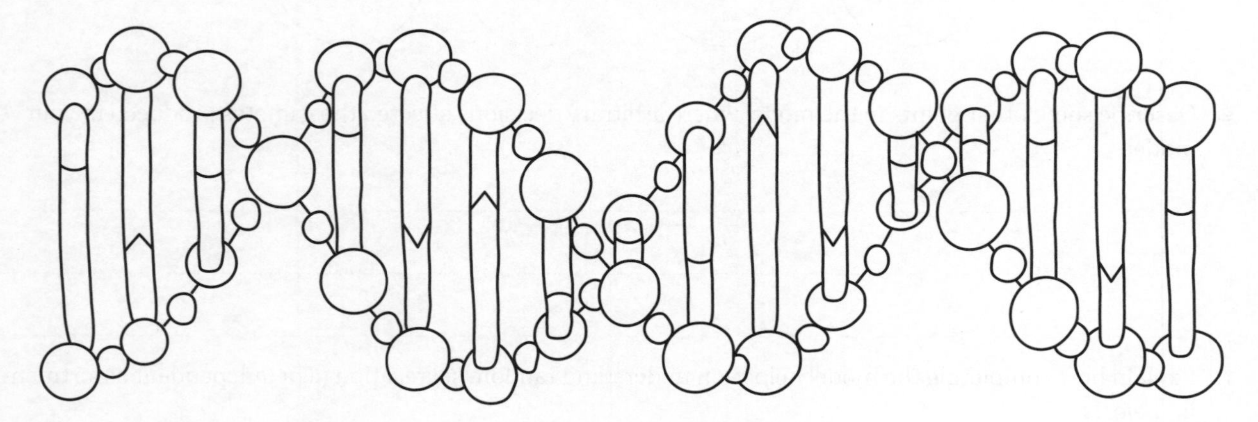

9780170449625

c Nucleotides easily join into a chain when a chemical bond forms between a phosphate and another sugar. A DNA molecule consists of two strands of nucleotides, with the nitrogenous bases facing each other and the phosphates and sugars on the outside. The bases form weak hydrogen bonds with each other. These weak bonds hold the two strands together but are easily broken. Adenine is attracted to thymine and guanine is attracted to cytosine. The bonds between phosphate and sugar are called covalent bonds and they are strong compared to hydrogen bonds. Strong chemical bonds are usually represented by solid lines whereas weak hydrogen bonds are usually represented by dotted lines.

Construct a chain of nucleotides down the page by drawing five nucleotides. Arrange the nucleotides with adenine (A) first, then two guanines (G), followed by cytosine (C) and then thymine (T). Draw in extra bonds between phosphates and sugars where necessary.

d Add the complementary strand of nucleotides to your diagram in part **c**. Leave a little space between the bases and draw in the hydrogen bonds between them. Label one hydrogen bond and one covalent bond.

e T forms two hydrogen bonds with A, and C forms three hydrogen bonds with G. T cannot form hydrogen bonds with any other base except A. The same is true for C and G. The nitrogenous bases presented in the key above indicate a jigsaw-like fit between the pairs of nitrogenous bases. Suggest why these nitrogenous bases are often represented like this when the chemicals are not actually shaped like this.

2 The double strands of DNA are twisted into a spiral, called a double helix. The twisting of DNA is the result of hydrophilic and hydrophobic interactions between the molecules that comprise DNA and the water in the cell. The nitrogenous bases are in the inner portion of the molecule and the twist reduces the contact between the hydrophobic nitrogenous bases and the water in the nucleus.

Explain how hydrophobia and hydrophilia contribute to the helical shape of the DNA molecule.

> **HINT**
>
> The verb *explain* requires you to consider cause and effect.

____

____

____

____

____

____

DNA replication

The double-stranded nature of the DNA molecule makes exact copying possible. Each strand acts as a template for new bases to attach according to the base-pairing rules: A with T, and G with C. Replication of DNA is a complex process and each step is controlled by one or more enzymes.

3 The following statements describe DNA replication but are presented in the wrong order. Re-write the statements into the table, matching them with the diagrams. One diagram needs to be supplemented by you.

- All pieces of DNA in the cell are in G_1 phase of the cell cycle and the cell is being prepared for DNA replication.
- The enzyme DNA polymerase helps nucleotides join on to the template strands of DNA according to the base pairing rules. The enzyme DNA ligase helps glue the phosphates and sugars together.
- S phase begins and the enzyme helicase unwinds the DNA helix.
- Two double-stranded DNA lengths are produced, each with one strand from the parent molecule and one new strand. DNA polymerase backtracks to 'proofread' and 'edit' the strand, correcting any base pair errors.
- There are many nucleotides floating free in the nucleus.
- The two double strands are wound back into helices. The DNA is checked again for errors before the cell enters mitosis.
- Helicase also unzips the DNA, separating the two strands. Other enzymes bind to the newly opened DNA and prevent it from zipping back together again.

	Diagram	Description of processes
a	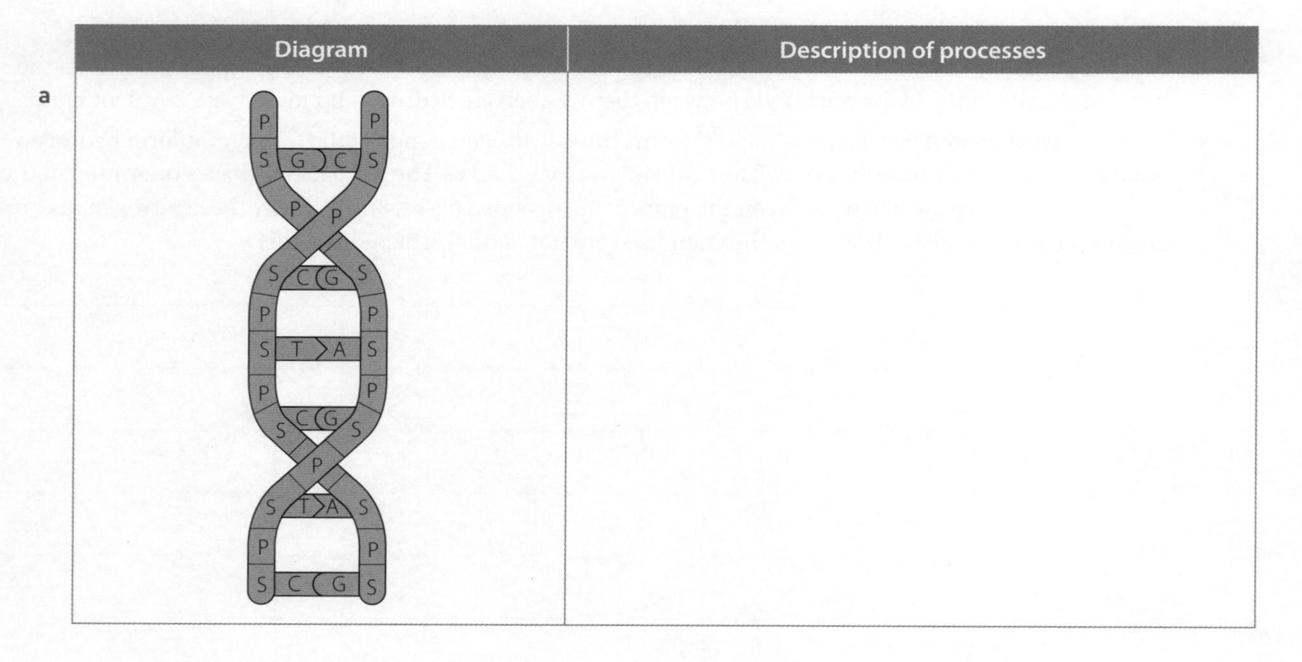	

	Diagram	Description of processes
b	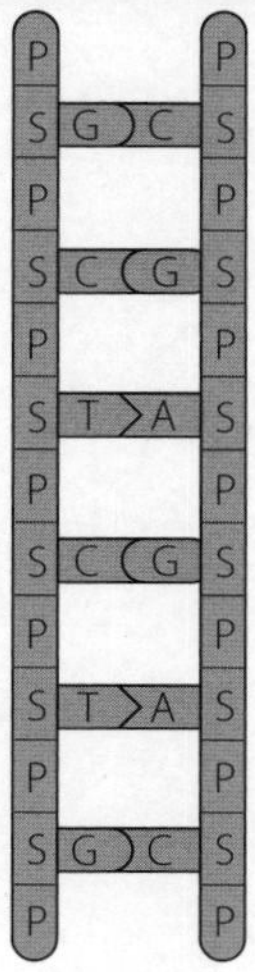 P P S G C S P P S C G S P P S T A S P P S C G S P P S T A S P P S G C S P P	
c	 P P S G C S P P S C G S P P S T A S P P S C G S P P S T A S P P S G C S P P	
d	 G S P A S P S T P C S P S C P S G P Draw these nucleotides onto the bottom part of the diagram in part **e**.	
e	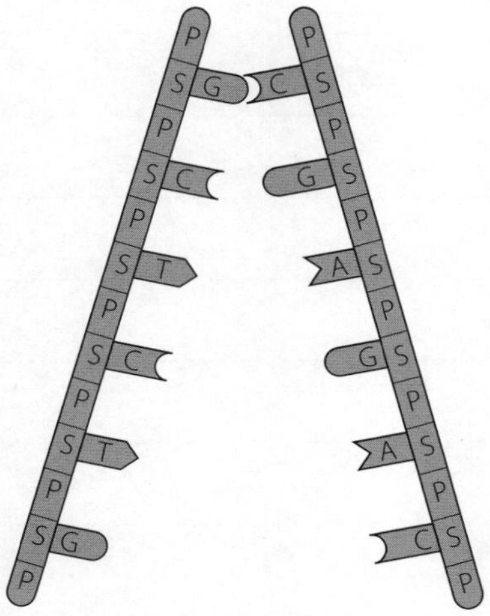 P P S G C S P P S C G S P P S T A S P P S C G S P P S T A S P P S G C S P P	

	Diagram	Description of processes
f	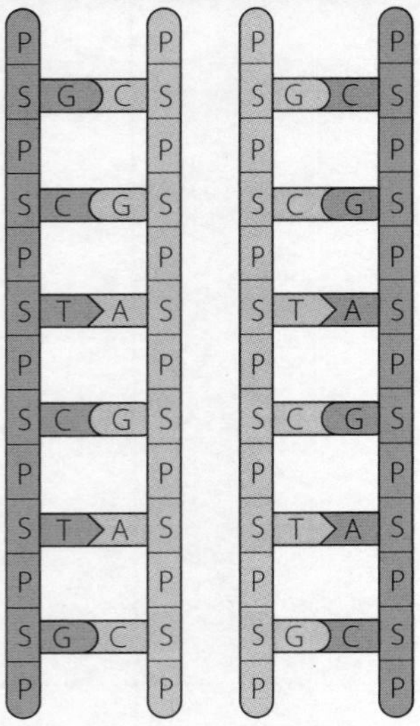	
g	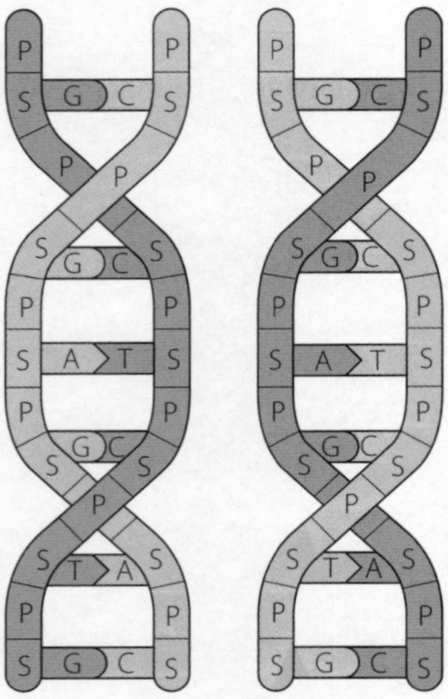	

4 Name three enzymes that are important in DNA replication.

5 Why is DNA replication important?

Cell division: the bigger picture

LEARNING GOAL

Assess the effect of the cell replication processes on the continuity of species.

Students were asked to brainstorm a question about cells and DNA. Here are their ideas:

So … in the BIGGER PICTURE … is exact replication of cells and DNA IMPORTANT?

YES	NO
… in MITOSIS, identical cells are made so an individual can grow.	… MEIOSIS never produces identical gametes because of CROSSING OVER and INDEPENDENT ASSORTMENT → variation in a population, even variation in a family.
… cancer cells are cells that don't divide normally → tumours.	… MEIOSIS can't replicate identical cells because the point is to make cells with ½ the normal number of chromosomes – so FERTILISATION can happen.
… mistakes during DNA replication (MUTATIONS) can lead to faulty genes; e.g. inability to make insulin (diabetes).	… mistakes (MUTATIONS) in DNA replication, mitosis and meiosis leads to new genes, new variations … maybe increased survival of the species.
… mistakes in MEIOSIS can lead to egg or sperm with extra or missing chromosomes; e.g. Down Syndrome, Turner's Syndrome.	

1 Use the students' ideas and other information in this chapter to assess the effect of the cell replication processes on the continuity of species.

HINT

The verb *assess* requires you to include points for, points against and a judgement.

2 Assess one of your classmates' responses to question **1** using the criteria in the table below.

Criteria	Thoroughly	Adequately	Limited	Missing
Response is structured to address the verb 'assess'				
Considers mitosis, meiosis and DNA replication				
Provides points for and points against exact replication				
Provides points for and points against non-exact replication				
Makes a clear link between cell replication processes and continuity of a species				
Provides a judgement				

3 DNA and polypeptide synthesis

INQUIRY QUESTION: WHY IS POLYPEPTIDE SYNTHESIS IMPORTANT?

WS 3.1 DNA across the kingdoms

LEARNING GOAL

Compare DNA in eukaryotes and prokaryotes.

DNA has the same chemical composition in all living things, and the DNA in both prokaryotic and eukaryotic cells must be replicated before cell division begins. The same enzymes uncoil, unzip and assist in attaching the nucleotides in prokaryotes and eukaryotes when DNA replicates.

However, prokaryotic and eukaryotic cells differ in some fundamental ways. Prokaryotic cells divide by binary fission whereas eukaryotic cells divide by mitosis. Prokaryotic DNA replication is relatively faster than that of eukaryotes. Prokaryotic cells usually have a single version of each gene, whereas it is normal for most eukaryotes to have two versions, or alleles, of each gene.

1 a Explain why most eukaryotic cells have two alleles of each gene, but prokaryotes have only one copy of each gene.

HINT

The verb *explain* requires you to consider cause and effect.

____

____

____

b What is the significance of the fact that DNA structure, molecular building blocks such as amino acids and some molecular processes are the same across all kingdoms?

____

____

____

2 Analyse the images, then complete the text using words from the word bank.

a

Science Photo Library/Dennis Kunkel Microscopy

b

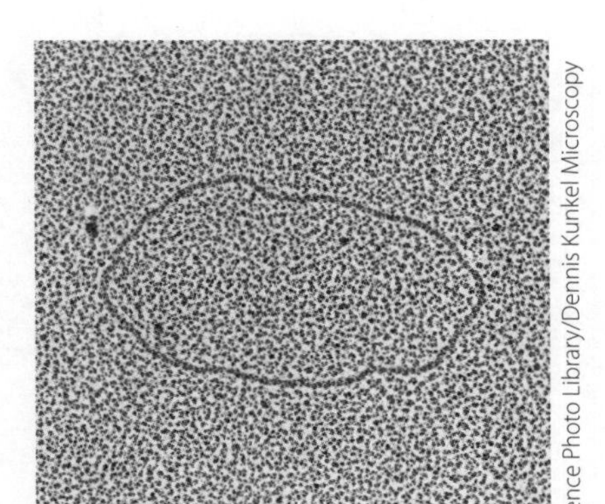

Science Photo Library/Dennis Kunkel Microscopy

c

Replication origin

Replication begins

Replication completed

2 circular daughter DNA molecules

Electron micrograph of a prokaryotic chromosome in a a relaxed state and b a supercoiled state. c The process of chromosome replication

linear	supercoiled	plasmids	proteins
nucleus	mitochondria	single	non-chromosomal
base	cytoplasm	conjugation	advantage

The DNA of prokaryotes and eukaryotes differs in its packaging and quantity. Eukaryotic DNA is located inside the ____________ and each chromosome consists of one piece of ____________ DNA supercoiled around histone ____________. Eukaryotes typically have many chromosomes with many ____________ pairs. Prokaryotic cells have two forms of DNA floating in the ____________: a ____________ chromosome and one or more small circular pieces of DNA. The chromosomal DNA is a loop in its relaxed state but is ____________ around dense proteins most of the time. Prokaryotic cells can have one or more small rings of ____________ DNA, called plasmids, floating separately in the cytoplasm. The genes on ____________ code for features that are not essential to the survival of the cell, but can confer a selective ____________, such as resistance to antibiotics. Plasmids replicate independently of the chromosome and can be swapped between prokaryotic cells in a process called ____________. Plasmids are not present in eukaryotes but non-chromosomal DNA is found in ____________ and chloroplasts.

3 Complete the table to summarise the crucial similarities and differences between the genetic material in prokaryotic and eukaryotic cells.

Features	Prokaryotic cells	Eukaryotic cells
DNA structure		
Shape of DNA pieces		
Supercoiling		
Sketch of chromosome		
Location of chromosomes		
Number of chromosomes		
Non-chromosomal DNA		
Role of DNA		

4 If DNA is the instruction book for making more DNA, what are the ingredients?

____

WS 3.2 Proteins are quite complicated

STUDENT BOOK
Pages 137–42

LEARNING GOAL

Investigate the structure and function of proteins in living things.

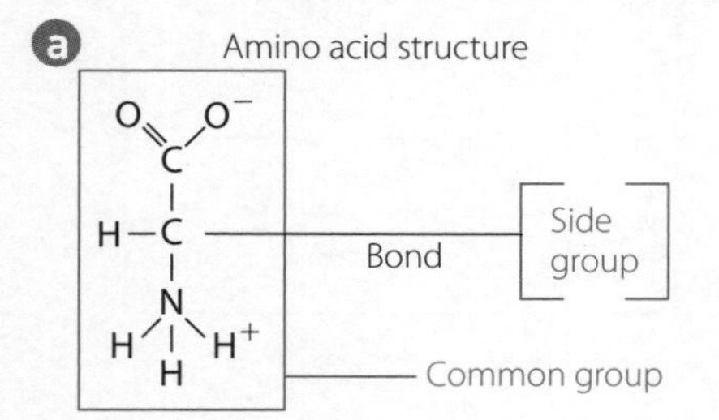

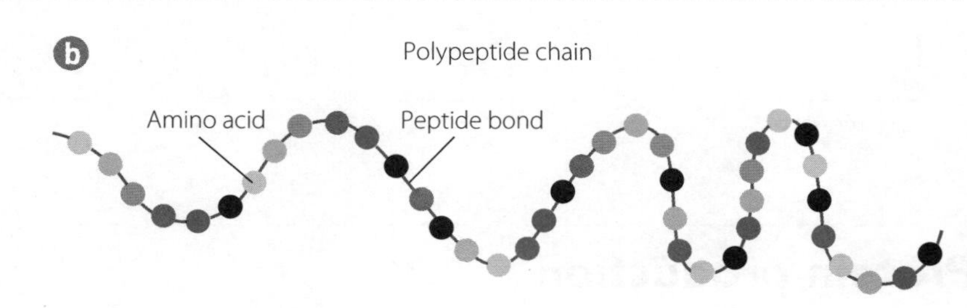

Polypeptide chains

The side group on an amino acid is what makes each amino acid different from the others. No matter which kingdom it belongs to, every organism can utilise the same 20 amino acids.

Amino acids join to make chains called polypeptide chains. An amino acid can be hydrophilic, hydrophobic or neither. This property causes some of the molecules to adjust themselves in a certain direction so, when they make a polypeptide chain, the chain will begin to twist and turn as it is being formed. This gives proteins a three-dimensional structure.

Proteins are crucial components of the structure and function of cell membranes. At the tissue level, they have a great variety of uses in multicellular organisms.

1 a Name the elements found in all amino acids.

____

b Distinguish between hydrophobic and hydrophilic molecules.

____

____

c Suggest the origin of the term *polypeptide*. Justify your answer.

____

____

2 Use secondary sources to investigate the following proteins found in mammals, then complete the table. Use words from the word bank for the examples.

Ferritin	Keratin	Amylase	Myosin
Insulin	Haemoglobin	Immunoglobin G	

Type of protein	Example	What this protein does
Antibody		
Enzyme		
Structural		
Functional		

→

Type of protein	Example	What this protein does
Storage		
Contractile		
Hormonal		

Protein production

One or more polypeptide chains are processed and folded to produce the final protein. For example, insulin begins as a single polypeptide chain of 86 amino acids called proinsulin, produced in the ribosomes in the cells of the pancreas. Processing involves the chain coiling and strong chemical bonds forming from one cysteine amino acid to another. Enzymes cut the chain in two places, making three sections – A, B and C. This cutting is called cleavage. The C section is discarded. The A chain (21 amino acids) and B chain (30 amino acids) undergo further processing and folding. Another cysteine–cysteine bond forms and the protein takes on its final three-dimensional shape. The correct shape makes it biologically active and the protein is then called insulin.

HINT

Cysteine is an amino acid and cytosine is a nitrogenous base found in DNA. Do not confuse them!

3 Use the text to label the features and annotate the stages in the processing of the polypeptide that becomes the protein insulin.

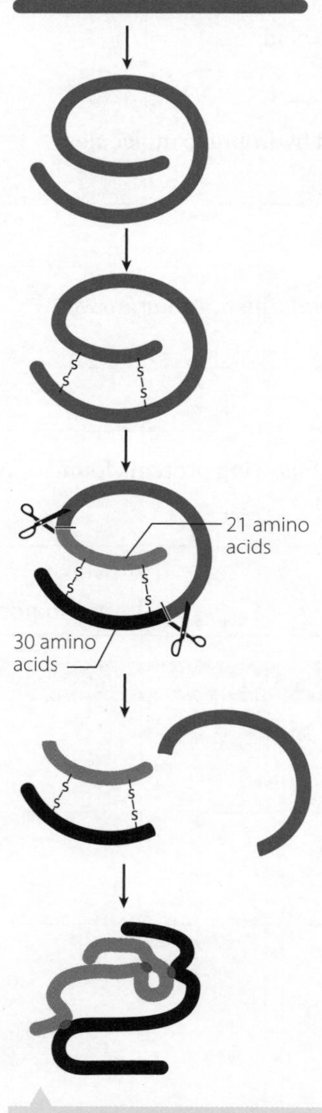

Polypeptide to protein – insulin

Protein shape

The shape of a protein is critical to its function because it determines whether the protein can interact with other molecules. One wrong amino acid in a protein can change the internal interactions between the protein's amino acids, which in turn may alter the shape of the protein. If the protein is subject to changes in temperature or pH, it can denature (lose its shape), even though the amino acid sequence does not change under these conditions.

4 Recall how denaturation of an enzyme changes the interaction between enzyme and substrate. Include detail about protein structure to explain why denaturation affects an enzyme's function. You may include diagrams in your answer.

WS 3.3

The universal code: the details of protein synthesis

STUDENT BOOK
Pages 120–8

LEARNING GOAL

Analyse the details of transcription and translation.

The role of genes across all kingdoms is to provide the instructions for joining amino acids to make a polypeptide. Polypeptide synthesis involves DNA and another nucleic acid, RNA, and the ribosomes in the cytoplasm. Further processing of polypeptides makes functional proteins.

1 a What are the ingredients for polypeptides?

____

b Which molecules and cellular structures are involved in polypeptide synthesis?

____

____

c Which organelles in eukaryotic cells are likely to transport polypeptides, and process and package them into proteins?

____

2 Use secondary sources to compare the features of DNA and RNA.

Features	DNA	mRNA	tRNA
Shape	Double helix		
Strands	Double		
Sugar	Deoxyribose		
Bases	A, T, C, G		
Location	Stays in nucleus		

Transcription

Transcription begins when RNA polymerase unzips the DNA in the region of a gene. This exposes the bases. The aim is to copy the order of the bases from the DNA. RNA nucleotides with complementary bases join together based on the order of the exposed bases of the template strand.

3 a Complete the following passage with the words from the word bank.

copy	messenger	transcription	cytoplasm
ribosomes	big	nuclear	

The factories for making proteins are the ________________ in the cytoplasm. DNA molecules are too ________________ to go through the pores in the ________________ membrane to the ribosomes with the instructions for making protein, so a ________________ of the gene's instructions needs to be made and sent to the ribosome via a ________________. Making the copy is called ________________, and the messenger is called mRNA. mRNA then leaves the nucleus and goes to a ribosome in the ________________.

b Use the text in part **a** to label the features and annotate the diagram below about transcription.

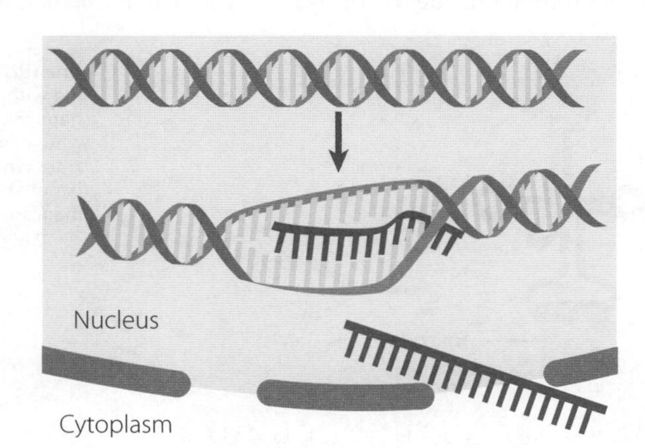

Translation

A codon is a group of three consecutive bases on a nucleic acid molecule. In 1963 the genetic code was deciphered for a bacterium, *Escherichia coli* (*E. coli*). Scientists worked out that most of the 64 codons were linked to the 20 amino acids used to build proteins. Shortly after, they realised the same code was shared by all living things. The table below lists which mRNA codons cause each amino acid to join the polypeptide chain during translation.

First base	Second base U		Second base C		Second base A		Second base G		Third base
U	UUU	Phe	UCU	Ser	UAU	Tyr	UGU	Cys	U
	UUC		UCC		UAC		UGC		C
	UUA	Leu	UCA		UAA	Stop	UGA	Stop	A
	UUG		UCG		UAG	Stop	UGG	Trp	G
C	CUU	Leu	CCU	Pro	CAU	His	CGU	Arg	U
	CUC		CCC		CAC		CGC		C
	CUA		CCA		CAA	Gln	CGA		A
	CUG		CCG		CAG		CGG		G
A	AUU	Ile	ACU	Thr	AAU	Asn	AGU	Ser	U
	AUC		ACC		AAC		AGC		C
	AUA		ACA		AAA	Lys	AGA	Arg	A
	AUG	Met/Start	ACG		AAG		AGG		G
G	GUU	Val	GCU	Ala	GAU	Asp	GGU	Gly	U
	GUC		GCC		GAC		GGC		C
	GUA		GCA		GAA	Glu	GGA		A
	GUG		GCG		GAG		GGG		G

Amino acid key

Ala	= alanine
Arg	= arginine
Asn	= asparagine
Asp	= aspartic acid
Cys	= cysteine
Gln	= glutamine
Glu	= glutamic acid
Gly	= glycine
His	= histidine
Ile	= isoleucine
Leu	= leucine
Lys	= lysine
Met	= methionine
Phe	= phenylalanine
Pro	= proline
Ser	= serine
Thr	= threonine
Trp	= tryptophan
Tyr	= tyrosine
Val	= valine

4 **a** Four different mRNA codons code for proline. Write out the codons.

b Identify the amino acid that matches each of these mRNA codons: GCU, AAU, CGC, GAU

c Which glutamic acid and valine codons are most similar? Justify your answer.

5 Using the table gives the impression that mRNA is all that is needed to make a polypeptide. The process of putting the amino acids in a chain in the correct order is complex and involves transfer RNA (tRNA). Codons on mRNA and anticodons on tRNA interact inside the ribosome as shown in the diagram below.

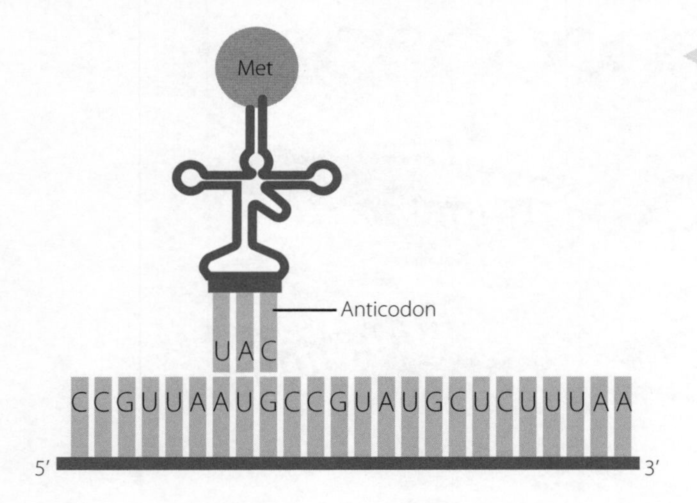

Using an example, distinguish between a codon and an anticodon.

6 Messenger RNA is made of nucleotides, as are DNA and genes. But proteins are made of amino acids; they are like a different language. A translator is needed that 'knows about' nucleotides *and* amino acids. Transfer RNA (tRNA) is involved in this translation. A tRNA molecule has three exposed nucleotides and one amino acid attached. The mRNA travels to a ribosome and the message is read, then tRNA molecules arrive and translate the nucleotide language into an amino acid language. Amino acids are joined to make a polypeptide chain.

Use the text to create an annotated diagram about translation.

> **HINT**
>
> Try starting with something similar to the diagram in Question 3b.

7 Insulin is a protein necessary for the control of glucose levels in the blood. In most animals, including humans, a single gene for insulin is found. The human *INS* gene is located on chromosome 11. The DNA strand below displays just 30 bases of the INS gene.

Coding strand

CAC CTG TGC GGC TCA CAC CTG GTG GAA GCT

Template strand

a Fill in the bases on the template strand of the insulin gene above.

b Write out the codons of the mRNA strand that would be made from this piece of DNA.

____

c Use the mRNA conversion table to work out the polypeptide chain made from the DNA in part **b**.

____

d Use amino acid abbreviations and anticodons to label the tRNA molecules required for the first amino acids in the polypeptide in part **c**.

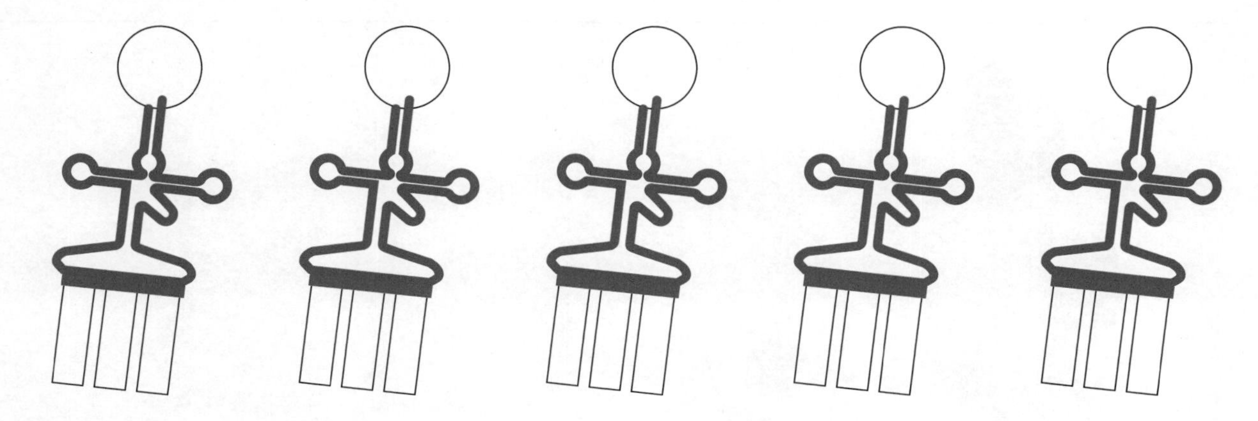

e You have been working with only a small portion of the INS gene and therefore producing only a portion of the final protein. Locate the section of the insulin protein translated from the DNA sequence given.

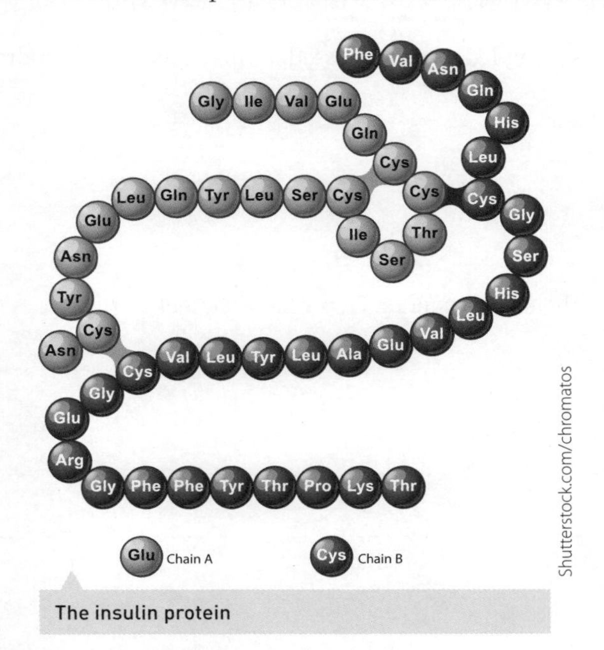

The insulin protein

f Calculate the percentage of the protein you have translated.

____

WS 3.4 DNA is not the whole story

STUDENT BOOK
Pages 128–36

LEARNING GOAL

Assess how genes and environment affect phenotypic expression.

Genes are expressed when the information encoded in the DNA gives instructions for the growth and functioning of the individual. The expression of genes contributes to the individual's observable traits, called the phenotype. Phenotypic variations can be a result of the effect of the environment on the expression and function of genes influencing the trait. Environmental factors can determine which genes are turned on or off. Several scenarios will be described to illustrate ways that environmental factors can affect the observed phenotype, without a change in genotype.

1 Consider the scenario below.

> The *C* gene is responsible for making black pigment in the fur of Himalayan rabbits. The expression of this is controlled by temperature. The fur pigment is made consistently between 15°C and 25°C and the gene is never expressed above 35°C. The rabbit's body heat will keep the *C* gene partly inactive even if it lives in a cool climate. In the central parts of the rabbit's body, where it is warmest, no pigments are produced and fur colour will be white. In the rabbit's extremities – the ears, tip of the nose, feet and tail – where the temperature is much lower, the gene is turned on and actively produces pigment. If the rabbit lives in a climate above 30°C, it will be white all over, and the *C* gene is never expressed.

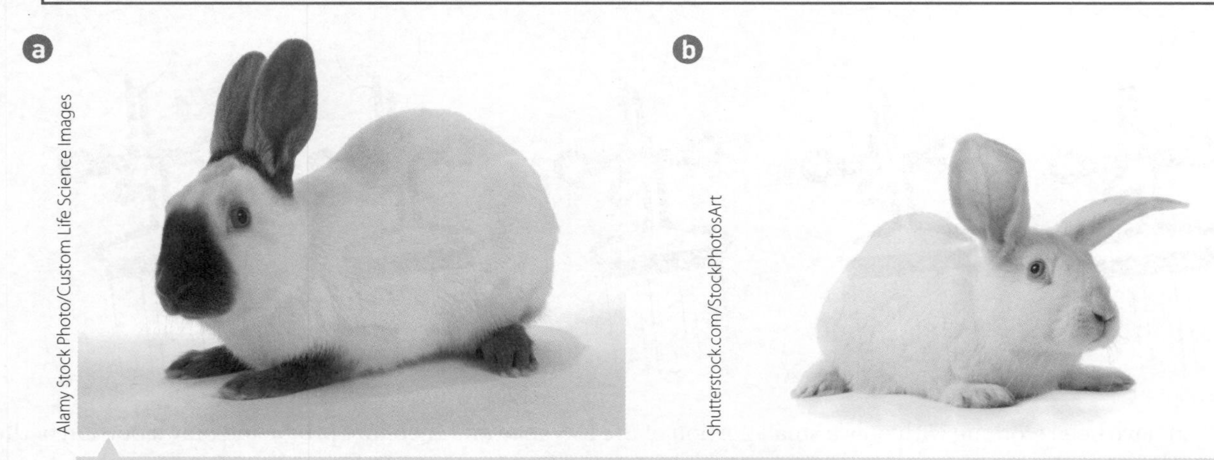

a Rabbit reared at 20°C or less; **b** Rabbit reared at temperatures above 30°C

Consider how the *C* gene is intended to be expressed in this scenario and address the following questions.

a Expected phenotype:

____

b Environmental factor:

____

c Resulting phenotype if the environmental factor has an effect:

____

d Explanation:

____

____

____

9780170449625

e A breeder wanted to see whether she could get her Himalayan rabbits to have black fur all over. She was thinking about keeping the rabbits in a refrigerated room at 15°C. Do you think this experiment will be successful? Propose reasons to support your judgement.

____

____

____

____

____

____

2 The sex of birds, mammals, most snakes and lizards is determined by sex chromosomes, and their genotype determines phenotype. Consider the scenario below.

> The Australian bearded dragon lizard has male and female sex chromosomes but also has a temperature-induced sex reversal at the embryo stage. The male and female phenotypes are easily distinguished, and the genotype can be tested in the lab. Laboratory studies showed that high incubation temperature during egg development reverses genotypic male embryos into phenotypic females, yet they retain the male genotype. Sex reversed females in the wild probably come from warm nests. The hypothesis supported by the current data is that there are two genes that are turned on under heat stress that override the sex chromosomes and trigger sex reversal.

Consider how the sex chromosomes (genotypic sex) are intended to be expressed and answer the following questions.

a Expected phenotype:

____

b Environmental factor:

____

c Resulting phenotype if the environmental factor has an effect:

____

d Explanation:

____

____

____

3 Analyse the graphs of data collected from laboratory experiments where bearded dragon eggs were incubated at increasing temperatures.

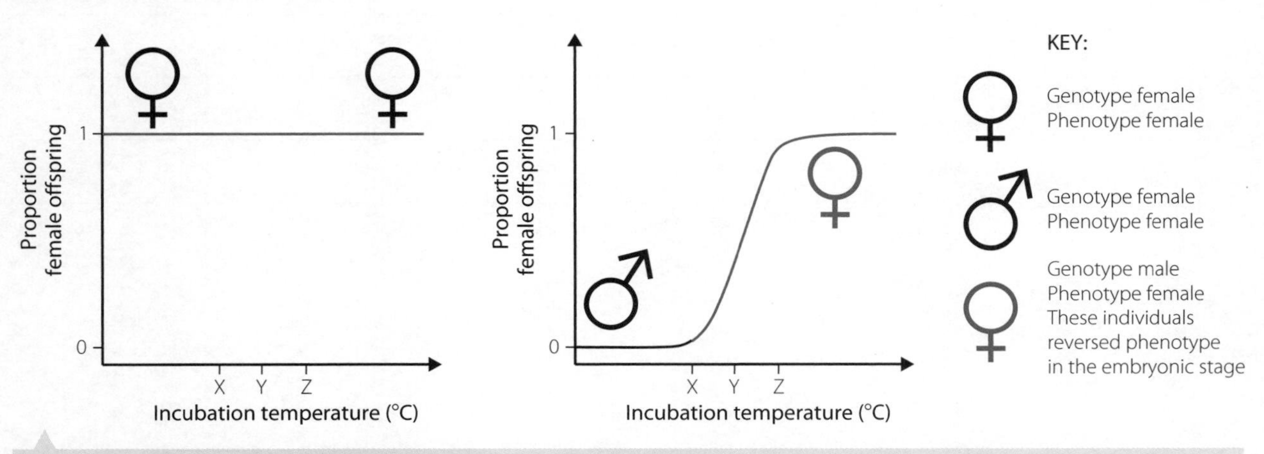

Effect of incubation temperature of embryos and proportion of female bearded dragons in the population.

a What does the graph on the left say about genotypic females?

b What does the graph on the right say about males?

c What does the position of the graph at temperature Y say about the population of males?

d For embryos incubated at temperature Z, make a statement about the makeup of the population.

4 Consider the scenario below.

> Phenylketonuria (commonly known as PKU) is an inherited disorder that increases the levels of the amino acid phenylalanine (phe) in the blood. Infants appear normal until they are a few months old. As phe levels build up, these children develop permanent intellectual disability. If babies are tested for the PKU gene at birth with a simple blood test, they can begin treatment immediately. Brain damage is largely prevented when children with the PKU gene are placed early and continuously on a low-phenylalanine diet. The PKU phenotype associated with having the gene is not expressed.

Use the genetic disorder phenylketonuria as an example to demonstrate that environment can affect phenotype.

4 Genetic variation

INQUIRY QUESTION: HOW CAN THE GENETIC SIMILARITIES AND DIFFERENCES WITHIN AND BETWEEN SPECIES BE COMPARED?

WS 4.1 Meiosis and variation

STUDENT BOOK
Pages 151–5

LEARNING GOAL

Explain how meiosis results in increased genetic diversity of a species.

The key to survival for any species is genetic diversity. If a species has sufficient genetic diversity, it will be more resilient to changes in the environment and, therefore, the more likely it is to survive and reproduce.

1 Define genetic diversity.

2 Meiosis is one of the key drivers in genetic diversity and variations within populations of multicellular organisms. The image at the right shows a cell towards the end of meiosis I.

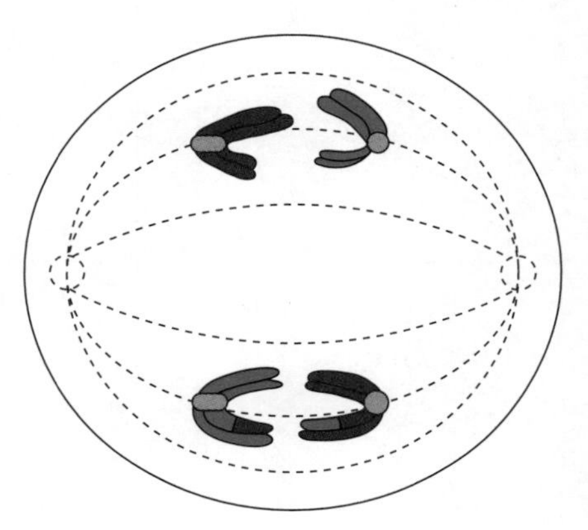

a In the circles below, draw the haploid daughter cells produced at the end of meiosis II.

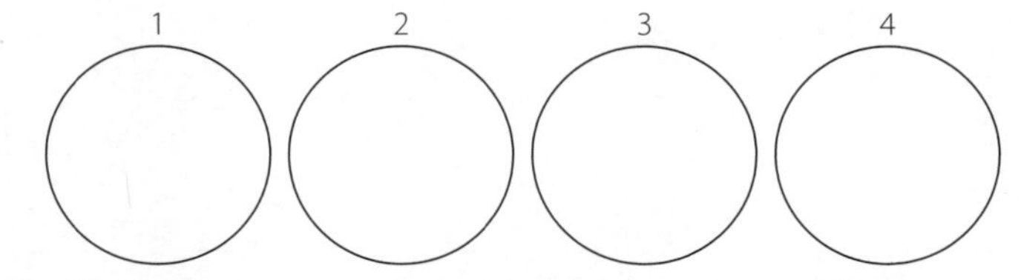

b Which of the gametes formed is likely to be the most genetically diverse?

c Justify your answer to part **b**.

9780170449625

WS 4.2 Inheritance patterns

STUDENT BOOK
Pages 160–2

LEARNING GOAL

Predict the outcomes of monohybrid genetic crosses using Punnett squares.

Organisms that reproduce by sexual reproduction have body cells that contain two homologous pairs of each chromosome. This is referred to as diploid, and is represented by the symbol $2n$. Each chromosome has many sections known as genes that determine specific traits in the organism. Pairs of genes are known as alleles. One allele may be dominant over the other, so the trait that it represents will be expressed over the less dominant, or recessive, allele.

1 Complete the table by writing the correct term or definition in each row.

Term	Definition
Gene	
	A variation of a gene on homologous chromosomes
Dominant gene	
	A gene that will only be expressed in the absence of a dominant gene
	The observable expression of a trait resulting from an organism's genotype and its environment
Genotype	
Heterozygous	
	Two of the same genes are inherited from an organism's parents
Codominant	
Incomplete dominance	

2 Gregor Mendel used pea plants to study inheritance. One of the variables in his study was the colour of the pea plant flowers, which were purple or white. When Mendel crossed purple flowers with white flowers, the offspring flowers were purple.

In this diagram, P is used to represent the dominant allele, and p is used to represent the recessive allele. Complete the table with reference to the information above.

Generation

P: PP Purple flowers × pp White flowers

F_1: All purple flowers Pp

F_2: Purple flowers : White flowers 3 : 1

Term	Example
Dominant phenotype	
Recessive phenotype	
Homozygous dominant genotype	
Heterozygous genotype	

HINT

When you are answering a question relating to alleles, genotypes or crosses, you can draw a Punnett square and refer to it in your answer to help you to structure your response and convey your understanding.

3 a The allele for freckles is dominant (F). The Punnett square below shows the crossing of a homozygous recessive parent with a heterozygous parent. Complete the Punnett square and determine the chance that their offspring will have freckles.

Parent 2

Parent 1		F	f
	f		
	f		

With freckles: ______

Without freckles: ______

b The Punnett square below shows the same trait with a homozygous dominant parent crossed with a homozygous recessive parent. Complete the Punnett square and determine the chance that their offspring will have freckles.

Parent 2

Parent 1		F	F
	f		
	f		

With freckles: ______

Without freckles: ______

4 A broccoli farmer noticed that some of his plants produced short stalks. Customers preferred this short-stalked broccoli because there was less waste. He decided to cross-pollinate two of the short-stalked broccoli plants and germinated the seeds that were produced. He found that the next generation of broccoli plants produced around 75% short-stalked plants with 25% of the plants still showing regular stalks.

Using a Punnett square, explain the results of the farmer's cross-pollination and determine whether the allele for short stalks is dominant or recessive.

HINT

The verb *explain* requires you to consider cause and effect.

5 Suggest one disadvantage of sexual reproduction for crops in which farmers identify favourable traits.

WS 4.3 Codominance and incomplete dominance

STUDENT BOOK Pages 174–5

LEARNING GOAL

Compare the processes of codominance and incomplete dominance.

Not all alleles fit binary dominant or recessive categories. In the genes of some organisms, pairs of alleles do not show dominance of one allele over the other. This is known as codominance or incomplete dominance, depending on how the phenotype is expressed.

1 Use a table to compare codominance and incomplete dominance.

HINT

The verb *compare* requires you to look for similarities and differences.

2 Explain the type of inheritance that is shown in the diagram below.

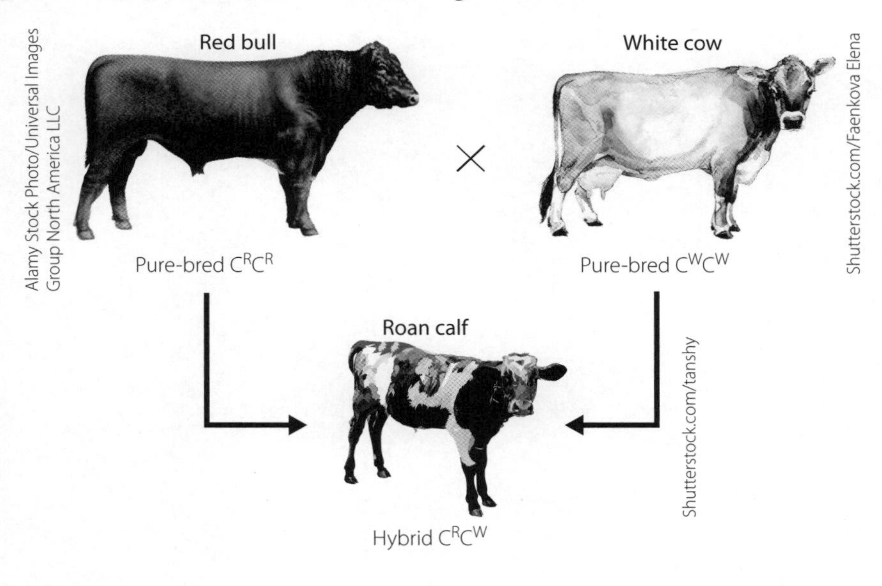

3 Two pure breed tulips were used in a genetic experiment. One tulip had red flowers and the other had white flowers. The two tulips were crossed and all of the first generation (F_1) produced had pink flowers. The F_1 tulips were then crossed with each other. The offspring produced in the next generation (F_2) displayed the following flower colours: 30 red, 65 pink, 33 white.

a Draw a Punnett square to explain the first cross. Use the following letters to represent the alleles: T^R = red, T^W = white.

b The 30 red tulips from the F_2 were crossed with 30 of the pink tulips also from the F_2. The cross produced 120 offspring (F_3). Calculate the phenotype(s) and the likely number of each phenotype expressed in the F_3.

WS 4.4 Pedigrees

STUDENT BOOK
Pages 166–7

LEARNING GOAL

Construct and analyse genetic pedigree charts to track the expression of a genetic trait through multiple generations.

Pedigree charts can be used as a visual representation to easily track genetic traits through multiple generations of the same family. Pedigree charts are commonly used for genetic counselling in order to advise parents on the likelihood of passing on potentially harmful genes to their children.

1 Draw the correct pedigree chart symbols in the table.

Symbol	Meaning
	Male – unaffected
	Male – affected
	Male – carrier
	Female – unaffected
	Female – affected
	Female – carrier

2 Justify this statement: 'If two affected parents produce offspring who lack the trait being studied, then the trait is dominant.' Use Punnett squares in your response.

3 Below is a pedigree chart tracing a genetic disease through a family.

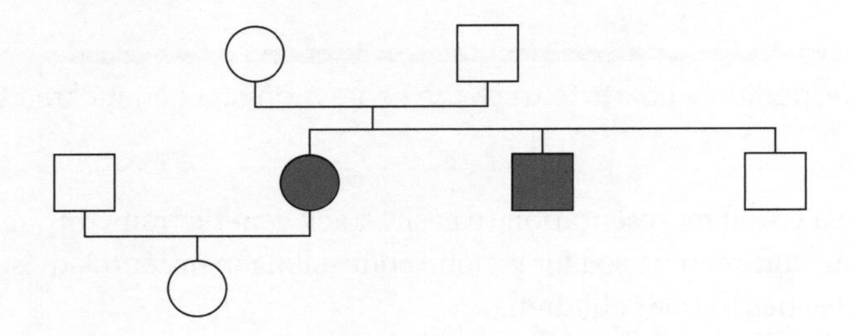

a Annotate the pedigree chart to deduce all possible genotypes of each individual using the letter 'g' to represent the allele.

b Explain how you can determine whether the disease is caused by a recessive or a dominant gene.

4 Hannah is a carrier of sickle cell anaemia. Hannah and Steven have three children: 1 boy (Joe) and 2 girls (Rachel and Jane). Rachel has sickle cell anaemia. Joe marries a woman who does not have sickle cell anaemia. They have one male child who has sickle cell anaemia.

a Draw a pedigree chart to indicate the family tree described above.

b Identify whether sickle cell anaemia is a dominant or a recessive trait.

c Joe and his wife were quite shocked to have a child with sickle cell anaemia. Using a Punnett square, calculate the probability that their next child will have sickle cell anaemia.

9780170449625

WS 4.5 Sex-linked pedigrees

LEARNING GOAL

Predict the outcome of sex-linked genetic trait crosses.

Haemophilia is a rare genetic condition that makes blood clot very slowly. This prevents scabs from forming and can result in people losing a lot of blood or bruising easily. One of the genes that control blood clotting is found on the X chromosome. X^H represents an X chromosome with the dominant allele for normal blood clotting. X^h represents an X chromosome with the recessive allele that causes blood to clot slowly. The Y chromosome is small and does not have the gene for blood clotting.

1 Identify the genotype of a male and a female with haemophilia.

2 Identify the genotype of a female that is a carrier but does not express haemophilia.

3 Explain how a male can have haemophilia when the condition has not been seen in his parents or grandparents.

4 Below is a pedigree chart tracking haemophilia over several generations.

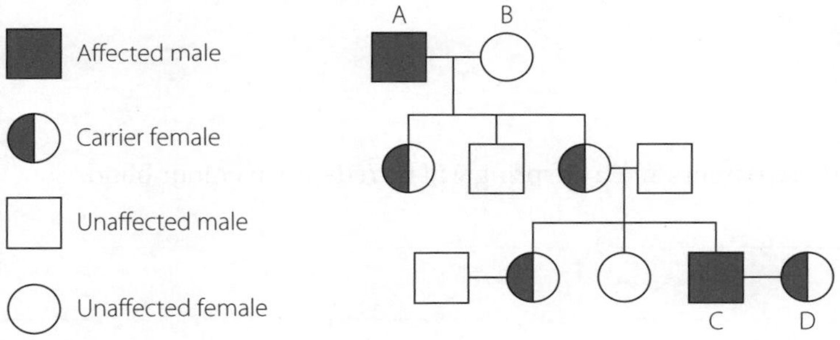

Affected male

Carrier female

Unaffected male

Unaffected female

A B

C D

a Identify the genotype of person A.

b Explain why both the female offspring of person A and person B are carriers of haemophilia.

c Calculate the probability that the offspring of person C and person D will have haemophilia. Use a Punnett square in your answer.

5 Red–green colour blindness is an X-linked recessive trait. Imagine that a woman has red–green colour blindness but her male partner does not.

a Calculate the probability that their female offspring will be red–green colour blind. Use a Punnett square.

b Explain why all of these parents' male offspring will be red–green colour blind.

____

____

____

____

6 Sex-linked traits can also be dominant. Below is an image of a sex-linked dominant trait pedigree chart.

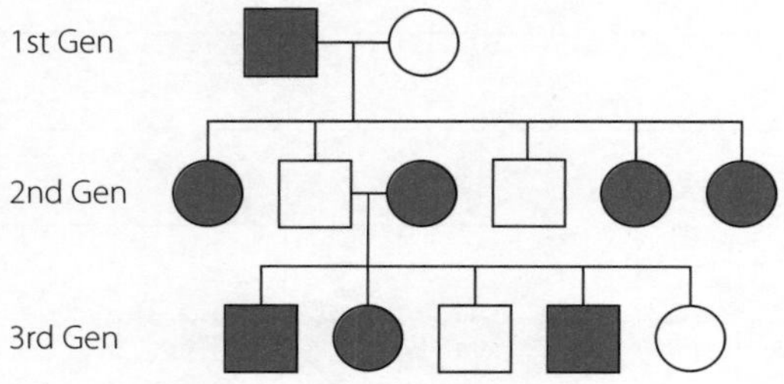

1st Gen

2nd Gen

3rd Gen

a Identify the genotype of the male and female in the first generation using 'D' to indicate dominant genes and 'd' to indicate recessive genes.

b Explain the pattern of inheritance shown in the second generation.

WS 4.6 Multiple alleles

STUDENT BOOK
Page 177

LEARNING GOAL

Predict the phenotype expressed when traits are coded for by multiple alleles.

Some traits are coded for by more than two sets of genes. An example of this in humans is blood type. There are three alleles that are associated with blood type. These lead to different antigens on the surface of red blood cells. The possible alleles are:

1. Allele I^A – leads to the production of antigen A
2. Allele I^B – leads to the production of antigen B
3. Allele I^I – leads to the production of neither antigen A nor B

The table shows the possible genotypes and phenotypes that can arise.

Alleles	Genotype	Molecular markers present	Phenotype (blood group)	Dominance or codominance
I^AI^A	Homozygous	A	A	–
I^BI^B	Homozygous	B	B	–
I^AI^B	Heterozygous	AB	AB	Codominance
I^AI^I	Heterozygous	A	A	Dominance
I^BI^I	Heterozygous	B	B	Dominance
I^II^I	Homozygous	O	O	–

1 Identify the genotype of a person with AB blood type. ______________________

2 Indicate the blood group of people with the following genotypes.

a I^AI^I ______________________

b I^BI^B ______________________

c I^II^I ______________________

3 Within a family, Parent 1 has blood type A and Parent 2 has blood type B. Draw each possibility of their genotype crosses in Punnett squares. Include the likely percentage of phenotypes in their offspring.

9780170449625

4 The table shows the antigens and the antibodies found in the blood of different blood group types.

- Antigens are protein markers that allow cells to be recognised by the body as self or foreign.
- Antibodies detect and bind to corresponding antigens to neutralise or destroy foreign cells.

	Group A	Group B	Group AB	Group O
Red blood cell showing ABO antigens	A	B	AB	O
Antibodies in plasma	Anti-B	Anti-A	None	Anti-A and Anti-B

If a patient is given the wrong blood type, the body will react and attack the foreign blood, resulting in a life-threatening condition.

Using the information in the table, explain which blood group type could safely be given to anybody and therefore is classed as a universal donor.

____

____

____

5 The placenta in placental mammals separates the blood of the mother and the offspring. Explain why this is important for the survival of both the mother and the offspring.

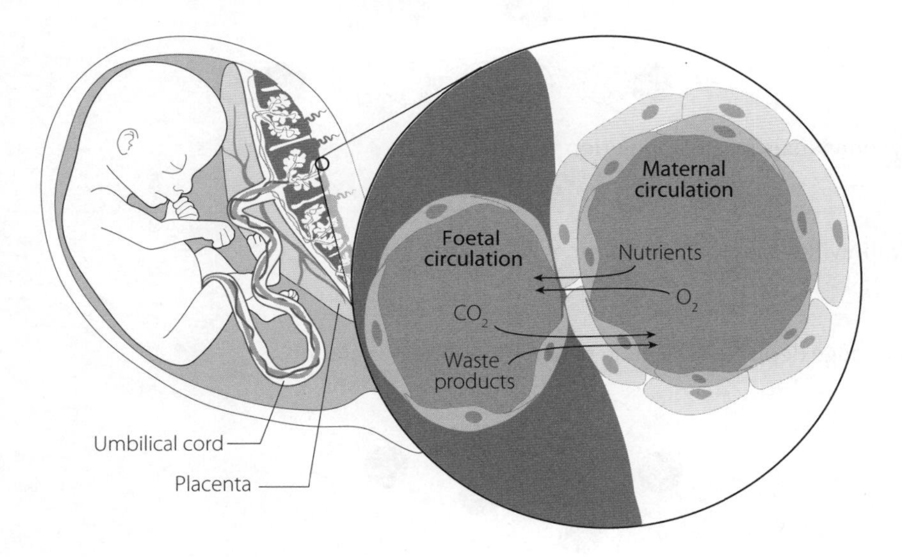

____

____

____

____

____

____

____

____

5 Inheritance patterns in a population

INQUIRY QUESTION: CAN POPULATION GENETIC PATTERNS BE PREDICTED WITH ANY ACCURACY?

STUDENT BOOK
Pages 201, 252

WS 5.1 Hardy–Weinberg calculation

LEARNING GOAL

Use the Hardy–Weinberg equation to predict the frequency of alleles found in a population.

1 Define the term *gene pool*.

2 In the population depicted below, the pink circles indicate a recessive allele for red fur and the unshaded circles represent a dominant allele for black fur. The pairs indicate an individual.

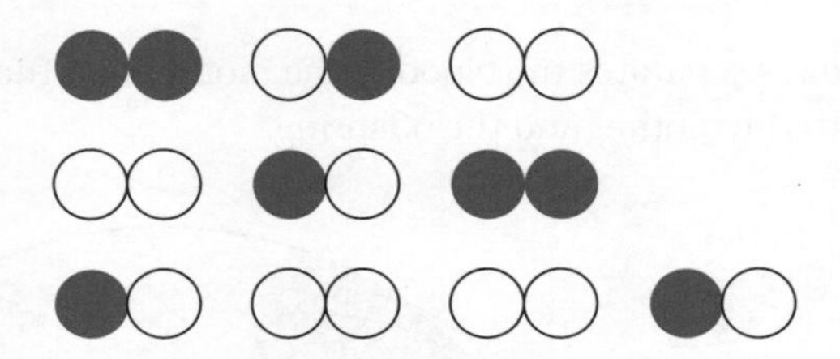

a How many individuals in the population will have red fur?

b How many individuals in the population will have black fur?

c Below is an image depicting the gene pool of the same population.

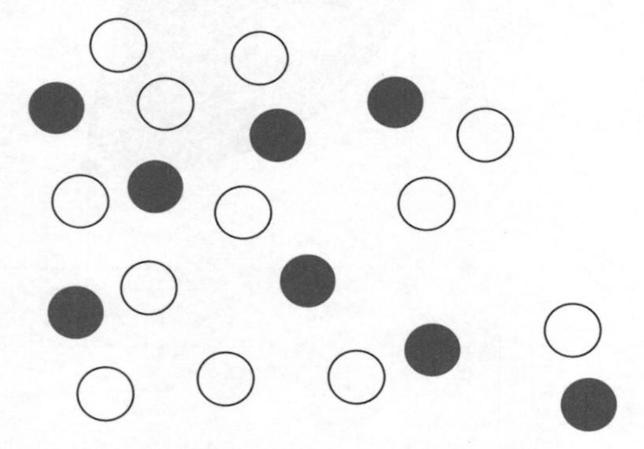

i How many recessive red fur alleles are there?

ii How many dominant black fur alleles are there?

9780170449625

d In order to calculate the Hardy–Weinberg equilibrium, we first need to calculate the allele frequency in the population.

Dominant allele (p): $p = \dfrac{\text{Number of dominant allele}}{\text{Total number of alleles}}$

Recessive allele (q): $q = \dfrac{\text{Number of recessive allele}}{\text{Total number of alleles}}$

Use these equations to calculate the p and q values in this population.

e Now you have your p and q values, you can use the Hardy–Weinberg equation to determine allele frequency in a population.

$p^2 + 2pq + q^2 = 1$

Input the values from part **d** into the equation.

3 Twenty per cent of organisms in a population of 40 individuals have 6 toes. Having 6 toes is recessive to having 5 toes.

a What percentage of the population has 5 toes?

b What are the p and q values for the population?

c What percentage of the population is heterozygous for the 6-toe trait?

DNA profiling and sequencing

STUDENT BOOK
Pages 197–9

LEARNING GOAL

Describe the processes used in DNA profiling and sequencing.

1 A sample of DNA is collected from a crime scene. There are two suspects for the crime, who have both consented to provide a sample of DNA.

Draw a flow diagram with annotated images to describe the process of DNA profiling in this scenario. Use the words from the word bank in your diagram.

restriction enzyme	gel electrophoresis	unknown sample
DNA ladder	suspect sample	polymerase chain reaction (PCR)

9780170449625

2 Complete the passage using the words from the word bank.

heating	sequence	complementary	shorter
isolating	primer	terminate	laser beam
quickly	DNA polymerase	lengths	
nucleotides	dye	electrophoresis	

The Sanger method of sequencing DNA begins with ____________ the DNA from the cells of the target organism. This is followed by sequencing reactions where the DNA strand is separated into single strands by ____________. A ____________ binds to the start of a strand of DNA and ____________ is used to build a complementary strand using free ____________. Chain-terminating nucleotides attach to their ____________ bases and prevent any further nucleotides attaching, and so ____________ the chain of DNA. This results in many different ____________ of DNA strands being produced, each with a ____________ label from the chain-terminating nucleotide. ____________ is used to push the different length strands through a capillary gel tube. The ____________ strands move more ____________ and emerge from the tube before the longer strands. A ____________ at the end of the tube detects the specific dye on each strand and identifies the base nucleotide sequence. The computer then analyses the order in which the complementary nucleotides pass through the laser beam and produces a ____________ of DNA for the original template strand.

WS 5.3 Genetics for conservation

STUDENT BOOK
Pages 202–5

LEARNING GOAL

Describe the use of population genetics in conservation efforts.

1 Using an example, describe the importance of genetic variation within a population when facing changing environmental conditions.

____

____

____

____

____

____

2 The two images below show varying sizes of populations. The different colours represent varying traits within the populations.

Population A

Population B

a Which population is more genetically diverse?

____

9780170449625

b The International Union for Conservation of Nature is an organisation that produces a Red List that classifies the conservation status of different species. Traditionally, this status is calculated on population size and distribution. Using the populations above as examples, explain why it is important for genetic diversity to also be assessed when generating a conservation status.

3 Research the Hemmersbach Rhino Force Cryovault project and describe how it is assisting the genetic diversity of endangered rhinoceros populations.

4 The minimum viable population (MVP) is the number of individuals of a species that can survive and maintain a healthy population in the wild without being affected by the genetic defects associated with inbreeding. There are several different methods that can be used to calculate MVP, taking into account many factors. Extinction debt is a term used to describe a species that is likely to become extinct even though there may be several individuals left in the wild. Explain why a species that falls below the MVP threshold may be classed as experiencing an extinction debt or 'inevitable extinction'.

WS 5.4 Genetics for disease and human evolution

STUDENT BOOK Pages 211–19

LEARNING GOALS

Describe the use of population genetics in determining inheritance of a disease.

Describe the use of population genetics to explore human evolution.

1 There are two main human migratory theories: the multiregional hypothesis and the replacement, or Out of Africa, hypothesis. The graph below depicts one of these theories.

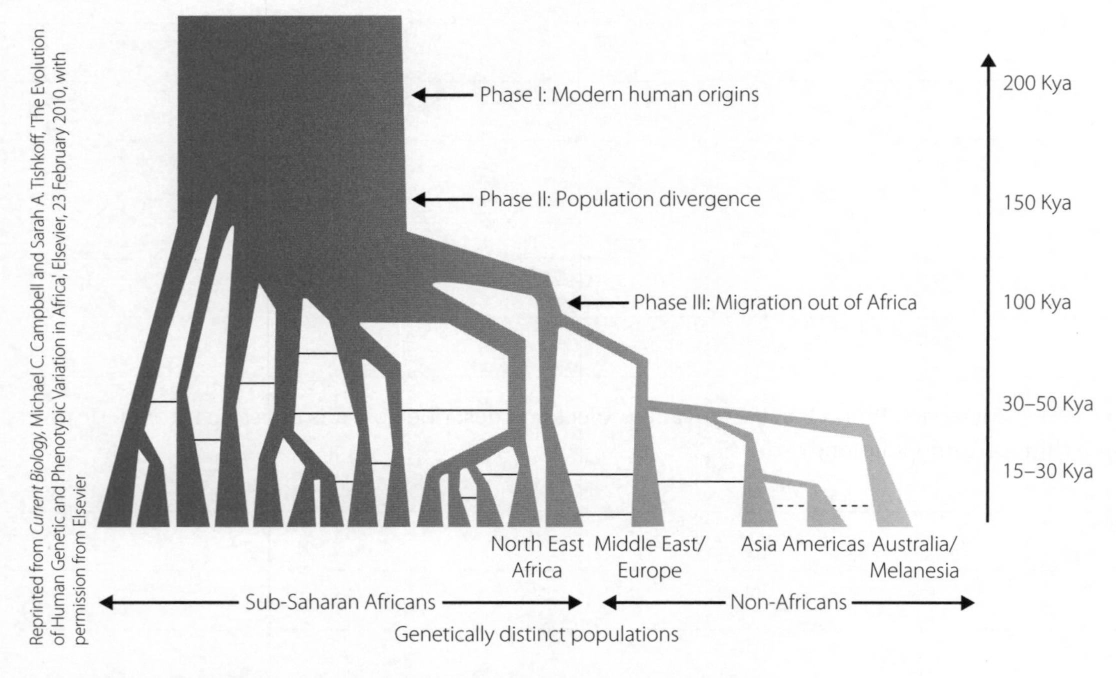

a Identify the theory the graph represents and justify your choice.

____

____

____

____

b Suggest what the lines in between each branch of the graph represent.

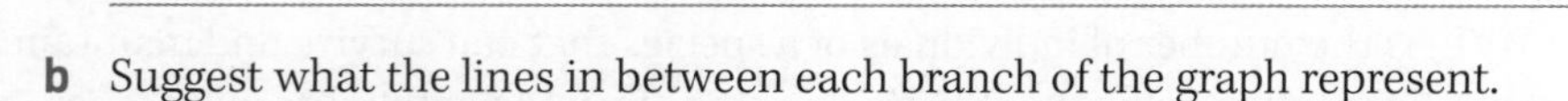

____

____

9780170449625

c Evaluate the statement: 'Sub-Saharan Africans are more genetically diverse than the rest of humanity combined.'

HINT

The verb *evaluate* requires you to make a judgement based on criteria or to determine the value of something.

2 Huntington's disease is a rare neurodegenerative disease found in humans. Around 5–7 people per 100 000 on mainland Australia have the disease; however, Tasmania has a higher proportion of around 12–13 people per 100 000 with the disease. One explanation for this is the fact that Tasmanians with Huntington's disease can be traced back to a common ancestor.

a Use secondary sources to describe Huntington's disease.

b A heterosexual couple with a history of Huntington's disease in their respective families is planning to have a child. Explain how genetic counselling could be used to advise the couple in their decision.

Module 5: Checking understanding

1 The gamete plays an important role in sexual reproduction because it carries:

A genetic information from both parents.

B half the genetic information of the parent.

C all of the genetic information of the parent.

D double the genetic information of the parent.

2 External fertilisation is typical of organisms that:

A are terrestrial.

B demonstrate a high degree of parental care.

C live in aquatic environments.

D release their sperm and egg at different times.

3 Protein synthesis involves:

A transcription and then replication.

B replication and then transcription.

C translation and then transcription.

D transcription and then translation.

4 Which part of the flower contains the embryo of the offspring? Justify your answer by explaining the process of fertilisation in a flowering plant.

5 Draw lines to match each term with an appropriate definition.

	Term		Definition
a	Double helix	**i**	A sugar
b	Nucleotides	**ii**	A type of weak bond between base pairs that holds the double helix together
c	Deoxyribose	**iii**	There are four kinds and they form specific pairs
d	Hydrogen bond	**iv**	Monomers that make up DNA
e	Nitrogenous bases	**v**	Bonds with thymine
f	Adenine	**vi**	Bonds with guanine
g	Cytosine	**vii**	Two strands of nucleotides twisted around each other

6 Identify two proteins and their functions in tissues.

7 Use the table to describe DNA replication and protein synthesis.

Process	DNA replication	Protein synthesis
When it occurs		
Type of cell involved in this process		
Main steps involved		
Result		

8 Use the table to compare mitosis and meiosis.

	Differences	Similarities
Mitosis		
Meiosis		

9 Distinguish between DNA and mRNA.

____

____

____

____

____

____

10 Below is a pedigree chart highlighting a dominant genetic disease passed from one generation to the next.

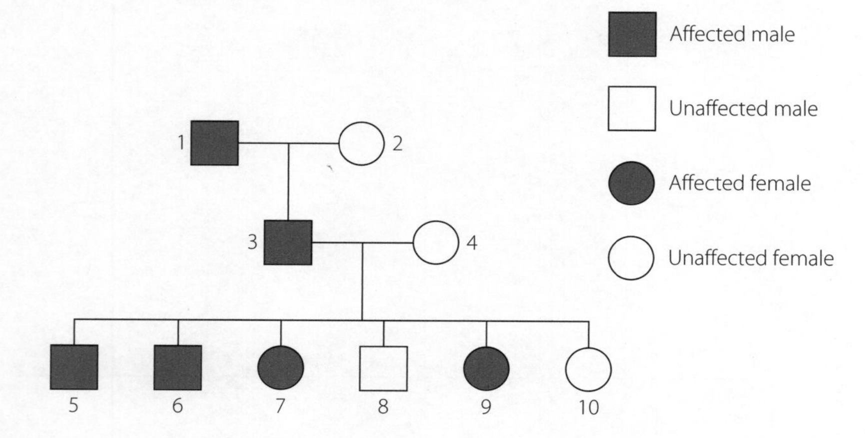

a Using the pedigree on the previous page, state the genotype of male 3 and justify your choice.

b Explain whether the gene is sex-linked.

11 Complete the table to show all possible genotypes for each phenotype.

Blood type	Genotypes
A	I^AI^i,
B	I^BI^B,
AB	
O	

12 Explain how an allele could be found in the majority of a population but the phenotypic expression of the allele is rare.

13 Explain how DNA profiling of endangered species could assist in breeding programs.

MODULE SIX »
GENETIC CHANGE

Reviewing prior knowledge

1 Use the words in the word bank to complete the paragraph.

codon	amino acids	polypeptide
nucleotides	translated	protein

DNA codes for ______________ using a sequence of three ______________. A group of three nucleotides is referred to as a ______________. During protein synthesis, the DNA segment is transcribed and ______________. Many amino acids are joined together to form ______________ chains. These polypeptide chains then form proteins in the cell. If a single nucleotide in a codon is changed, the resulting amino acid may be altered and this would affect the entire polypeptide chain and the ______________ that is produced.

2 a Use the labels in the word bank to complete the diagram of the peppered moth and its genetic makeup.

genotype	heterozygous	phenotype
alleles	homozygous	homozygous

Black
Black
Pepper
DD
Dd
dd
D D
D d
d d

b Identify whether the black genotype is dominant or recessive.

____

c The black allele was a mutation in peppered moths that was first reported around 1848 in Manchester, United Kingdom. Around this time, soot from factories started to darken the usually grey/white bark of trees in the area. Describe how this mutation was evolutionarily beneficial.

____

____

____

____

____

____

9780170449625

3 Which of the following is the process of choosing parent organisms for the characteristics that are wanted in their offspring?

A Active selection

B Reproductive selection

C Selective breeding

D Breeding selection

4 Describe the location of DNA in a prokaryotic cell.

A A linear chromosome and plasmids

B A single chromosome and a loop in the mitochondria

C A single circular chromosome and one or more plasmids

D Several circular chromosomes and a nucleoid

5 Describe the link between DNA, genes and polypeptides.

6 Complete the table to compare mitosis and meiosis.

Cell division	Role	Outcome
Mitosis		
Meiosis		

7 Create a diagram to identify the male and female reproductive structures in a flower.

8 Define the term ***biodiversity***.

9 Why is bioethics important in medicine?

6 Mutation

INQUIRY QUESTION: HOW DOES MUTATION INTRODUCE NEW ALLELES INTO A POPULATION?

WS 6.1 Mutagens

STUDENT BOOK
Pages 228–33

LEARNING GOAL

Identify and describe a range of mutagens.

1 Classify the following mutagens as *electromagnetic radiation*, *chemical* or *naturally occurring*.

Mutagen	Classification	Mutagen	Classification
Sunlight		Viruses	
Cigarettes		Transposons	
Processed meat		X-rays	

2 UV light can lead to a specific type of point mutation of two adjacent pyrimidine bases (cytosine or thymine), known as a dimer. This mutation causes the pyrimidine bases to bond together with covalent bonds, preventing hydrogen bonds forming with complementary bases and disrupting DNA replication. This process is shown in the diagram below.

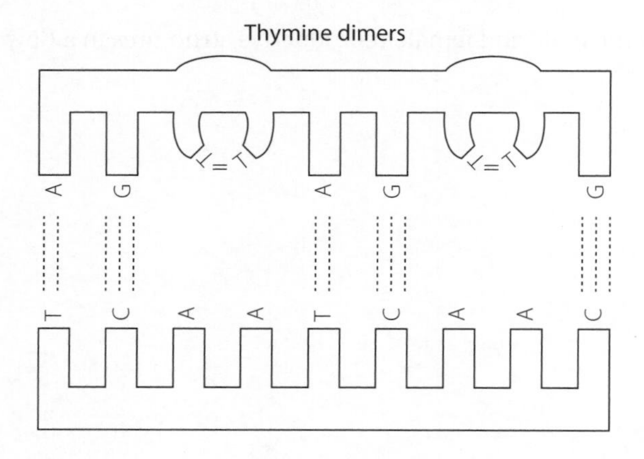

a Explain, using your knowledge of enzymes, why thymine dimers would prevent DNA from being transcribed.

HINT
The verb *explain* requires you to consider cause and effect.

____

____

____

____

9780170449625

b Thymine dimers can be repaired naturally in two different ways. Research and describe the following two processes using annotated diagrams.

Process	Description	Diagram
Photoreactivation		
Nucleotide excision repair		

c From your research, explain why photoreactivation is not possible in humans.

WS 6.2 Types of mutations

STUDENT BOOK
Pages 236–8

LEARNING GOAL

Describe types of point mutations and their results in an organism.

1 In many flowers, colour is regulated by a series of pigment enzymes. One such pathway is shown below.

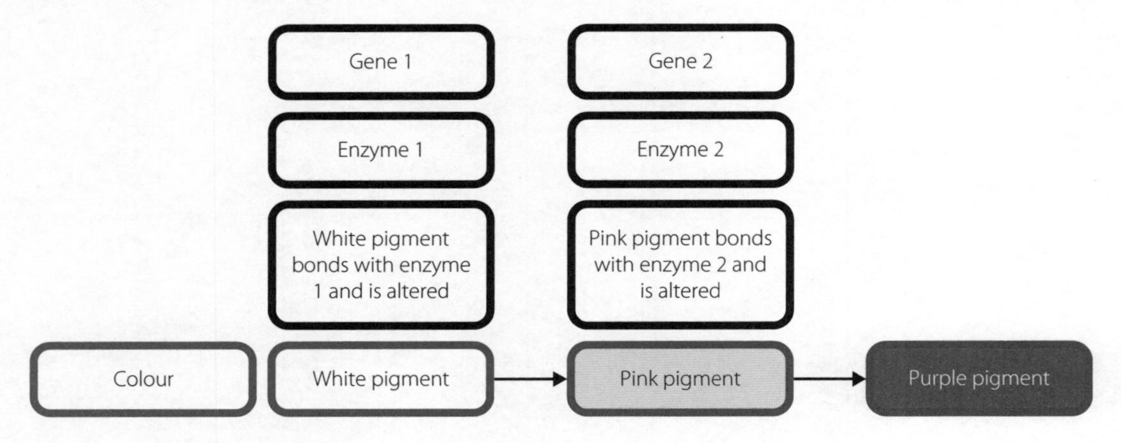

a A substitution point mutation occurs in gene 1. Explain how this could affect enzyme 1 and its active site.

____

____

____

____

____

____

____

____

b The image at the right shows the chromosome containing gene 1 and gene 2.

Describe the effects of the following types of mutation on flower colour.

Gene 1
Gene 2

i Frameshift in gene 2

____

____

____

____

ii Chromosomal inversion

____

____

____

____

9780170449625

2 Below is a normal sequence of DNA and its corresponding polypeptide.

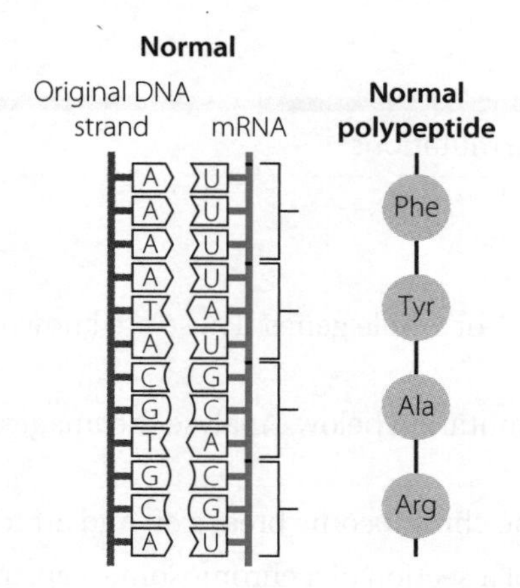

Identify the following types of point mutations and explain how the polypeptide has been affected.

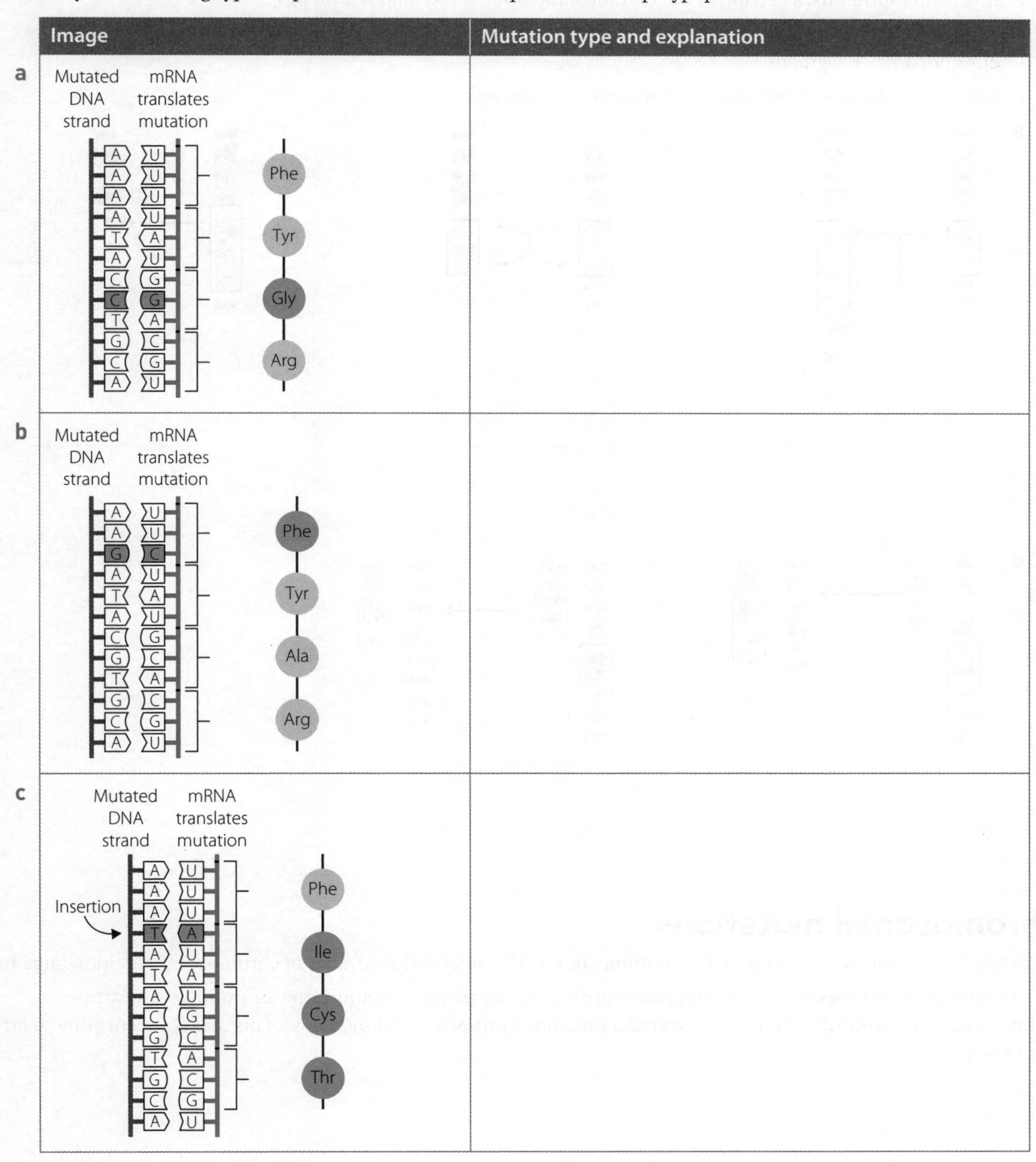

	Image	Mutation type and explanation
a	Mutated DNA strand: A A A A T A C C T G C A mRNA translates mutation: U U U U A U G G A C G U Phe – Tyr – Gly – Arg	
b	Mutated DNA strand: A A G A T A C G T G C A mRNA translates mutation: U U C U A U G C A C G U Phe – Tyr – Ala – Arg	
c	Mutated DNA strand: A A A T A T A C G T G C A Insertion mRNA translates mutation: U U U A U A U G C A C G U Phe – Ile – Cys – Thr	

Gene and chromosomal mutations

LEARNING GOAL

Describe types of gene and chromosomal mutations.

Gene mutations

Mutations can affect large segments of DNA or whole genes. These are known as block mutations and there are several types that can occur.

1 Read the descriptions of each type of mutation below. Analyse the images and write the name of each type of mutation below each image.

- Insertion mutations: a section of one chromosome breaks off and attaches to a different chromosome
- Duplication mutation: replication of a section of a chromosome, causing multiple copies of the same genes on that chromosome
- Inversion mutations: a section of the chromosome rotates 180° and reattaches
- Translocation mutations: whole chromosomes or a segment become attached or exchanged with another chromosome or segment
- Deletion mutations: a section of chromosome is removed

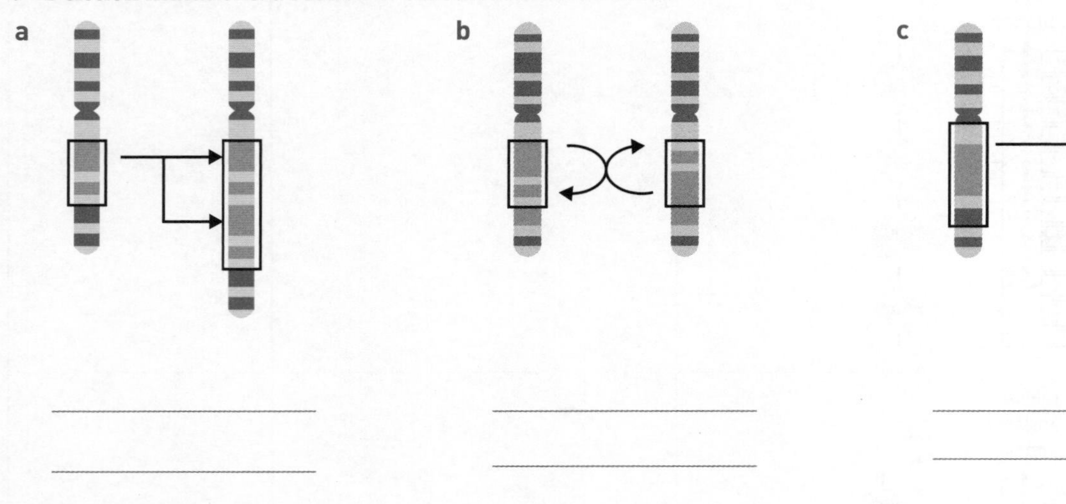

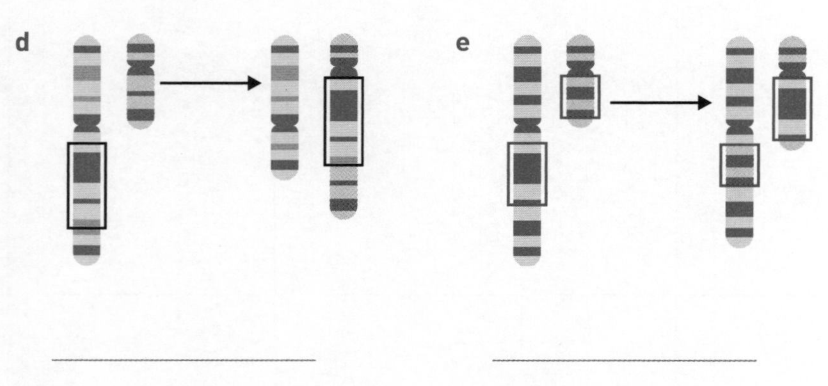

Chromosomal mutations

Occasionally, mutations can involve entire chromosomes. This affects the number of chromosomes an individual has.

Aneuploidy is the presence of an abnormal number of a particular chromosome; for example, an extra chromosome (e.g. trisomy, with three copies) or a missing chromosome (monosomy). This usually occurs due to errors in meiosis.

9780170449625

2 Below is a diagram outlining the process of non-disjunction that can result in aneuploidy.

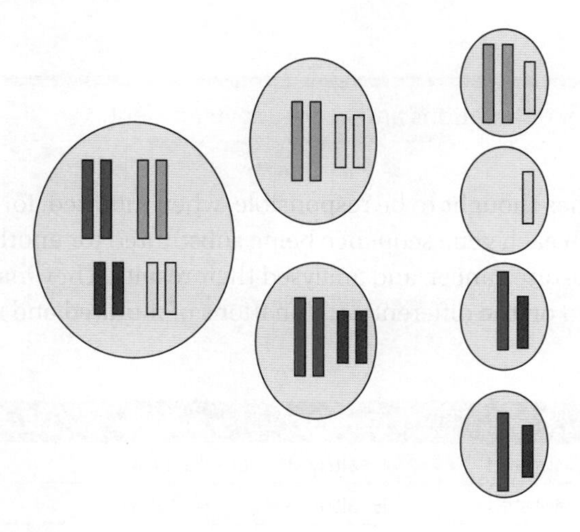

Use the diagram and your understanding of meiosis to explain when non-disjunction occurred in this case, and how an organism could inherit an extra chromosome.

3 Using secondary sources, research the following types of chromosomal mutations and complete the table.

Type of mutation	Description of mutation	Description and example of syndrome in humans	Chromosome affected
Trisomy			
Monosomy			
Sex-linked trisomy			

WS 6.4 Mutations and cancer

STUDENT BOOK
Pages 241–3

LEARNING GOAL

Analyse information relating to gene mutations and cancer development.

Scientists isolated four somatic genes thought to be responsible, when mutated, for prostate cancer in men. The mutation involves one nucleotide in each gene sequence being substituted for another. The same scientists examined healthy people and people with prostate cancer, and analysed their results. They managed to calculate the likelihood of a person developing cancer based on the different combinations of mutated and non-mutated genes. Their results are shown below.

Person	Gene 1	Gene 2	Gene 3	Gene 4	Likelihood of developing cancer (%)
1	Healthy	Mutated	Healthy	Healthy	0
2	Mutated	Mutated	Healthy	Mutated	30
3	Mutated	Healthy	Mutated	Mutated	90

1 Describe the type of mutation being studied.

2 Explain whether this mutation could be inherited.

3 Identify the gene mutation combination most likely to result in cancer.

4 From the data above, explain which gene the scientists can discount as a cause for cancer.

5 Explain why a silent mutation in gene 2 does not increase the likelihood of developing cancer.

9780170449625

6 The scientists have been granted funding to develop a drug that will prevent the mutation of one gene. Justify the gene that they should focus their research on.

Genetic drift and genetic bottle necks

STUDENT BOOK Pages 248–9

LEARNING GOAL

Describe and model the processes of genetic drift, founder effect and genetic bottlenecks.

Allele frequency in a population is not always driven by natural selection or favourable characteristics. Natural disasters or random accidents can alter the ratios of allele numbers in a short amount of time. This can then result in a permanent change in a population.

1 Use the diagram to explain how favourable characteristics may not be in the majority of a population.

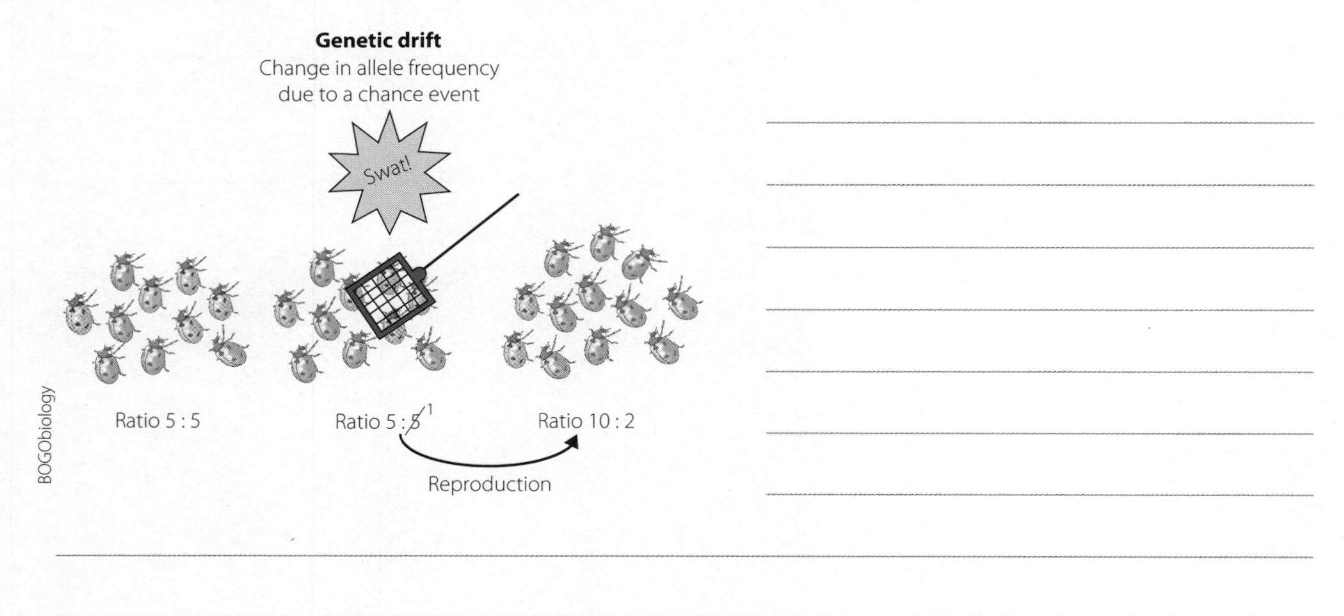

2 Use the words from the word bank to complete the paragraph.

isolated	genetic drift	founder	alleles	colonised

Islands are perfect examples of ________________ areas that can be ________________ by a small group of individuals. This small group is known as the ________________ population. This new population is unlikely to have an equal range and frequency of ________________ compared to the original population. This change in the frequency of alleles can result in ________________.

3 Below is a model used to compare the founder effect and a genetic bottleneck.

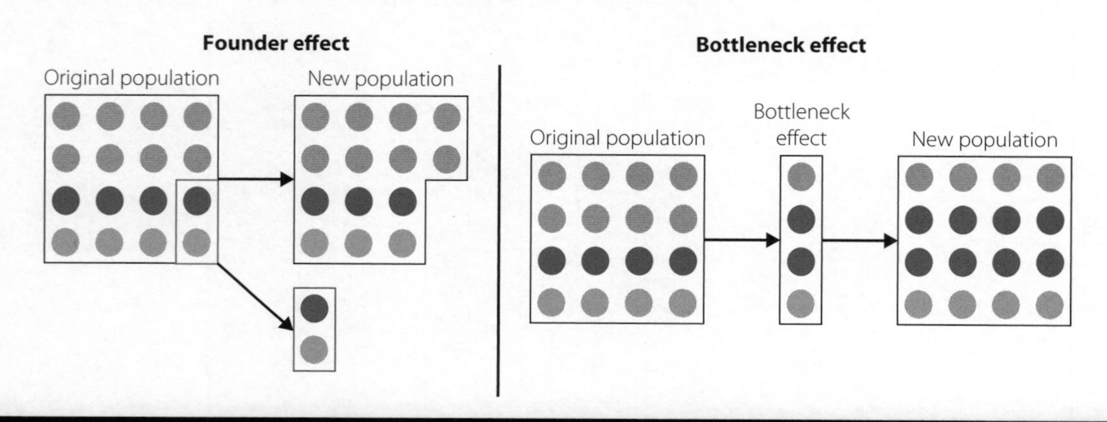

Assess the effectiveness of the models on the previous page at correctly comparing the two genetic processes.

HINT

The verb *assess* requires you to include points for, points against and a judgement.

4 Using the following equipment, design an activity that could effectively model the process of a genetic bottleneck. Write a detailed method for your design.

Shutterstock.com/Bogdan Vacarciuc

Lollies of the same shape but different colours

Shutterstock.com/urfin

A conical flask

Remember to include the following stages in your activity.

- Original population allele frequency
- Event causing a genetic bottleneck
- New reduced population allele frequency
- Future allele frequency after genetic bottleneck

7 Biotechnology

INQUIRY QUESTION: HOW DO GENETIC TECHNIQUES AFFECT EARTH'S BIODIVERSITY?

STUDENT BOOK
Pages 259–61, 272–3, 276–7, 282–3

What is biotechnology?

LEARNING GOALS

Consolidate the definition of biotechnology.

Consider social, ethical and biodiversity implications.

Biotechnology defined

'Biological materials (*bio*) as tools (*technology*)' is a literal breakdown of the word *biotechnology*. A common misconception is that biotechnology is relatively new and involves futuristic laboratories and factories. Biotechnology is not new; this becomes more obvious when we expand the definition of the word. Ancient and traditional biotechnologies that utilised microorganisms include making beverages such as wine and beer, and foods such as bread, yoghurt and cheese. Agriculture fits the definition of biotechnology and began at least 10 000 years ago when people began manipulating plant and animal breeding, rather than just hunting and gathering from the wild.

> Biotechnology: the use of living organisms to make or modify a product; to improve animals or plants for human use; to utilise microorganisms for specific purposes

1 Which of the following examples fit the definition of biotechnology? Circle Yes or No.

a Collecting Native hop (*Dodonaea viscosa*) from a forest, boiling the thick, leathery leaves and applying it to relieve earache. Yes / No

b Collecting berries and tubers from a forest for food. Yes / No

c Storing milk in the stomach of a calf, finding that the milk turns into cheese, and that the cheese lasts longer than milk. Yes / No

d Tracking and killing a goanna, then cooking and eating it. Yes / No

e Building a fence around a group of wild sheep and killing them for food over a few months. Yes / No

f Using yeast to ferment the sugars in barley into an alcoholic drink. Yes / No

g Collecting seeds from wild plants, then grinding them into flour to make bread. Yes / No

h Collecting seeds from wild grasses, then planting them and collecting the seeds the following year. Yes / No

i Using bacteria to break down chemical pollution. Yes / No

j Altering the genes of crops, so that they are more nutritious, or grow more efficiently. Yes / No

k Bringing a stud bull from another farm to mate with your herd of females. Yes / No

l Yeast growing on the skins of grapes. Yes / No

m It is found that a useful glue could be extracted from the skin of a frog. Scientists are working on a synthetic version of the glue to avoid using living frogs. Yes / No

An example of biotechnology

Possum skin cloaks were traditionally made and worn by Aboriginal peoples of south-eastern Australia for many thousands of years. The cloaks were worn for warmth, but also had several other uses: as baby carriers, as coverings at night, as drums in ceremonies and for burial. To produce a possum skin cloak, possum pelts were stretched out and dried, then a bone was used to pierce holes in the edges of the skin. Pelts were then sewn together using thread made either from plant fibre or the tendons of either a kangaroo or an emu. The inside of the cloak was decorated with designs etched into the leather using mussel and oyster shells, and bone and stone tools. Sap collected from the black

9780170449625

wattle tree was mixed with ground ochre, a pigmented clay, and painted onto the designs. Cloaks were secured for wearing using bone or wooden pins, or echidna quills.

Aboriginal peoples say that wearing a cloak creates a connection to their culture and is an emotional experience. Traditionally, possum cloaks are buried with their owners and, until recently, few existed outside of museum collections. Revitalised interest in possum skin cloaks among Aboriginal communities has led to the establishment of cloak-making workshops, which have created opportunities for passing on stories and knowledge, and have strengthened relationships across the multiple generations involved. Approximately 100 possum skin cloaks are now held in Aboriginal and Torres Strait Islander communities, and are used in cultural ceremonies. The Peter MacCallum Cancer Centre in Melbourne has a possum skin cloak available for use by Aboriginal and Torres Strait Islander people when going through cancer treatment.

In Australia, possums are a protected species, but they are classified as feral pests in New Zealand. Cured skins can be imported from New Zealand to make modern possum cloaks.

2 a Examine the process for making a possum skin cloak. Which aspects fit the definition of biotechnology?

b Discuss the social and ethical implications of making modern possum skin cloaks.

HINT

When considering the social implications of an action, think about how people in a society or group are affected by the action. Ethical implications focus on whether the action is right or wrong and do not need to focus only on people. Discussing implications means looking at the positive and negative effects of the action.

Positive social implications	Negative social implications
Positive ethical implications	**Negative ethical implications**

Biotechnology in farming

Humans have been controlling the reproduction of other species for thousands of years. Early farmers would select the best ram (more wool, more meat, more docile) to mate with the flock, or select seeds from the best wheat plants to sow for next year's crop.

3 Complete the table by matching the terms from the word bank with their appropriate definitions.

Variety	Primary producer	Crossing	Cultivation
Domestication	Strain	Species	Breed

	Term	Meaning
a		An individual or company carrying on a business such as animal farming, plant cultivation, fishing, pearling, tree farming or felling
b		The process of breeding with the intention to create hybrid offspring between breeds, varieties or strains
c		A group of organisms that can breed with each other and produce offspring that also produce offspring
d		Taming an animal and keeping it as a pet or on a farm; a sustained and multigenerational relationship where humans influence the care and reproduction of the animals
e		Collecting wild seeds, saving seed from last year's crop, planting and caring for the crop; using artificial pollination to choose the parents of next year's seeds; cloning plants to increase the quantity of a desired variety
f		A stock of animals or plants of the same species having a distinctive appearance and typically having been developed by deliberate selection; usually individuals are homozygous for the definitive features
g		A stock of animals or plants of the same species, having a distinctive appearance and typically having been developed by deliberate selection; usually individuals are homozygous for the definitive features
h		A sub-group of microbes of the same species, having some distinctive features and genetic variation

4 Investigate the following sheep breeds in Australia. Identify the features of each breed that were selected in creating each breed.

Merino	Dorper	Corriedale	East Friesland

5 The first European colonists in Australia quickly began growing crops and raising animal herds. They brought wheat seeds from England, which were not suited to the Australian climate and suffered from local fungal diseases called rusts.

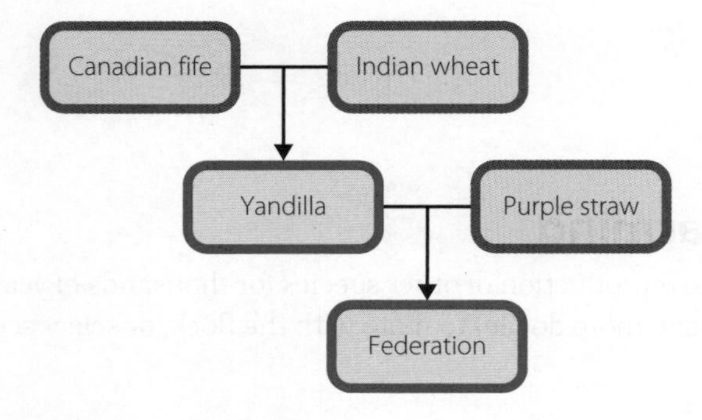

The diagram shows five varieties of wheat. All are the same species, but each variety had its own advantages and disadvantages for growing in Australia. Yandilla and Federation were new varieties created by selective breeding in Australia at the end of the 1800s. Investigate the characteristics of each variety and justify the breeding program that led to the Federation variety.

Hybrids

Crossing between varieties creates hybrids. Successful hybridisation can lead to hybrid vigour – increased strength and better health in the hybrid (heterozygous) individuals than is found in the homozygous parents. Sometimes hybrids display advantages that are not present in either of the parental varieties, because of the compounded effects of their new gene combinations, giving new phenotypes.

6 The advantages of hybridisation have been stated. Describe the disadvantages of hybridisation in agricultural plants and explain the effect these have on the primary producer.

HINT

Consider using a Punnett square in your answer, showing a cross between two heterozygous parents.

WS 7.2 Insulin case study

LEARNING GOAL

Analyse past, present and future biotechnology associated with insulin production.

Classical insulin manufacture

For centuries, half of all people who developed Type 1 diabetes died within two years; more than 90% were dead within five years. Insulin was identified as the hormone that Type 1 diabetics lacked in 1922. Fortunately, pork- and beef-derived insulin is nearly identical to human insulin. Starting with the pancreases of pigs and cattle, a pharmaceutical company extracted, purified and marketed insulin. Daily injections of insulin became the therapy for those with Type 1 diabetes.

1 a Insulin is a pharmaceutical product. Justify its production as biotechnology.

b Assess the importance of the discovery of insulin, and knowledge about its pharmaceutical production, to society.

HINT

The verb *assess* requires you to include points for, points against and a judgement.

c Bovine (beef) and porcine (pig) pancreases are a by-product of the meat industry. Some religious groups do not permit the use of any drugs, dressings or implants that contain or are derived from animals; others focus on material only from pigs or cows. In emergency or starvation situations, those rules could be waived for followers of some religions and, if disease sufferers were lacking reasonable alternatives, these products could be allowed. Analyse the impact that religious rules or personal preference might have on a decision to use insulin.

Modern insulin manufacture

In 1982, insulin made by genetically modified bacteria became a viable alternative to insulin sourced from animals. Production involves recombinant DNA technology – incorporating a gene from one organism into the DNA of another. There are currently three brands of recombinant insulin: Humulin and Insuman, produced by bacteria (*Escherichia coli*), and Novolin, produced by yeast (*Saccharomyces cerevisiae*).

The process to make recombinant insulin

1. Cut human *INS* gene encoding the polypeptide proinsulin from chromosome 11.
2. Insert *INS* gene into a plasmid (DNA vector) and into a host cell.
3. Make a culture of bacterial or yeast host cells containing the DNA vector.
4. Control culture and fermentation conditions to optimise yields.
5. Proinsulin polypeptide is converted to insulin protein by further processing.
6. Multi-step purification of human insulin from the culture of cells.
7. Crystallise insulin.
8. Modify into different insulin products: slow release, fast acting etc.

2 a Recall the difference between proinsulin and insulin.

b Distinguish between *DNA vector* and *host cell.*

c The diagram below represents just one part of the process described. Which step is illustrated?

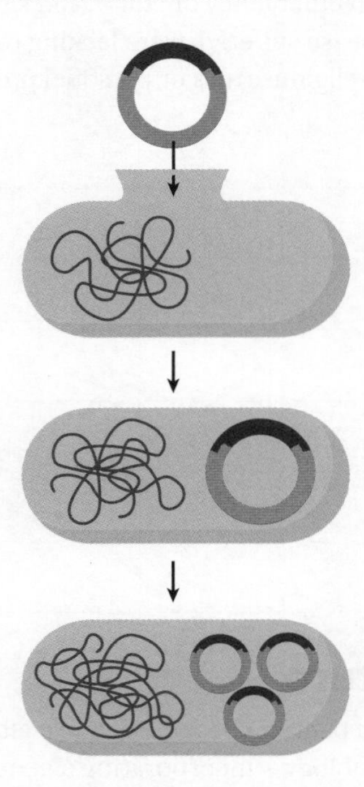

d Annotate the diagram in part **c** as thoroughly as possible. Include labels based on your knowledge of cells from previous worksheets.

e Describe what should be drawn as the next part in the diagram. Make a sketch of your prediction.

f Yeast cells can also incorporate bacterial plasmids, making them an alternative host to bacterial hosts. The producers of recombinant insulin have found that yeast is more capable than bacteria at completing the processing of the polypeptide proinsulin into functional insulin. Suggest why this might be the case.

Restriction enzymes and recombinant insulin

DNA ligase enzymes were described in the 1960s – they repair DNA by helping nucleotides 'glue' together during DNA replication. Recombinant DNA technology became a possibility once restriction enzymes were also discovered. Restriction enzymes are produced by bacteria and act like scissors, cutting up DNA. Restriction enzymes do not cut randomly; they cut at specific DNA sequences, which is one of the key features that make them suitable for DNA manipulation. There are many different restriction enzymes.

3 a Explain why the discovery of restriction enzymes was crucial to the development of recombinant insulin.

b Describe how restriction enzymes and ligases could be used to make the vector plasmid in the diagram on the previous page.

4 a An incomplete model to produce a recombinant plasmid is shown below. The events taking place at each stage are described in the table. Use the instructions on the following page to annotate the diagram.

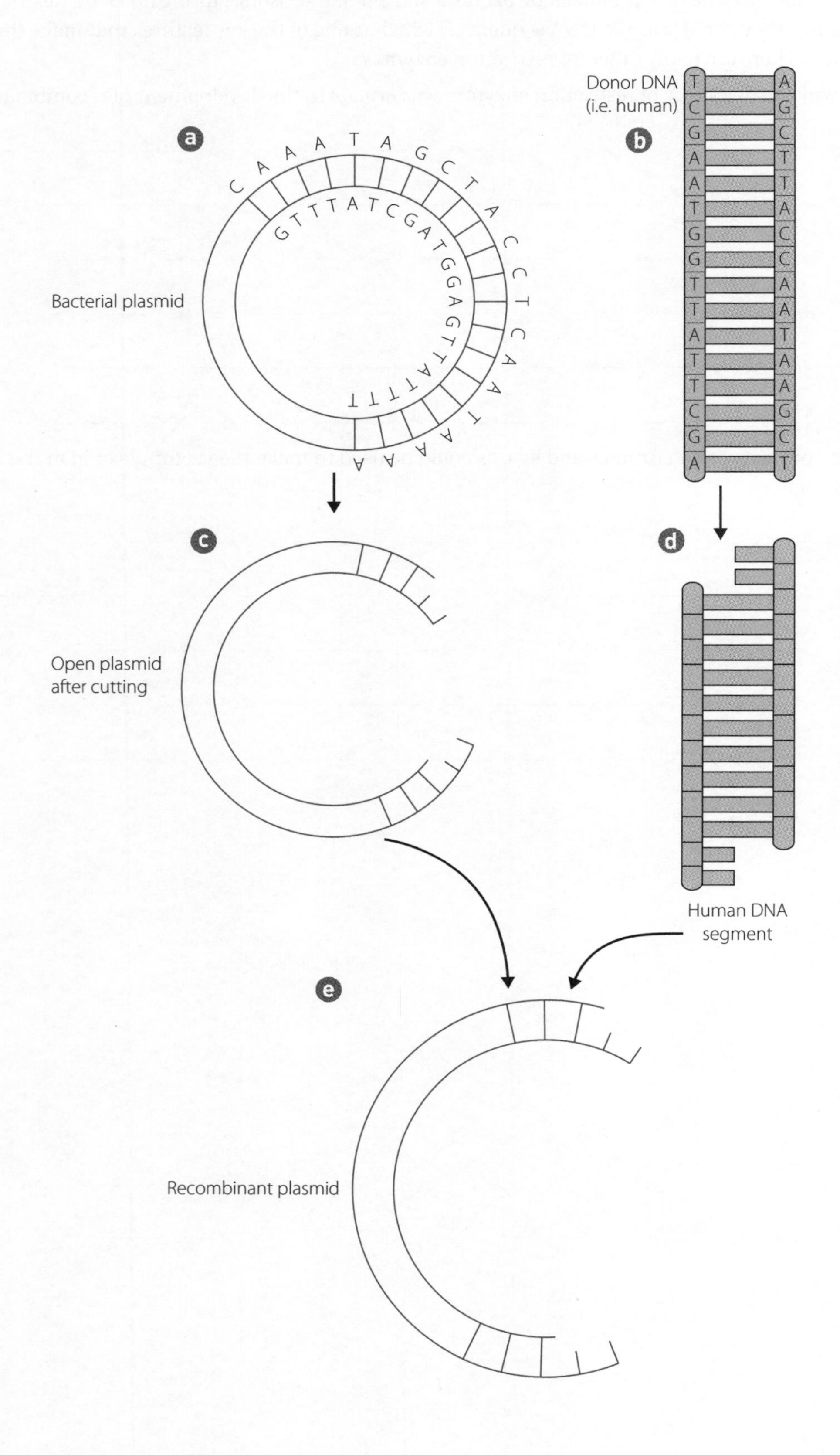

Events taking place at each stage	Instructions for completing the model
Part A	
• A restriction enzyme is added to a tube containing plasmids. • The enzyme binds to the sequence AGCT. • The enzyme makes a cut in the DNA between the adjacent A and G on the same side of the DNA strand.	• This enzyme travels clockwise around the plasmid looking for the sequence AGCT. • Highlight these letters on the plasmid. • Now search the inner bases of the plasmid in an anticlockwise direction. • Highlight the letters AGCT. • Draw a cutting line between the A and the G in these two highlighted sequences like this:
Part B	
• The same restriction enzyme is used to cut the donor DNA in two places to remove a whole gene.	• Scan upward on the left-hand DNA strand looking for AGCT and highlight the sequence (there will be two). • Scan downward on the right-hand DNA strand looking for AGCT and highlight the sequence (there will be two). • Draw a cutting line between the A and the G in these highlighted sequences.
Part C	
• The restriction enzyme has left two bases unpaired on each end of the DNA strand in the plasmid.	• Transfer the letters of each base on to the cut plasmid. • Label the sticky ends.
Part D	
• The restriction enzyme has left two bases unpaired on each end of the donor DNA segment (a gene).	• Transfer the letters of each base onto the human DNA segment (some bases have been left out to make this step less onerous). • Label the sticky ends.
Part E	
• Open plasmids and donor DNA segments are mixed in test tubes. • Sticky ends are attracted to each other and the donor DNA segment becomes incorporated into the bacterial plasmid. • Ligase enzymes help the nucleotides join on both sides of the DNA strands. • The result is called a recombinant plasmid.	• Draw the human DNA segment into the opening in the plasmid. • Transfer the letters of each base onto the recombinant plasmid. • Annotate any other details onto the model.

b Why was the donor DNA segment gene easy to insert into the plasmid? Use evidence from the model in your answer.

c The hypothetical restriction enzyme in this model finds the sequence AGCT and cuts between A and G. What is the effect of using a different restriction enzyme that finds the sequence TAT and cuts between the first T and the A?

d Evaluate the model used in this question for genetically modifying a bacterium.

> **HINT**
>
> The verb *evaluate* requires you to make a judgement based on criteria or to determine the value of something.

Future directions of research

Fermentation of genetically modified single-celled organisms has been the principle source of insulin for four decades. There is potential for other genetically modified options, such as using a plant-based system. Theoretically, seeds or leaves could produce high levels of proinsulin. Once the plants were genetically altered, cuttings or tissue culture could be used to clone the plants. This would be low cost reproduction. Seeds could act as long-term storage of recombinant insulin until required.

5 Summarise the theoretical benefits of recombinant insulin from plant-based sources outlined in the text, adding any others you can think of. Suggest disadvantages of plant-based insulin production.

Benefits of using recombinant plants to produce human insulin	Disadvantages of using recombinant plants to produce human insulin

6 Recently published peer-reviewed research documents one new area of enquiry: gene editing relating to diabetes. Stem cells were produced from a patient with a rare, genetic form of insulin-dependent diabetes called Wolfram syndrome. Researchers re-programmed the stem cells and turned them into pancreatic cells. They then used the gene-editing tool CRISPR-Cas9 to correct the genetic defect in the stem cells.

Next, they implanted the edited and non-edited pancreatic cells into laboratory mice with diabetes. The mice receiving the edited cells were able to produce their own insulin and were not diabetic in the six months they were monitored. The other mice remained diabetic.

> **HINT**
>
> CRISPR-Cas9 is a new gene editing tool that can edit a mutation and convert a DNA sequence to a more favourable sequence. It is very precise.

a Identify the two main areas of biotechnology research involved in this story.

b What does the term *peer-reviewed* mean?

c Using CRISPR-Cas9 technology on human patients would be very costly. If it becomes available to the general population for editing mutations, it is likely that only wealthy people will have access to the technology. What are the implications for society?

d Studies using mice are common in scientific research and have revealed important scientific information. Identify three arguments for and against using mice in research.

e The research about gene editing relating to diabetes was published in the scientific journal *Science Translational Medicine* on 22 April 2020, in volume 12, issue number 540. The title was 'Gene-edited human stem cell-derived β cells from a patient with monogenic diabetes reverse pre-existing diabetes in mice'. The authors were Kristina G. Maxwell, Punn Augsornworawat, Leonardo Velazco-Cruz, Michelle H. Kim, Rie Asada, Nathaniel J. Hogrebe, Shuntaro Morikawa, Fumihiko Urano and Jeffrey R. Millman.

There are various systems for referencing an article. A common one is Harvard referencing. Use the Harvard referencing system to show how this paper should appear in a reference list. If you are not familiar with the system, use secondary sources to find out about it.

HINT

Harvard referencing is an internationally recognised way to correctly format reference material.

9780170449625

8 Genetic technologies

INQUIRY QUESTION: DOES ARTIFICIAL MANIPULATION OF DNA HAVE THE POTENTIAL TO CHANGE POPULATIONS FOREVER?

WS 8.1 Uses and advantages of current genetic technologies

LEARNING GOALS

- Outline uses of genetic technologies.
- Compare advantages of reproductive technologies.

Humans have been using genetic technologies for a long time. Selective breeding of plants and animals has been practised for thousands of years, and there is evidence of artificial pollination from 870 BCE. Artificial insemination was first employed in the early 1700s and has been used commercially since the 1980s. More recent technologies include in vitro fertilisation (IVF), cloning and DNA recombinant technologies.

1 Outline the uses of cloning. Include the phrases from the word bank in your answer.

gene cloning	whole-organism cloning	reproductive technology	cellular

2 Draw a table to compare the advantages of selective breeding and IVF.

Processes and outcomes of artificial insemination and artificial pollination

STUDENT BOOK
Pages 283–5

LEARNING GOALS

Outline the processes of artificial insemination and artificial pollination.

Distinguish between artificial insemination and artificial pollination.

The techniques of artificial insemination and artificial pollination have a long history. Today, both of these reproductive technologies are regularly used in agricultural production around the world.

1 Draw a flow chart to summarise the process of artificial insemination.

HINT

A flow chart requires you to put information in a series of small boxes and use arrows to indicate the flow of information.

2 Use a diagram to explain the process of artificial pollination.

9780170449625

3 Create a table to distinguish between the outcomes of artificial insemination and artificial pollination.

LEARNING GOALS

- Compare whole-organism cloning and gene cloning.
- Outline the impact of cloning on genetic diversity.
- Assess the reliability of secondary sources.

Rather than relying on selective breeding, cloning is a technology that allows scientists to select for a particular trait. Commonly used forms of cloning include molecular cloning, gene cloning, cell cloning and whole-organism cloning, also known as reproductive cloning.

1 Compare gene cloning and whole-organism cloning.

> **HINT**
> The verb *compare* requires you to look for similarities and differences. Use a table or Venn diagram.

2 The film *Jurassic Park* was based on the premise of using cloning to revive extinct organisms. De-extinction, or bringing back extinct organisms, is an idea that is gaining increasing popularity among some scientists.

Assess the effectiveness of using cloning as a tool for bringing extinct organisms to life.

> **HINT**
> In your answer include a description of the applicable cloning technique, an outline of the advantages and limitations of using cloning in this situation and a clear judgement.

9780170449625

Techniques and applications of recombinant DNA technology

STUDENT BOOK Pages 293–8

LEARNING GOALS

Describe the process of recombinant DNA technology.

Outline the application of transgenics in medical applications.

1 The process of recombinant DNA technology involves cutting DNA at particular locations.

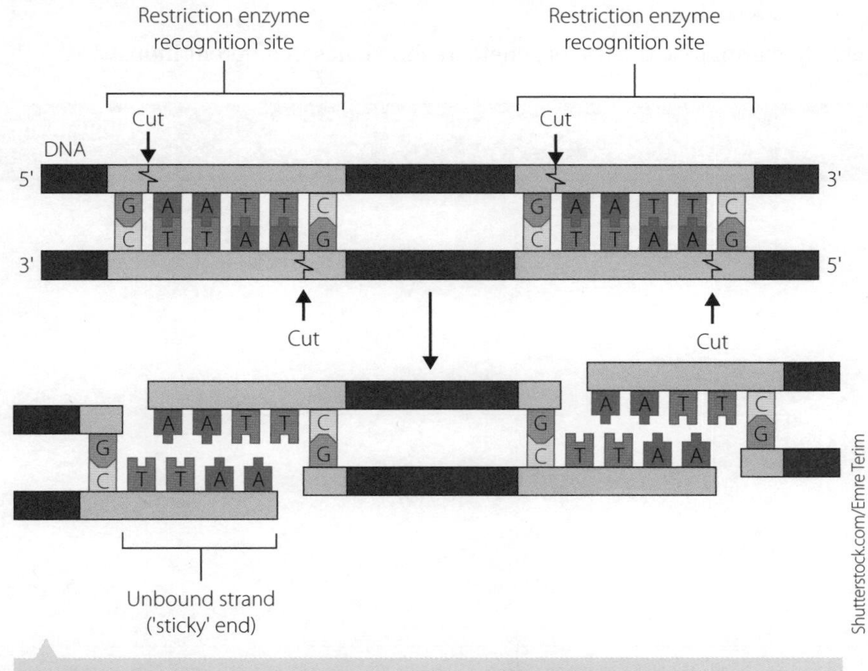

Restriction enzymes play an important role in recombinant DNA technology.

With reference to the diagram, explain the role of restriction enzymes and sticky ends in recombinant DNA technology.

2 Gavin drew the flow chart below to describe how transgenic mice are produced and used in the medical industry. Provide feedback to Gavin on his diagram. Correct any mistakes and provide suggestions for improvements.

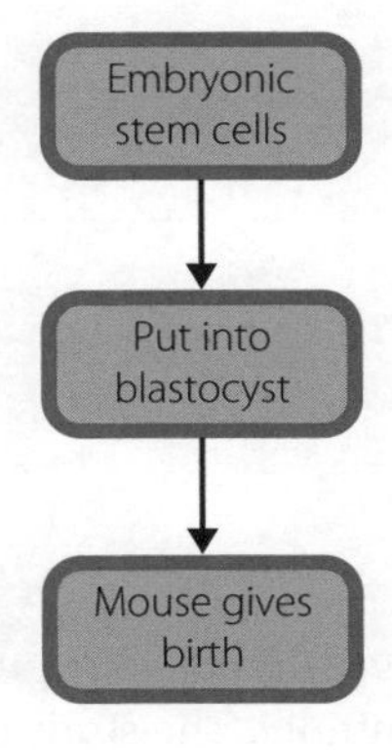

Genetic technologies in agricultural, medical and industrial applications

STUDENT BOOK
Pages 299–301

LEARNING GOALS

Discuss the application of genetic technologies in agricultural, medical and industrial applications.

Interpret data relevant to the industrial application of genetic technology.

Genetic technologies such as recombinant DNA technology and the production of transgenic organisms are growing increasingly popular in a range of industries.

1 Complete the table to summarise the use of genetic technologies in different industries.

Industry	Genetic technology example	Benefit/s of the named technology	Limitation/s of the named technology
Agricultural			
Medical			
Industrial			

2 Researchers in Saudi Arabia have created a gene catalogue where the complete DNA sequence for quinoa was recorded. Quinoa is a popular agricultural plant, but the tall, thin stalks are susceptible to damage during storms. The recording of the gene catalogue for quinoa has the potential to allow farmers to grow quinoa more tolerant of extreme climate events.

Use secondary sources to outline an example that explains how a gene catalogue may benefit medicine.

3 In 2011, the journal *Applied and Environmental Microbiology* published a paper on bioremediation; that is, the use of biological processes to break down pollutants. The study examined the ability of various strains of fungal species to break down the synthetic polymer polyester polyurethane. Researchers prepared solutions of liquid polymer and added the fungi (see photo below). As the polymer was degraded, the liquid turned from white to clear. The optical density of the suspended polymer was measured in order to determine the effectiveness of the fungi in breaking down the polymer. The graph below shows the optical density of the suspended polymer over 16 days, when various species of fungi are added.

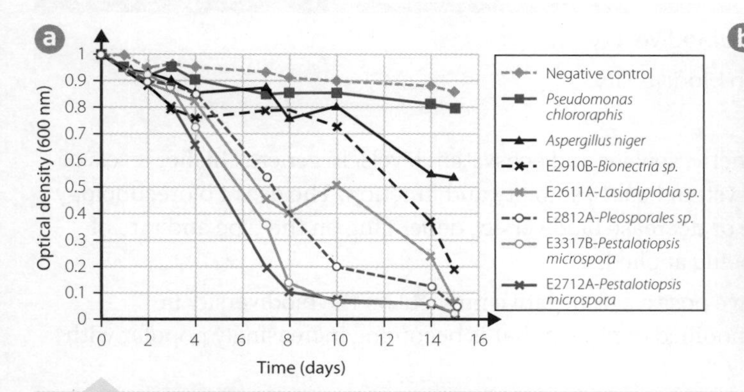

a Degradation of the polymer over 16 days by various fungi species. b As the fungi degrade the polymer the solution turns from white to clear. Optical density measures are used to determine the effectiveness of degradation by the fungi.

a Identify a possible hypothesis for the investigation described.

b Identify one controlled variable for this investigation.

c Explain why the variable you identified in part **b** is required for the investigation to be valid.

d With reference to the graph, suggest which strain of fungal species would be most useful for the bioremediation of polyester polyurethane pollution in waterways.

HINT

Include specific reference to data from the graph in your answer.

e Use the information provided to propose one possible direction of future research for this project.

Biotechnology in agriculture: effects on biodiversity

LEARNING GOALS

Explain how biotechnology can increase and decrease biodiversity.
Interpret data relevant to the impact of agriculture on biodiversity.

The term *biodiversity* encompasses variation at the genetic, species and ecosystem levels. In general, higher levels of biodiversity tend to correlate with higher levels of ecosystem health, stability and function. The use of biotechnology in the agricultural setting has the potential to increase or decrease biodiversity, depending on the type and rate of application, and the scale at which the technology is being applied.

1 The planting of genetically modified crops can have positive or negative impacts on the biodiversity in surrounding ecosystems. Bt corn is a genetically modified organism that is becoming increasingly popular with farmers (see graph below).

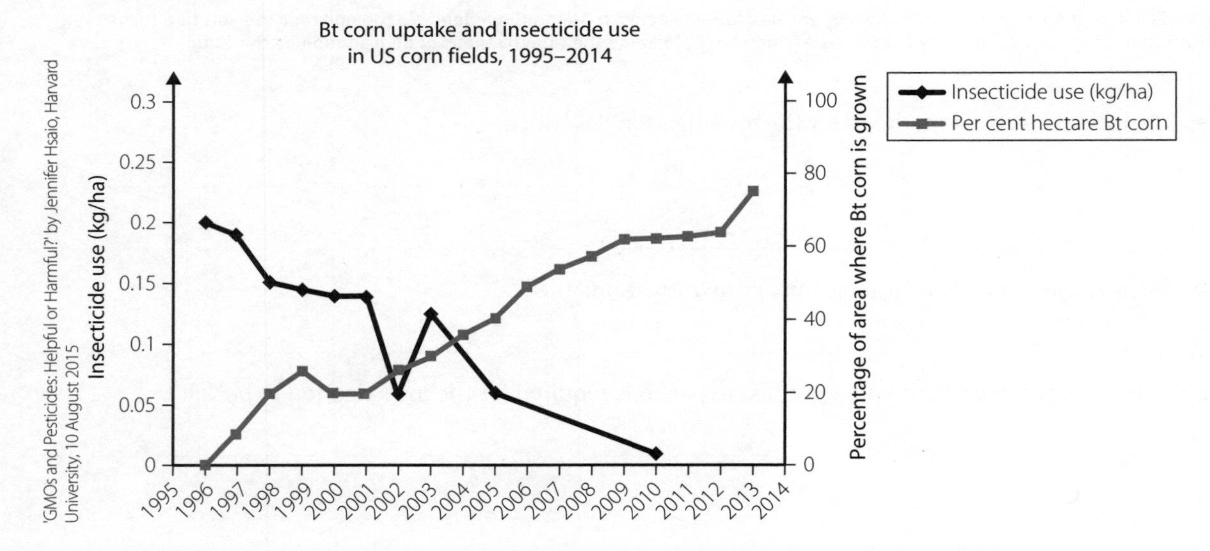

Use the information provided to predict the outcome on biodiversity in the ecosystems where Bt corn is planted.

HINT

Include specific reference to data from the graph in your answer.

2 Biodiversity is a complex measure of ecosystem health that can be measured with a range of statistical tools. One simplified aspect of biodiversity is richness, and this may refer to the number of species, the number of Orders (meaning the taxonomic grouping), the number of ecosystems or the number of genes.

The table on the next page presents a subset of data collected from a study that aimed to measure the contribution of agricultural land to local biodiversity levels. In local areas, invertebrates were sampled from farms and nearby national parks to compare biodiversity levels between the two land use types. At least 97 per cent of all animals are invertebrates, so they are an effective group to measure when determining impacts on biodiversity. The number and types of invertebrates found in a location will change in response to impacts the land receives and, in general, activities that disturb the land will encourage invertebrates into a location as they take advantage of available niches. On the other hand, if land management techniques, such as the application of pesticides, are too intense then it is likely that invertebrate biodiversity will decrease.

In the study, three sites were sampled on each farm and the adjacent national park. The owner of Sunny Creek Farm, Lilli, had been growing genetically modified herbicide resistant wheat for five years, whereas Gerald at Windy Valley Farm grew non-genetically modified wheat.

9780170449625

Location	Site number	Invertebrate Order richness per site	Average invertebrate Order richness per location
Sunny Creek Farm	1	23	
Sunny Creek Farm	2	27	
Sunny Creek Farm	3	19	
Sunny Creek National Park	1	18	
Sunny Creek National Park	2	22	
Sunny Creek National Park	3	15	
Windy Valley Farm	1	29	
Windy Valley Farm	2	32	
Windy Valley Farm	3	27	
Windy Valley National Park	1	20	
Windy Valley National Park	2	15	
Windy Valley National Park	3	18	

a Calculate the average invertebrate richness for each location and record it in the table.

b Graph the average invertebrate richness per location on the grid below.

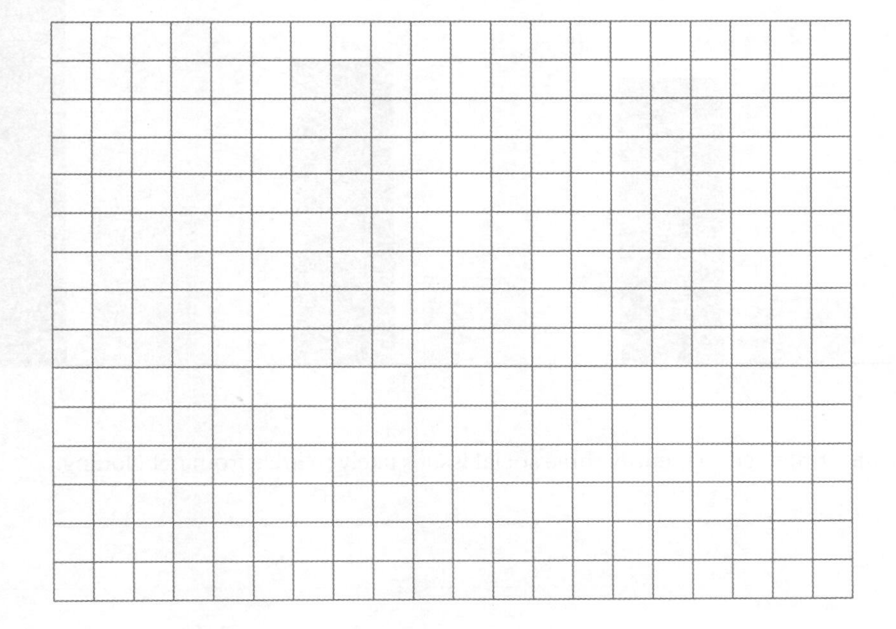

c Account for the trend shown in the graph.

Social, economic and cultural contexts in biotechnologies

STUDENT BOOK
Pages 305–6

LEARNING GOAL

Interpret secondary sources to assess the social and economic influences on whole-organism cloning.

There are only a few companies in the world that offer the service of pet cloning. Sinogene in China, ViaGen Pets in the USA and Sooam Biotech in South Korea are three of these companies. For around $100 000, pet owners can choose to have their cat, dog or horse cloned. The process involves collecting skin samples from the pet and sending it to the company for the extraction of viable cells. Cells are placed on a growth medium and harvested after two weeks. Two more animals are then needed: an egg donor and a surrogate mother. Under sedation, surgery is performed on these animals to first harvest eggs and then, once the egg has been prepared with the donor skin cells, up to 15 cloned embryos are injected into the uterus of the surrogate mother. Using this process, the chance of a successful pregnancy is approximately 40 per cent.

Cloning is a technology that has wide applications across a number of industries. Increased use of the technology, along with information available in the media, has influenced the way cloning is perceived by the general public. The graph below shows the public perception of cloning in the USA from 2002 to 2018.

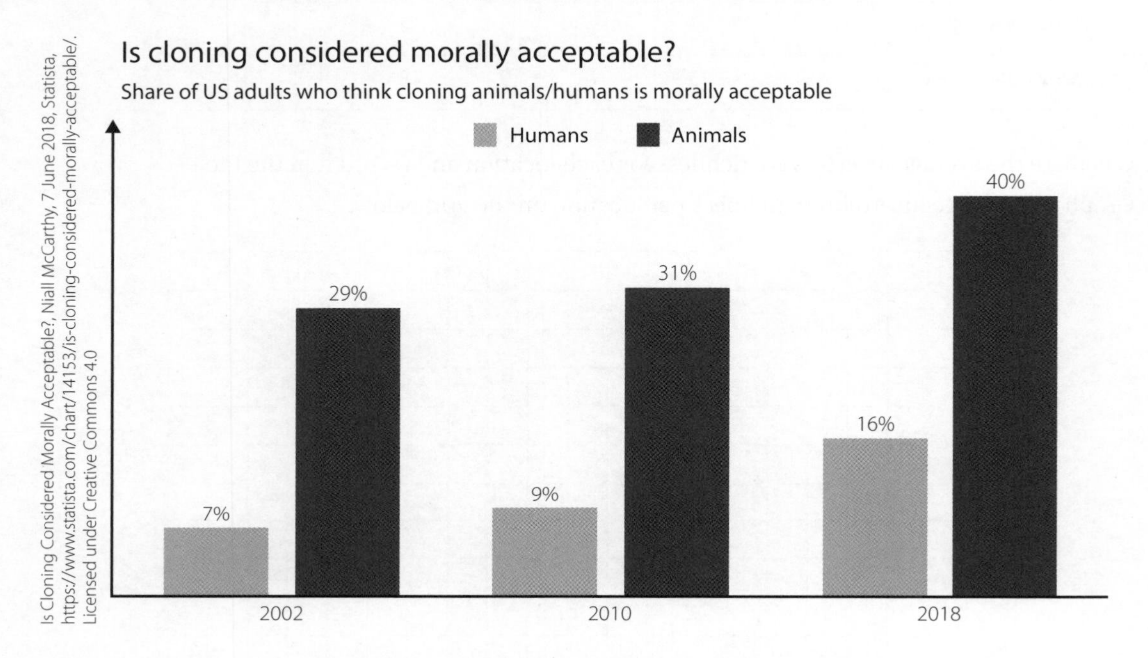

1 a Use the stimulus provided to identify three social issues likely to arise from pet cloning.

9780170449625

b Discuss one economic reason that may influence the size of the global pet cloning business.

c With reference to the stimulus material provided, predict what is likely to happen to the size of the pet cloning industry in five years' time.

HINT

Ensure you include specific reference to data and/or information included in the provided stimulus.

2 Outline the criteria required to assess the validity of a secondary source.

Module 6: Checking understanding

1 Explain how a point mutation could result in no phenotypic change.

2 Describe how the founder effect could result in a vast difference in allele frequencies between two populations of the same species.

HINT

You may like to include a diagram in your answer.

3 Bark from the Cinchona tree was chewed by Indigenous people in South America to ease fever associated with malaria. This type of biotechnology involves:

A the production of an antibiotic to cure fevers.

B fermentation.

C the use of medicinal plants.

D the genetic engineering of plants to make medicine.

4 What is the purpose of restriction enzymes in genetic modification?

5 Describe fermentation as it occurs in bread-, beer- and wine-making. Explain why fermentation is an example of a biotechnology.

6 Create a table to compare artificial insemination and artificial pollination.

7 Use an example to describe how biotechnology can decrease biodiversity in a population.

8 The process of xenotransplantation can ease the pressure on organ donation waiting lists. The graph below summarises the number of available organ donors and transplants conducted in the USA from 1991 to 2013.

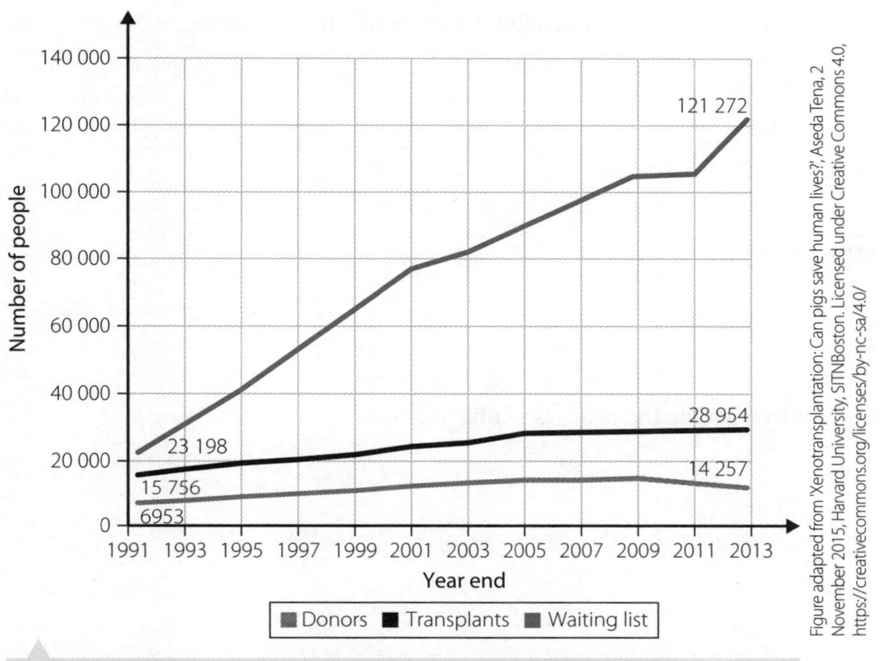

The number of organ donors (grey), the number of transplants conducted (black) and the number of people on the organ transplant waiting list (pink) in the USA from 1991 to 2013

a What is xenotransplantation? Include the terms from the word bank in your answer.

complementary	surface markers	proteins

b Using information from the graph, justify the use of xenotransplantation to produce human organs.

MODULE SEVEN »
INFECTIOUS DISEASE

Reviewing prior knowledge

1 Compare the terms *eukaryotic* and *prokaryotic*.

2 Use examples to distinguish between unicellular and multicellular organisms.

3 Define the term *disease*.

4 Use a flow diagram to describe the process of phagocytosis.

9780170449625

Causes of infectious disease

INQUIRY QUESTION: HOW ARE DISEASES TRANSMITTED?

WS 9.1 Classification of pathogens in plants and animals

STUDENT BOOK
Pages 314–31

LEARNING GOALS

Describe a range of plant and animal diseases caused by a variety of pathogens.
Distinguish between cellular and non-cellular pathogens.

A pathogen is an infectious agent that transmits disease from one host to another. Pathogenic organisms are varied and include unicellular and multicellular organisms. An understanding of these organisms is essential for treating infectious diseases and preventing further outbreaks.

1 Complete the table to describe a variety of pathogens.

Pathogen	Structure	Relative size	Example disease
Prion			
			Acquired immune deficiency syndrome (AIDS) caused by the human immunodeficiency virus (HIV)
	Unicellular prokaryotic organisms that have a cell wall		
Fungus			
	Unicellular eukaryotic organisms		
Macroparasite		~1 mm – 30 cm, although very wide range of sizes	

2 Distinguish between cellular and non-cellular pathogens using examples. Include the terms from the word bank in your answer.

living	non-living	protein	cell wall

3 Annotate the diagram to describe three diseases caused by pathogens in plants. In your answer, include the name of the pathogen, the name of the disease and the symptoms of the disease.

iStock.com/colematt

4 Some pathogens have complicated life cycles that require more than one host at different stages. Use secondary sources to describe the life cycle of one pathogen that inhabits at least two animals during its life cycle. In your answer, include the name of the pathogen, the names of the hosts, a description of the pathogen life stages and an explanation of how the pathogen is able to move between hosts.

Modes of transmission of infectious disease

LEARNING GOALS

- Describe direct and indirect modes of infectious disease transmission.
- Describe the transmission of dengue fever during an epidemic.
- Interpret data related to the transmission of dengue fever.

The term *transmission* refers to how a pathogen is spread from host to host. Understanding how diseases are spread between hosts plays an important role in being able to control the severity of disease outbreaks and protect the health of populations.

1 Identify the mode of transmission (direct or indirect) for each example given below.

	Transmission example	Mode of transmission
a	Shutterstock.com/ Dragana Gordic	
b	iStock.com/PeopleImages	
c	iStock.com/cometary	
d	Shutterstock.com/ dekazigzag	
e	iStock.com/frank600	

	Transmission example	Mode of transmission
f	iStock.com/AleMoraes244	
g	Shutterstock.com/Africa Studio	

2 a Use an example to distinguish between transmission of infectious disease via direct and indirect contact.

HINT

The verb *distinguish* asks you to look for differences.

b Conduct a peer review of part **a**. Ask a class member to mark your answer to part **a** using the following rubric.

Criteria	Circle the selected mark
• Includes an appropriate example of both direct and indirect transmission • Describes at least two differences between the modes of transmission	4
• Includes an appropriate example of both direct and indirect • Describes one difference between the modes of transmission OR • Includes an appropriate example of direct or indirect transmission • Describes two differences between the modes of transmission	3
• Describes both direct and indirect transmission OR • Describes direct transmission and provides an appropriate example OR • Describes indirect transmission and provides an appropriate example	2
• Provides some relevant information	1
Comment on one positive aspect of the answer:	
Provide a suggestion for improvement:	

9780170449625

3 An epidemic is an outbreak of an infectious disease in a community at a particular time. Multiple factors, including the mode of transmission, can work together to impact the severity of a disease epidemic.

a In 2016, many dengue fever epidemics occurred around the world. Use information you find in secondary sources to complete the table below.

Disease name	Dengue fever
Pathogen name and classification	
Disease transmission	
Symptoms	
Treatment	
Location of 2016 epidemic	
Number of cases and number of deaths during the 2016 epidemic	
Management and prevention strategies used to control the epidemic	

b Use Harvard referencing to cite one resource you used to complete part **a**.

> **HINT**
>
> Harvard referencing is an internationally recognised way to correctly format reference material.

c Assess the validity of the reference you identified in part **b**.

> **HINT**
>
> For a reference to be *valid*, the information should come from a qualified author and should be current. The information should also be accurate; i.e. consistent across multiple sources.

4 Many researchers have predicted that the transmission of dengue fever is likely to change with climate change. The data below are taken from a 2018 paper by Lee et al. entitled ***Potential effects of climate change on dengue transmission dynamics in Korea.*** Table 1 below shows the number of dengue fever cases in Korea each year between 2001 and 2008 and Table 2 shows the number of dengue fever cases per month from 2012–2016.

Table 1

Year	2001	2002	2003	2004	2005	2006	2007	2008
Cases	6	9	14	16	34	35	97	51
Year	2009	2010	2011	2012	2013	2014	2015	2016
Cases	59	125	72	149	252	154	255	313

Source: Lee H, Kim JE, Lee S, Lee CH (2018) Potential effects of climate change on dengue transmission dynamics in Korea. PLoS ONE 13(6): e0199205. https://doi.org/10.1371/journal.pone.0199205

Table 2

Year	Total	Jan	Feb	Mar	Apr	May	Jun	Jul	Aug	Sep	Oct	Nov	Dec
2012	149	4	7	5	6	3	9	20	26	24	18	16	11
2013	252	14	13	8	7	10	15	37	58	34	33	14	9
2014	165	13	8	5	6	10	14	32	26	11	18	17	5
2015	255	11	12	15	10	15	10	22	36	24	41	39	20
2016	313	32	36	32	23	18	22	39	42	24	13	18	14
Ave.	226.8	14.8	15.2	13	10.4	11.2	14	30	37.6	23.4	24.6	20.8	11.8

Source: Lee H, Kim JE, Lee S, Lee CH (2018) Potential effects of climate change on dengue transmission dynamics in Korea. PLoS ONE 13(6): e0199205. https://doi.org/10.1371/journal.pone.0199205

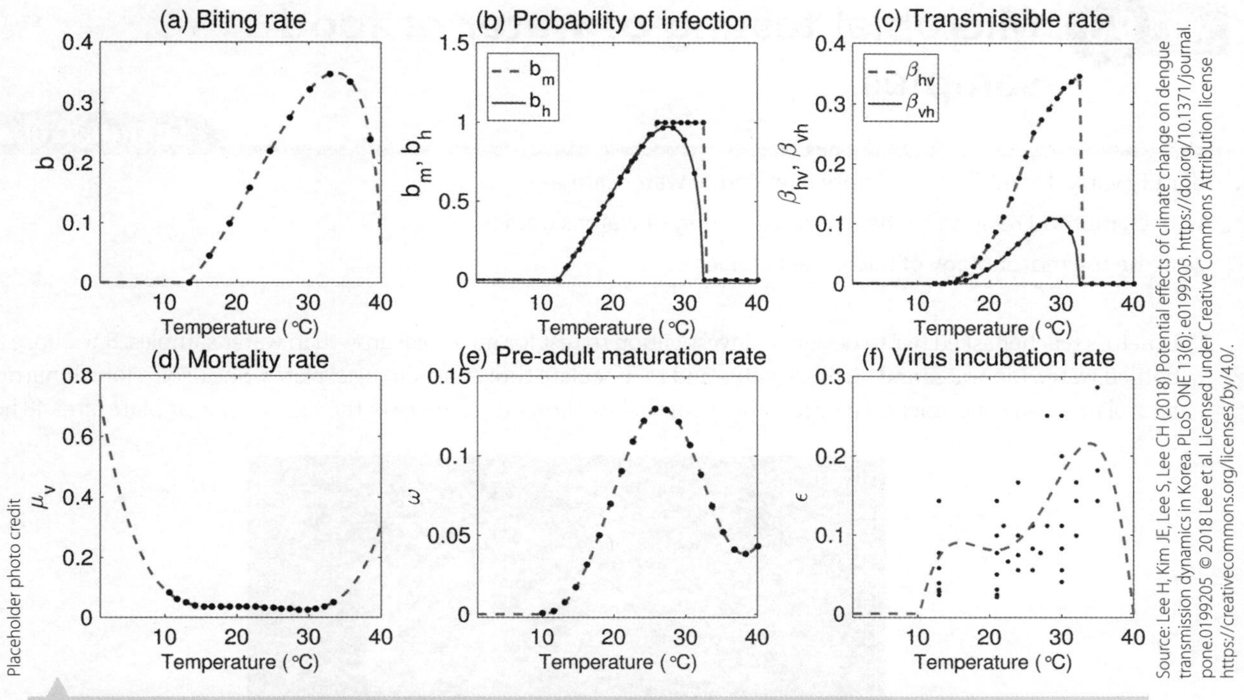

Source: Lee H, Kim JE, Lee S, Lee CH (2018) Potential effects of climate change on dengue transmission dynamics in Korea. PLoS ONE 13(6): e0199205. https://doi.org/10.1371/journal.pone.0199205 © 2018 Lee et al. Licensed under Creative Commons Attribution license https://creativecommons.org/licenses/by/4.0/.

Graphs a–f above show the influence of temperature on a number of factors that relate to the transmission of dengue via mosquitoes. The scales on the *y*-axes are referred to as 'parameters'.

a Describe the yearly trend suggested by the data. Include specific reference to data as part of your answer.

b Between 2012 and 2016, what time of year had the highest average number of cases of dengue fever in Korea?

c Use the data provided to describe the conditions most favourable for dengue transmission.

d In their paper, Lee et al. (2018) note that the climate of Korea has changed from warm temperate to subtropical; in other words, it is characterised by warmer, wetter conditions than previously experienced. While average summer temperatures are around 25°C, climate change predictions suggest continued increases in temperature into the foreseeable future.

Explain how these conditions will likely impact future dengue epidemics in Korea. In your answer, include specific reference to the stimulus provided.

Microbial testing of water or food samples

STUDENT BOOK
Page 337

LEARNING GOALS

Identify variables related to microbial testing of water samples.

Assess a method relating to the microbial testing of water samples.

Describe the morphology of microbial colonies.

1 Lynette's teacher asked her to design an investigation to test for microbial growth in water samples. She chose to test bottled water, tap water and distilled water, and to inoculate three nutrient agar plates. She used a fourth nutrient agar plate as an experimental control. The image below shows the growth of the tap water agar plate after 48 hours.

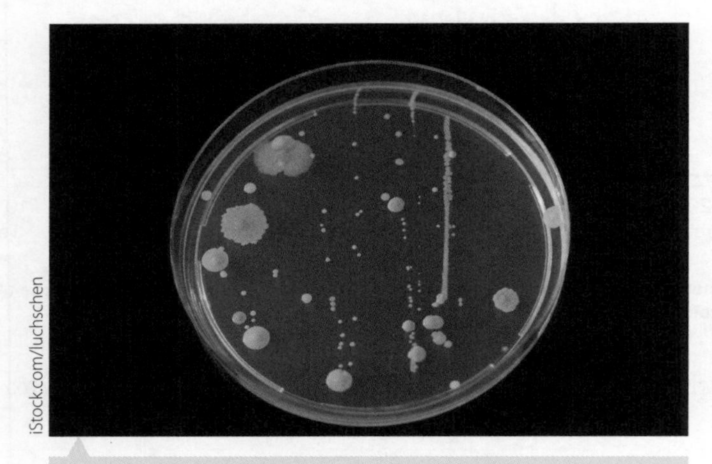

Microbial colonies growing on nutrient agar after 48 hours' growth time

a Write an appropriate aim for Lynette's investigation.

b Identify the dependent and independent variables for Lynette's investigation.

c Identify two controlled variables.

d Explain how controlling one of the variables you identified in part **c** ensures validity of the investigation.

e What treatment should be applied to the experimental control agar plate?

9780170449625

f What was the purpose of the experimental control?

g Justify the use of one control measure used to mitigate any risks from the investigation.

h As part of her investigation, Lynette was required to submit a method to her teacher. Her work is shown below.

1 Collect four nutrient agar plates.
2 Inoculate three agar plates with three different types of water: tap water, distilled water and bottled water.
3 Label the agar plates.
4 Place all agar plates in the incubator upside down at 25°C.
5 Count the colonies.

Use the rubric provided below to mark Lynette's method.

Criteria	Mark
Designs a valid experiment that: • outlines the sequential steps • identifies the independent variable and all relevant variables • includes quantities • includes a method of measuring and recording the dependent variable • includes repetition	5
Designs a valid experiment with one component missing	4
Designs an appropriate experiment that outlines steps and identifies variables	3
Outlines an experiment for investigating microbes in water but not in steps	2
Identifies a correct step in the method	1

i Selected mark: ____________

ii Explain why you selected that mark.

2 One way to gather results in a microbial growth investigation is to describe the morphology of the colonies that grow. The diagram below provides a summary of how to identify the morphological characteristics of a microbial colony.

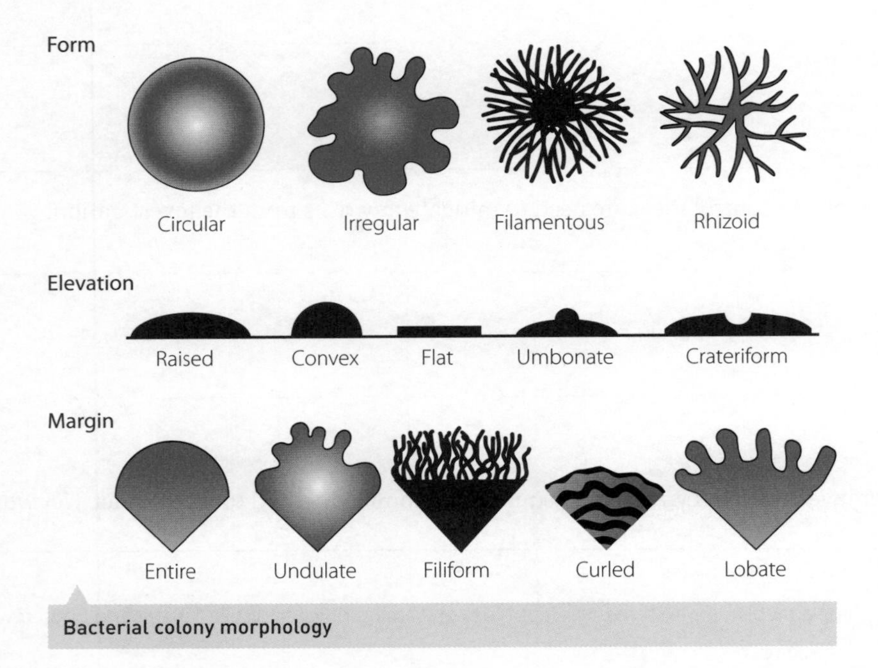

Bacterial colony morphology

Select two of the colonies from the tap water agar plate in question **1** and refer to the morphology diagram above to complete the following table.

Morphology	Colony 1	Colony 2
Form		
Elevation		
Margin		

Investigating the work of Koch and Pasteur

STUDENT BOOK
Pages 340–2

LEARNING GOALS

Describe Koch's postulates.

Describe Pasteur's swan neck flask investigation.

The work of Robert Koch (1843–1910) and Louis Pasteur (1822–1895) contributed significantly to microbiology and to our understanding of infectious disease. The technique of agar plating, for example, was invented by Koch, while one of Pasteur's contributions was the establishment of germ theory.

1 In 1984, Australian gastroenterologist Dr Barry Marshall applied the principles of Koch's postulates to prove *Helicobacter pylori* was responsible for causing gastritis and peptic ulcer disease. Dr Marshall and his colleague Robin Warren were awarded a Nobel prize for this work in 2005.

a Using information you gather from secondary sources, outline the steps Dr Marshall took to prove the link between *H. pylori* and gastritis.

b Dr Marshall is said to have fulfilled Koch's postulates by infecting a gerbil with *H. pylori* after he famously completed the experiment on himself. Explain why this step 'fulfilled' the postulates.

2 In 1859, Louis Pasteur designed his now famous swan-neck flask experiment, which allowed the spontaneous generation theory to be discredited. The design Pasteur used for this investigation is now often used to demonstrate the application of the scientific method.

For his investigation, Pasteur used two flasks, such as the one pictured below, to boil broth, after which he broke off one of the swan necks. After a while he observed microbial growth in the flask with the broken neck.

An early swan-neck flask similar to that used by Pasteur

a Identify the dependent and independent variables in Pasteur's investigation.

______

______

b Identify one variable Pasteur would have controlled.

______

c Explain why controlling the variable you identified in part **b** contributed to ensuring the investigation was valid.

______

______

______

______

d Write a hypothesis for Pasteur's investigation.

______

______

e What could Pasteur have done to ensure his investigation was reliable?

______

______

f Why does the swan-neck flask investigation discredit the spontaneous generation theory?

______

______

______

Causes of disease and effects on agricultural production

LEARNING GOAL

Describe causes of plant and animal diseases and their effects on agricultural production in Australia.

The agricultural industry contributed $58 billion to the Australian economy during the 2018–2019 financial year. Agricultural diseases not only impact the health of the plant and animal hosts, but can also reduce agricultural production and have large economic effects.

1 The Grains Research and Development Corporation published a report in 2009 by Murray and Brennan entitled *The Current and Potential Costs from Diseases of Wheat in Australia*. The table below is taken from this report and shows the ranked losses of wheat due to diseases in northern, southern and western regions of Australia, and in Australia as a whole.

Ranking of disease losses (1 = highest loss)

Disease	Northern		Southern		Western		Australia	
	Potential	Present	Potential	Present	Potential	Present	Potential	Present
Yellow spot	3	1	6	7	1	1	2	1
Stripe rust	1	2	2	1	5	5	1	2
Septoria nodorum	24	24	25	18	2	2	7	3
Crown rot	4	4	3	4	10	8	5	4
Root lesion *Pratylenchus neglectus*	9	6	4	5	3	3	6	5
Rhizoctonia barepatch	26	26	10	3	4	4	10	6
Cereal cyst nematode	26	26	1	2	12	17	3	7
Root lesion *Pratylenchus thornei*	5	3	13	10	15	14	9	8
Common bunt	11	14	5	12	9	12	11	16
Stem rust	2	9	7	21	7	10	4	17

Source: *The Current and Potential Costs from Diseases of Wheat in Australia*, Gordon M. Murray and John P. Brennan, Grains Research and Development Corporation, 2009

a Use the data in the table to identify the three diseases that currently cause the most loss in the Australian wheat industry.

b Select one of the diseases you identified in part **a** and use secondary sources to complete the table below.

Disease name	
Pathogen classification	
Pathogen name	
Disease transmission	
Symptoms	
Management and prevention strategies	

c The table below from Murray and Brennan (2009) shows the incidence of wheat diseases as a proportion of years when the disease occurs and as a proportion of the crop area when the disease develops. Use the information in the table to answer the questions that follow.

Incidence of wheat diseases as a proportion of years when disease occurs (%) and as a proportion of the crop area affected when the disease develops (%) in the GRDC regions and Australia

Disease	Northern		Southern		Western		Australia	
	Years	Area	Years	Area	Years	Area	Years	Area
Necrotrophic leaf fungi								
Bipolaris leaf spot	20.0	25.0	5.8	3.0	0.0	0.0	6.3	6.0
Wirrega blotch	0.0	0.0	48.1	0.5	0.0	0.0	21.0	0.2
Septoria tritici blotch	0.0	0.0	24.3	17.1	1.5	1.9	11.2	8.2
Septoria avenae blotch	0.0	0.0	2.0	2.0	0.0	0.0	0.9	0.9
Septoria nodorum blotch	18.6	23.2	5.1	4.3	66.7	100.0	30.8	43.8
Ring spot	0.0	0.0	69.7	24.7	20.0	25.0	38.0	20.2
Yellow spot	55.8	84.3	79.8	39.1	66.7	100.0	70.4	70.5
Biotrophic leaf fungi								
Powdery mildew	37.3	25.0	68.2	5.4	59.0	25.0	58.9	16.4
Stem rust	66.9	75.7	18.4	52.2	35.3	25.0	33.9	46.4
Leaf rust	66.9	53.2	30.4	56.5	24.7	25.0	35.1	44.0
Stripe rust	91.2	79.2	84.2	80.7	46.9	60.2	71.5	72.7
Downy mildew	20.0	25.0	3.0	0.2	4.4	5.5	6.7	6.8
Flag smut	38.6	25.0	25.7	8.4	29.1	25.0	29.4	17.7

→

Disease	Northern		Southern		Western		Australia	
	Years	Area	Years	Area	Years	Area	Years	Area
Root and crown fungi								
Foot rot	37.1	23.2	62.6	13.9	15.3	19.1	40.0	17.6
Crown rot	99.3	37.4	81.2	58.2	49.8	25.0	72.8	41.8
Take-all	37.1	23.2	74.9	32.1	49.1	25.0	58.1	27.8
Damping off/root rot	35.9	25.0	68.2	63.2	49.1	25.0	55.0	41.7
Rhizoctonia barepatch	0.0	0.0	76.0	50.5	60.8	65.9	56.0	46.8
Eyespot	0.0	0.0	11.9	9.5	0.0	0.0	5.2	4.1
Basal rot	0.0	0.0	2.0	4.0	0.0	0.0	0.9	1.8
Common root rot	83.0	43.4	69.9	65.6	4.7	5.9	47.9	39.0

Source: *The Current and Potential Costs from Diseases of Wheat in Australia*, Gordon M. Murray and John P. Brennan, Grains Research and Development Corporation, 2009.

i Which group of fungal diseases affects the largest area in Australia?

____

ii Which Australian region is least likely to be impacted by 'blotch' diseases?

____

iii Crown rot is a root and crown fungus. What proportion of years are impacted by this disease in Australia?

____

iv On average, what proportion of years in the Southern region are impacted by biotrophic leaf fungi?

____

Adaptations in pathogens that facilitate entry and transmission

LEARNING GOAL

Describe adaptations of different pathogens that facilitate entry into and transmission between hosts.

Pathogens and their hosts exist in close evolutionary battles. Pathogens aim to enter their host's tissues or cells and a host's defence mechanisms work hard to prevent the pathogen from gaining entry. Pathogenic organisms utilise adaptations to first adhere to host cells, followed by invasion of host tissues or cells. Adaptations are also utilised by pathogens to facilitate host-to-host transmission.

1 Use an example to distinguish between pathogen adhesion and invasion.

2 A range of strategies are used by pathogens to facilitate adhesion and invasion. Complete the table to describe some of these strategies.

Adhesion structure	Type of pathogen	Adhesion or invasion?	Labelled diagram and/or description: how does it work?
Surface proteins			
Pili			
Fimbriae			

9780170449625

Adhesion structure	Type of pathogen	Adhesion or invasion?	Labelled diagram and/or description: how does it work?
Microtubule protrusion			
Secretion of hydrolytic enzymes			

3 *Helicobacter pylori* is a bacterium that can cause gastritis in humans. Studies suggest that the shape of the bacterial cell is essential for the bacterium to establish persistent infection. Use information from secondary sources to annotate the diagram below to describe two physical features of this pathogen that allow it to evade host defence mechanisms and cause infection.

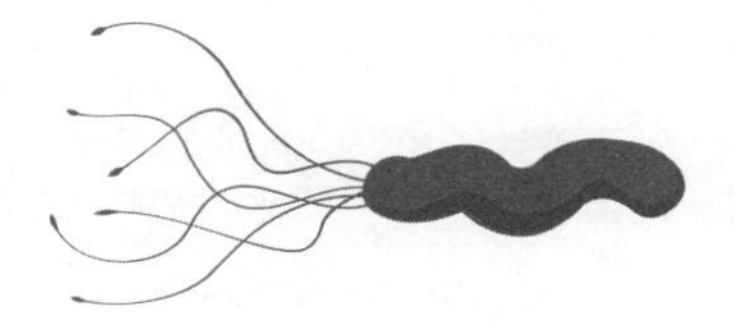

4 Use an example to describe the link between pathogen adaptations that facilitate transmission, and the mode of transmission of a pathogen between hosts.

____

____

____

____

____

____

10 Responses to pathogens

INQUIRY QUESTION: HOW DOES A PLANT OR AN ANIMAL RESPOND TO INFECTION?

WS 10.1 Plant responses to infection

STUDENT BOOK Pages 364–6

LEARNING GOAL

Describe the physical and chemical components of a plant's immune system.

1 Match each of the physical barriers in a plant's immune system to its description.

	Barrier
a	Lignification
b	Stomatal closure
c	Tylose formation
d	Cell wall
e	Waxy cuticle
f	Bark

	Description
i	Stops water from collecting on the surface of the plant and therefore any pathogen that needs water to proliferate. Water running off the leaf essentially cleans the leaf
ii	Thick cellulose outer layer of cells that limits the entry of pathogens into plants
iii	Waterproof and indigestible component of cell walls that thickens the cell walls of xylem vessels, preventing pathogen entry through xylem vessels
iv	Balloon-like swellings that block xylem vessels to prevent the spread of pathogens through the xylem
v	A thick outer layer of the trunk of a woody plant; contains many defensive chemicals such as terpenoids, alkaloids and hydrolytic enzymes to protect the plant from pathogens
vi	The gas exchange structures close to prevent any airborne pathogen entry into the plant from the outside environment

2 Below is a diagram depicting tylose formation in the xylem of a plant. Use the diagram and your knowledge of xylem to explain how this feature is necessary to prevent the spread of a water-living pathogens in plants but is also harmful to the plant.

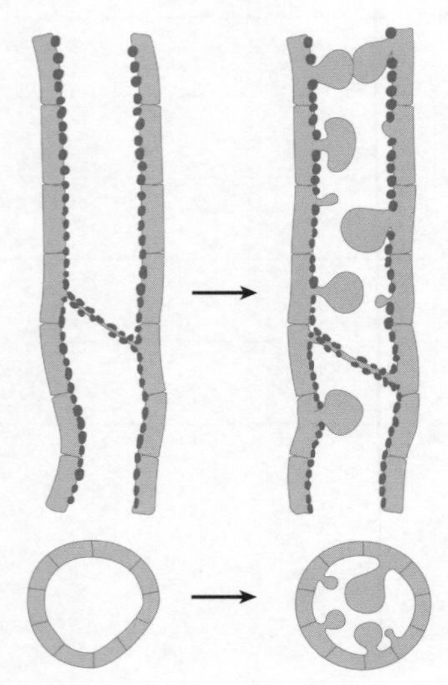

9780170449625

3 Using secondary sources, complete the table describing some of the chemical defences in the plant immune system.

Chemical plant defence	Named example	Mode of action
Hydrolytic enzymes		
Alkaloids		
		Found in mint and have antibiotic and antifungal properties
	Tannins	
	Defensins	

WS 10.2 Active immune response in plants

STUDENT BOOK
Pages 366–9

LEARNING GOAL

Explain the action of a plant's active immune response.

1 Plant pathogens can be broken down into two main groups: necrotrophs and biotrophs. Using secondary sources, complete the table describing these two groups of pathogens.

Term	Definition	Type(s) of pathogens
Biotroph		
Necrotroph		

2 Panama disease is a plant disease that affects bananas. You can find a case study about Panama disease on page 349 of the Student Book.

a Describe the symptoms of Panama disease.

b The xylem and phloem are damaged by Panama disease. Explain how this would be detrimental to the health of the plant.

3 Most of the banana plants in the region affected by Panama disease are all the same species and many are genetically identical. Explain how this type of monoculture is dangerous for farmers.

9780170449625

4 Below is a diagram outlining the hypersensitive response in the active immune system of plants. Correctly label each part of the diagram using the annotations listed.

- Cell wall increases in thickness, adding lignin and more cellulose to the structure, sealing the pathogen inside
- Fungal pathogen sends hyphae into the cell through cell wall
- Antimicrobial molecules inside the cell are triggered, signalling the cell wall to seal off the infected cell
- Healthy cell
- Infected cell dies and releases the signalling molecule methylsalicylic acid, which triggers healthy cells to produce antipathogenic chemicals
- Fungal pathogen attaches to the cell

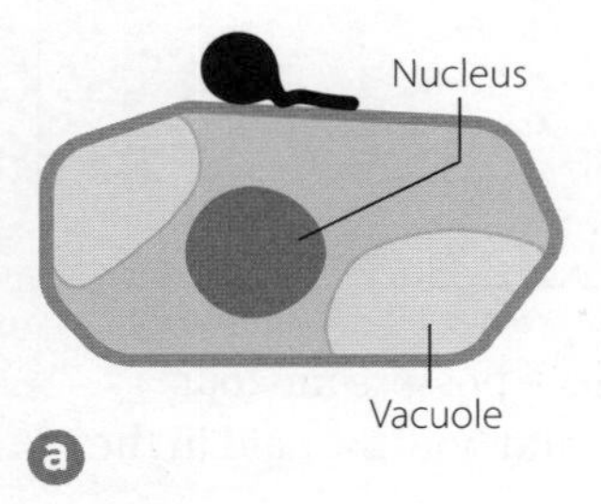

a

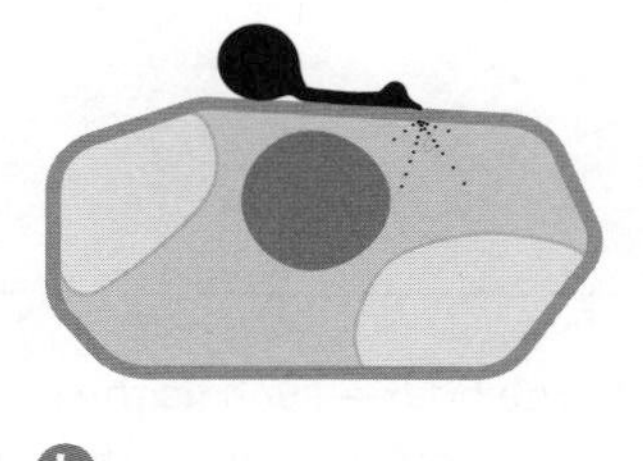

b

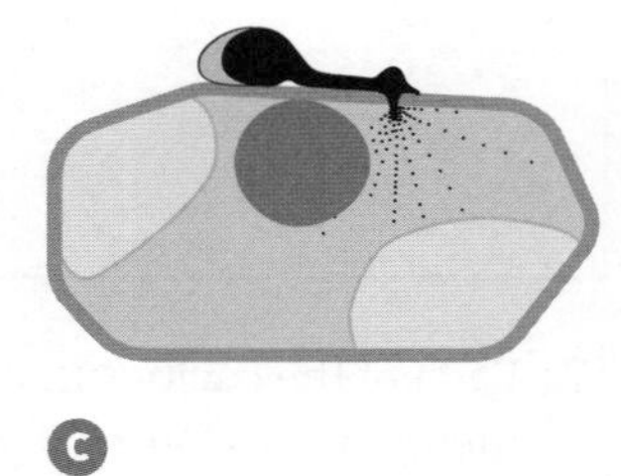

c

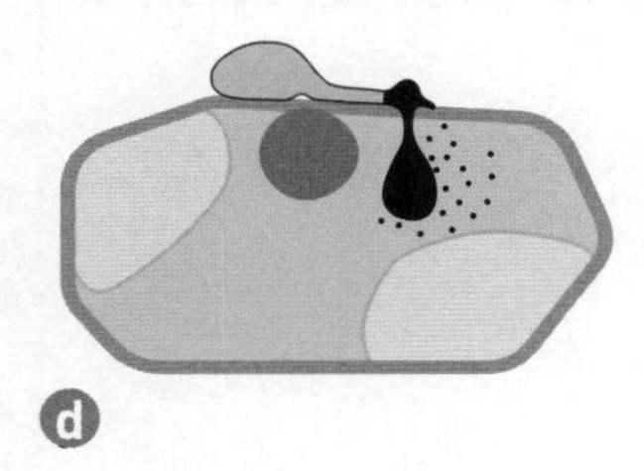

d

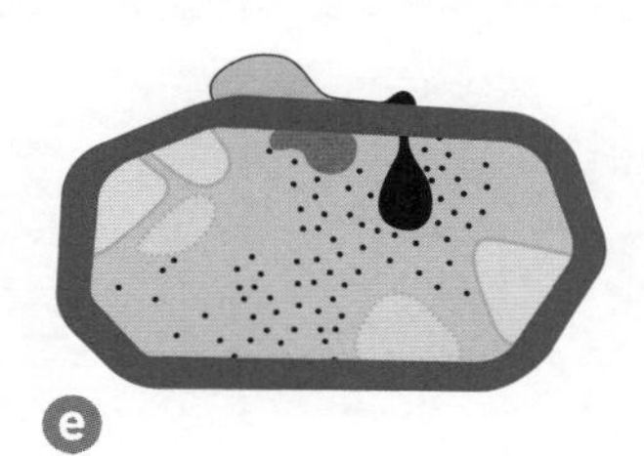

e

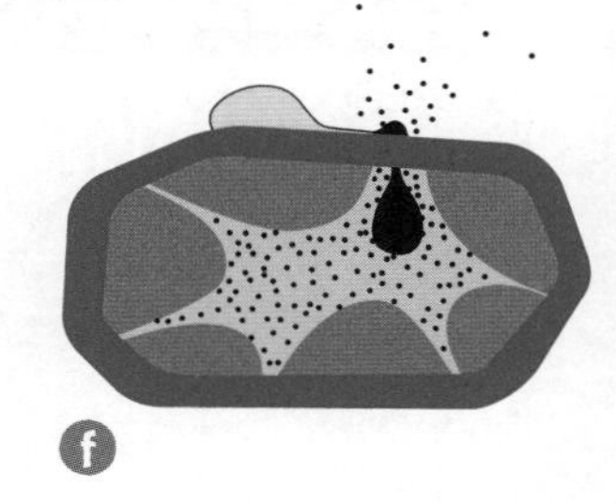

f

WS 10.3 Physical barriers to infection in humans

LEARNING GOAL

Describe the physical barriers found in humans that prevent infection.

1 Human skin is the first physical barrier that potential pathogens must pass through before entering the body. The skin has several mechanisms to prevent the entry of pathogens such as the 'bricks and mortar' arrangement of the stratum corneum.

Annotate the diagram below to indicate how shedding of the stratum corneum aids the immune system.

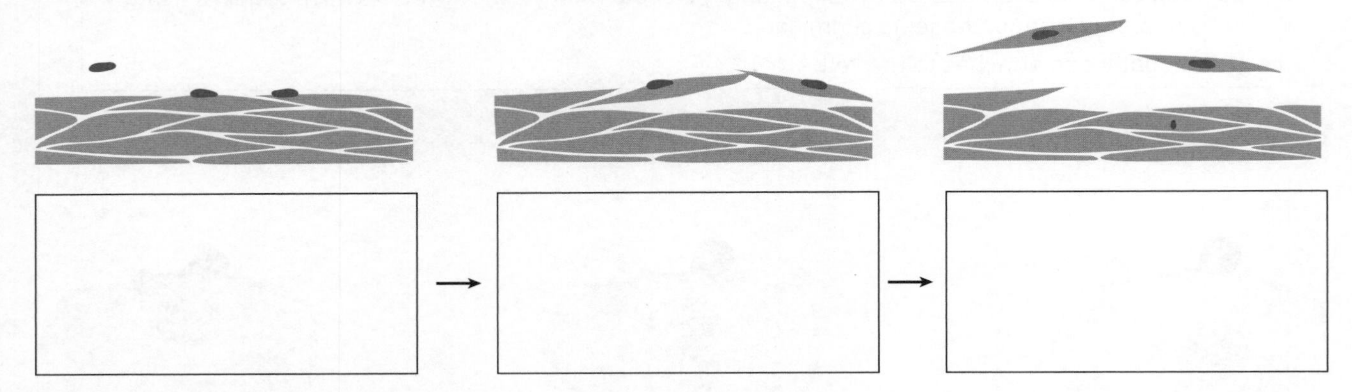

2 Each of the cavities that form an entry into the body – such as the eyes, mouth and anus – possess mucous membranes. Complete the table describing the components of the mucous membrane and how they aid in the immune response.

	Components of a mucous membrane	Diagram	Description
a	Cell junction		
b	Mucus		
c	Cilia		

9780170449625

3 Acne is a common skin condition, particularly with teenagers. Acne is an infection of the skin by the bacteria *Propionibacterium acnes*. Human skin is around 5.4–5.9 on the pH scale, making it slightly acidic. This is yet another part of its defence mechanisms. *Staphylococcus epidermidis* is a commensal bacterium that helps to outcompete and prevent the growth of other pathogenic bacteria on the skin.

a Draw a line graph using the data below.

pH	Percentage growth (%)	
	Staphylococcus epidermidis	*Propionibacterium acnes*
1	0	0
1.5	0	0
2	0	0
2.5	0	0
3	0	0
3.5	0	0
4	20	0
4.5	40	20
5	100	50
5.5	100	75
6	100	100
6.5	100	100
7	100	100
7.5	50	100
8	20	50
8.5	0	25
9	0	0

b Describe the term *commensalism* with reference to the passage on the previous page.

c Determine the optimum pH range for *Staphylococcus epidermidis*.

d Determine the optimum pH range for *Propionibacterium acnes*.

e Many people with acne resort to washing their faces regularly with soap. Regular soap usually ranges from pH 6–8. Some specialist acne brands produce face wash products that are slightly acidic. Explain why, when treating acne, it would be more beneficial to wash using a specialised product than to wash using regular soap.

WS 10.4 Chemical barriers to infection in humans

LEARNING GOAL

Describe the chemical barriers against infection in humans.

1 Complete the table describing the human body's chemical defences against infection.

	Location	How does it prevent infection?
Saliva		
Tears		
Urine		
Sebum and sweat		
Gastric juices		

2 Below is a column graph indicating the different stomach pH levels of different organisms.

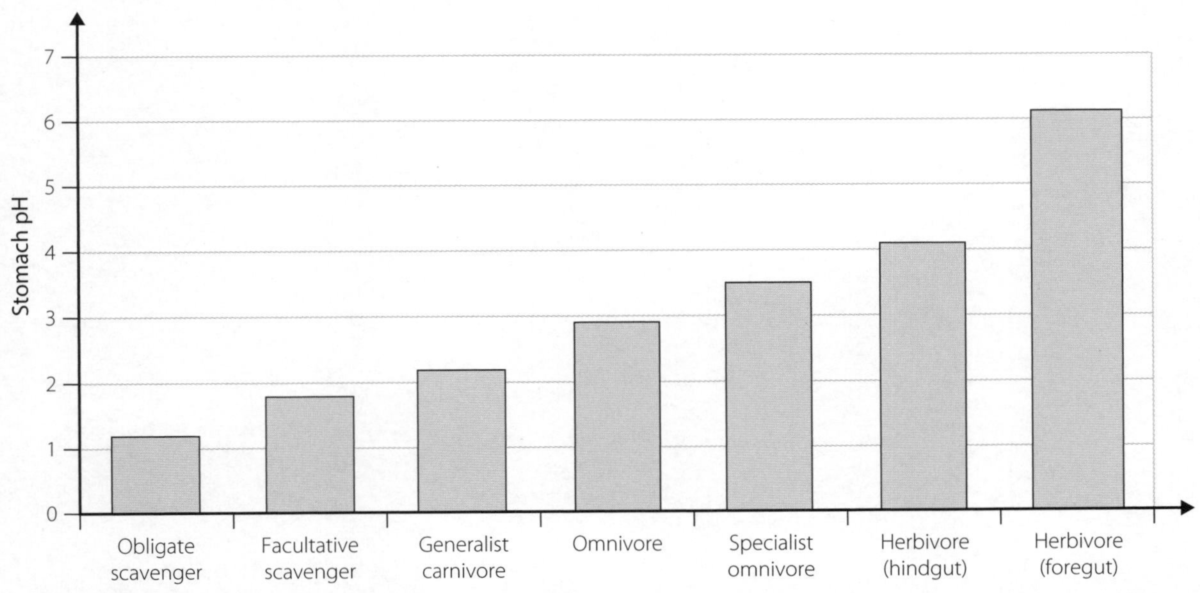

Source: Beasley DE, Koltz AM, Lambert JE, Fierer N, Dunn RR (2015) *The Evolution of Stomach Acidity and Its Relevance to the Human Microbiome*. PLoS ONE 10(7): e0134116. https://doi.org/10.1371/journal.pone.0134116. Licensed under Creative Commons 4.0.

a A vulture is an obligate scavenger. Suggest a reason why the stomach acid of a vulture has a very low pH.

b The stomach acid in a human is around pH 1.5. Although a plant-based diet can offer many benefits to humans with appropriate supplementation, vegans often use the argument that 'humans are not supposed to eat meat' to convince people to change their diets. Explain how the data above contradicts this statement.

WS 10.5 Innate immune responses in humans

LEARNING GOAL

Describe the inflammatory and fever response to infection.

When the body's physical and chemical barriers are breached, one of the next responses of the immune system is inflammation.

1 Use the text below to complete the flow chart of the process of inflammation. Draw a diagram of each stage.

- Phagocytes leave the blood and enter the interstitial spaces to fight invading pathogens.
- Blood vessels widen, increasing blood flow to the area.
- The walls of capillaries become more permeable, and fluid from the blood enters the interstitial space.
- The skin or other physical barrier is penetrated and pathogens enter the body.

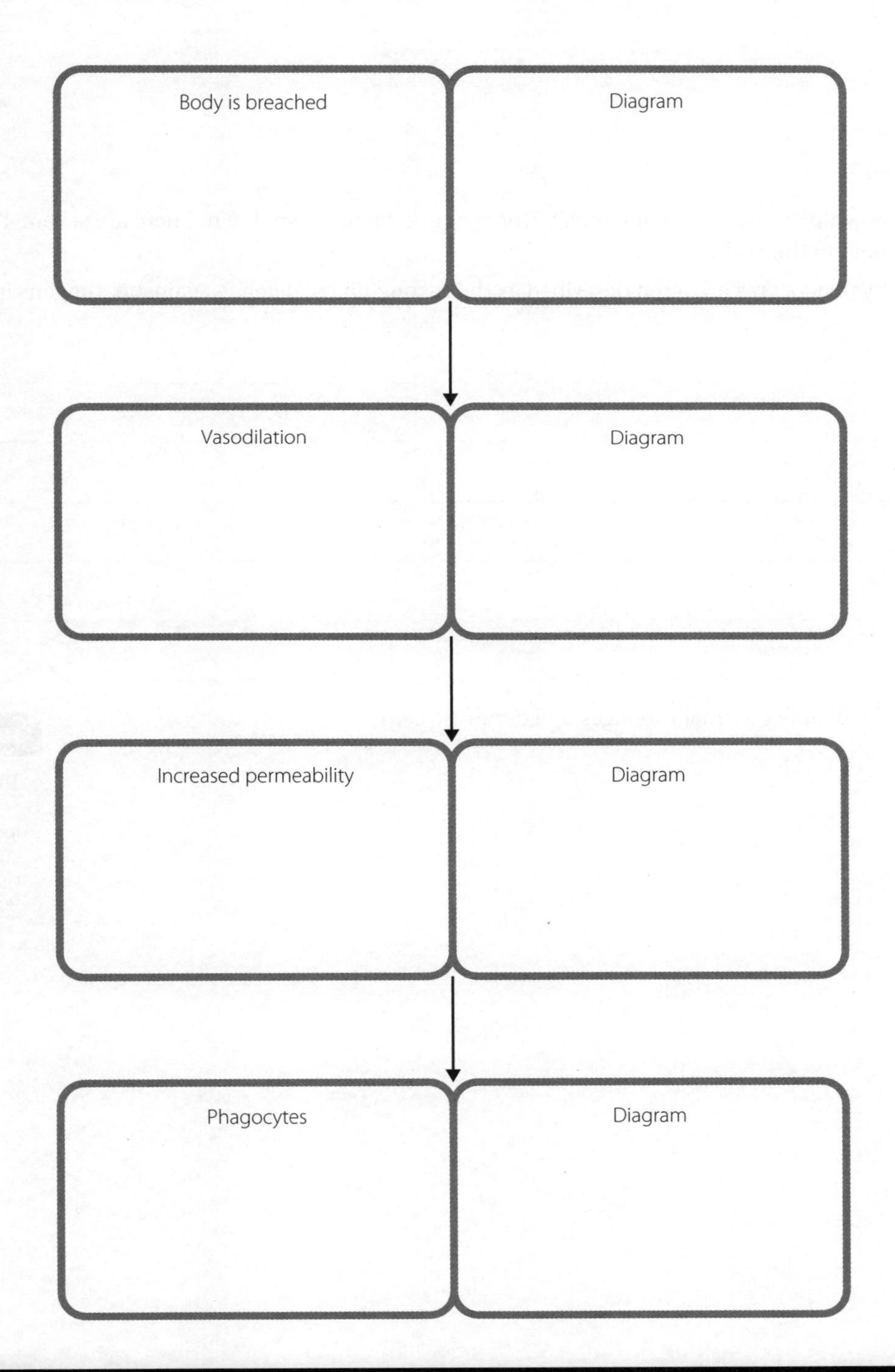

2 There are several symptoms of inflammation that a doctor looks for during a general examination. Complete the table to explain the processes in the body that lead to each of the following symptoms.

Symptom	Cause
Redness	
Heat	
Swelling	
Pain	

3 Phagocytes are white blood cells that engulf pathogens and break them down. There are several types of phagocytes found in the body.

a Explain why phagocytes are often described as the second line of defence against pathogens in the immune system.

b Compare the phagocytes macrophages and dendritic cells.

HINT

The verb *compare* requires you to look for similarities and differences. Use a table or Venn diagram.

c The diagram below shows the process of phagocytosis.

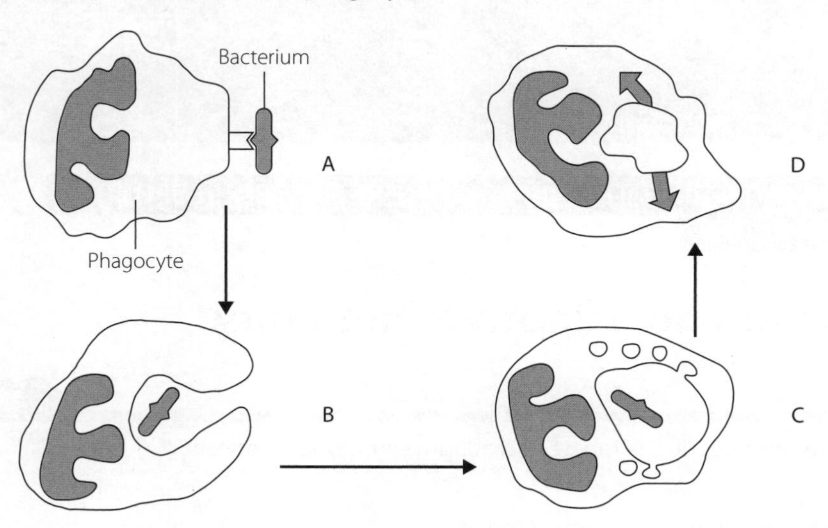

Describe what is happening at each stage.

4 Below is a flow diagram summarising the body's mechanism for fever.

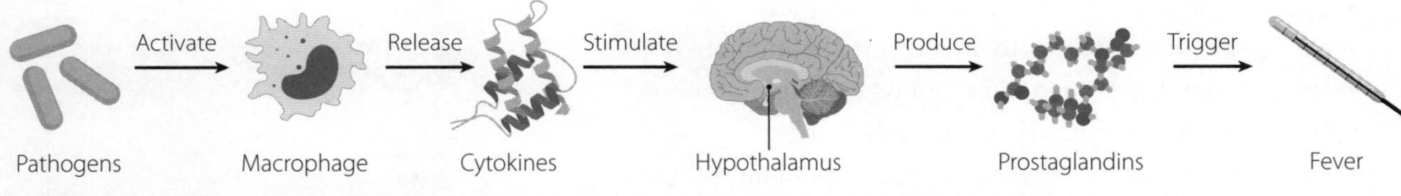

Source: 'Anti-Inflammatory, Antipyretic and Antibacterial Study of Kabasura Kudineer Choornam' by Devasia Neethu, KMCH College of Pharmacy, 2017

a Using the diagram, explain how macrophages stimulate fever.

b Explain, using your knowledge of enzymes, why changing the internal temperature of the body could be dangerous for the organism.

11 Immunity

INQUIRY QUESTION: HOW DOES THE HUMAN IMMUNE SYSTEM RESPOND TO EXPOSURE TO A PATHOGEN?

Innate and adaptive immunity

LEARNING GOAL

Compare the processes involved in the innate and adaptive immune response.

1 Use the words in the word bank to complete the passage.

adaptive	penetrate	inflammation	phagocytosis
innate	fever	chemical	physical

The ____________ immune system consists of ____________ and ____________ barriers that prevent the entry of pathogens into the body. It utilises non-specific cellular responses such as ____________ and biochemical responses such as ____________ and ____________. If a pathogen manages to ____________ or overwhelm the innate immune system, the ____________ immune response is the next and final line of defence against disease.

2 Compare adaptive and innate immunity using the phrases below.

- Targets a specific area of the body to protect
- Uses chemicals and cells to kill pathogens
- Can fight throughout the whole body at once
- Has a specific response to an individual pathogen
- Responds in the same generalist way to all pathogens
- Does not have a memory of past infections
- Remembers specific pathogens and how to fight them
- Present at birth
- Develops over a person's lifetime

Innate	Similarities	Adaptive

3 Any molecule that can be recognised by lymphocytes and triggers an immune response is known as an antigen. Most cellular pathogens, such as fungi, bacteria and parasites, have antigens made of protein on their surface. Antigens are also found on viruses and can even be classified as toxins, which some bacteria secrete.

Explain why a pathogen with no antigen would be extremely dangerous.

4 The adaptive immune system utilises two main types of cells: B cells and T cells. Complete the table describing each of these cell types.

Type of cell	Where do these cells develop and mature?	Types of cells found in the body	How does this cell destroy pathogens?	Humoral or cell-mediated?
B cell				
T cell				

WS 11.2 Humoral immune response

STUDENT BOOK
Pages 400–2

LEARNING GOAL

Explain the humoral response in active immunity.

During the development of B cells, many millions of variations of membrane-bound antibodies are generated, and each B cell has its own unique antibodies. Each one of these antibodies has the ability to bind to a specific type of antigen. This means that through a process of random generation a B cell in your body right now may have the antibody on its surface to destroy a pathogen that does not exist yet.

1 Explain why an antibody will only bind to one type of antigen.

2 As B cells mature in the bone marrow, only 10 per cent complete this process and are released into the blood. This is because B cell clones that present 'self' antibodies or 'autoantibodies' on their surface are destroyed. Explain why B cells with autoantibodies must be destroyed to maintain the health of the organism.

9780170449625

3 Use the following labels to annotate the diagram below, which describes the humoral primary response to a pathogen.

- Antigen
- Naive B cell
- Surface antibody binding to matching antigen
- Activated B cell
- Formation of many clones
- Plasma cells created to produce many antibodies
- Memory cell created to respond to future infections
- Antibodies produced to destroy antigen-containing pathogen

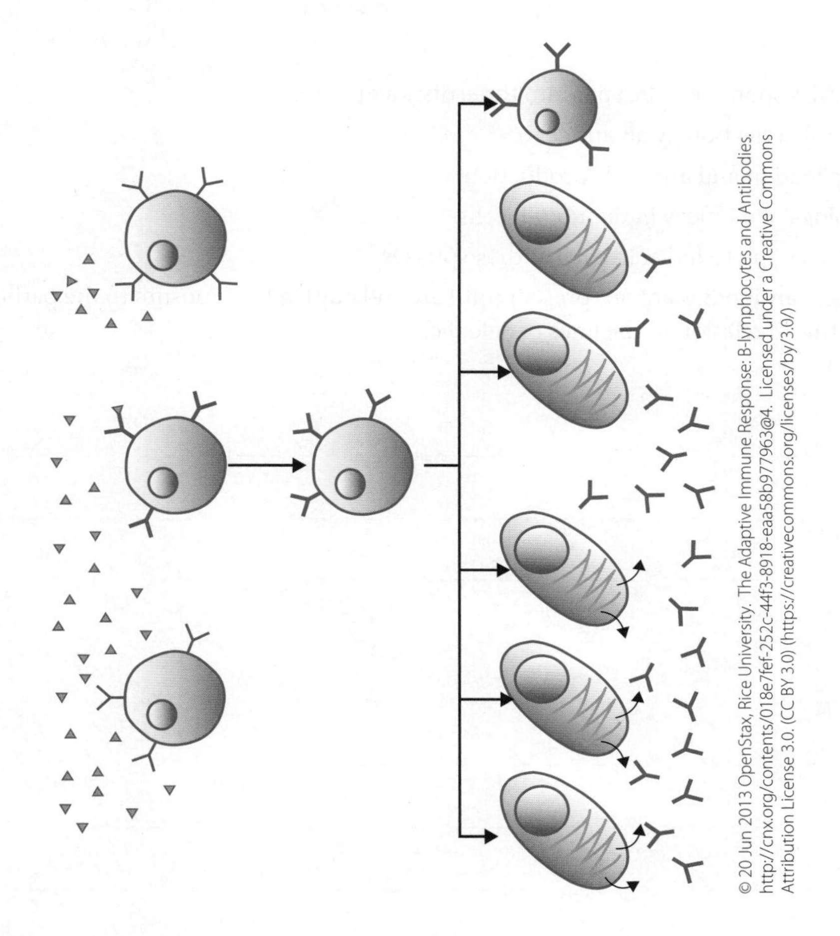

4 Discuss why memory B cells are important in protecting the body from future infection.

5 The graph below shows the level of antibodies found in the blood after exposure to a pathogen.

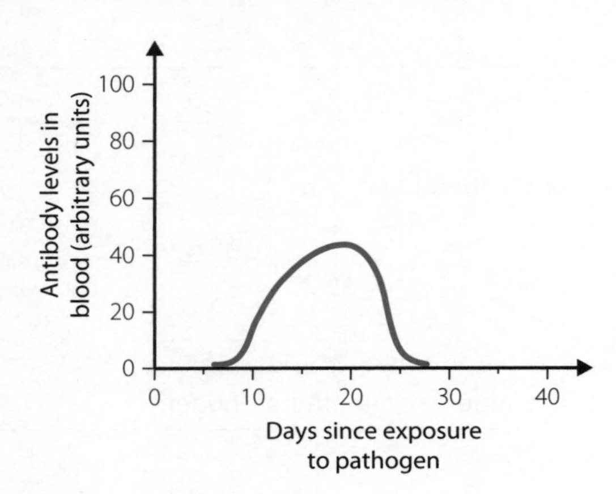

a The humoral response can be split into three distinct phases.

1: Naive B cell activation by an antigen.

2: B cell replication and antibody production

3: Immunological memory in memory B cells

Annotate the graph to indicate each of these phases.

b On the graph, antibodies are not present until around day 6 after exposure to the pathogen. Describe the reason for this in relation to the lines of defence.

6 Below is a graph showing antibody quantity after first and second exposure to an antigen.

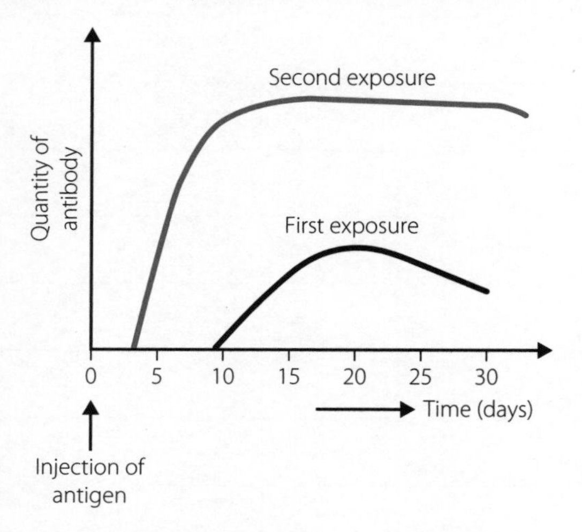

9780170449625

a Account for the trends shown in the graph.

> **HINT**
>
> When you are asked to *account for* a trend, you must describe the trend and then give a possible explanation for the trend. It is always best practice to use data from any graph in your answer.

b Compare the first exposure to the second exposure.

> **HINT**
>
> A response to a *compare* question must have similarities and differences, but does not need an explanation. It can be written as a table.

WS 11.3 Antibodies

LEARNING GOAL

Describe antibody activity in fighting infection.

1 Complete the table describing antibody function by drawing a diagram of each description.

Process description	Diagram of process
Neutralisation of bacterial toxins: antibodies bind to bacterial toxins, blocking the action of the toxin	
Agglutination: antibodies bind antigens on the surface of the cell and form antigen–antibody complexes, activating phagocytes and the complement cascade, leading to antigen/cell destruction	
Opsonisation: bound antibodies tag pathogens for destruction by phagocytes	

9780170449625

2 The adaptive immune system develops over a person's lifetime as they are exposed to more pathogens. A mother's milk contains a substance known as colostrum, which has many of the mother's antibodies within it.

a Describe why colostrum is important for the health of newborn babies.

b Explain why a baby will not become immune to the diseases its mother has had through drinking colostrum.

WS 11.4 Cell-mediated immunity

STUDENT BOOK
Pages 404–6

LEARNING GOAL

Explain cell-mediated active immunity.

The innate and adaptive immune systems do not function independently. They work together to protect the organism from infection. Below is a diagram that shows part of this interaction.

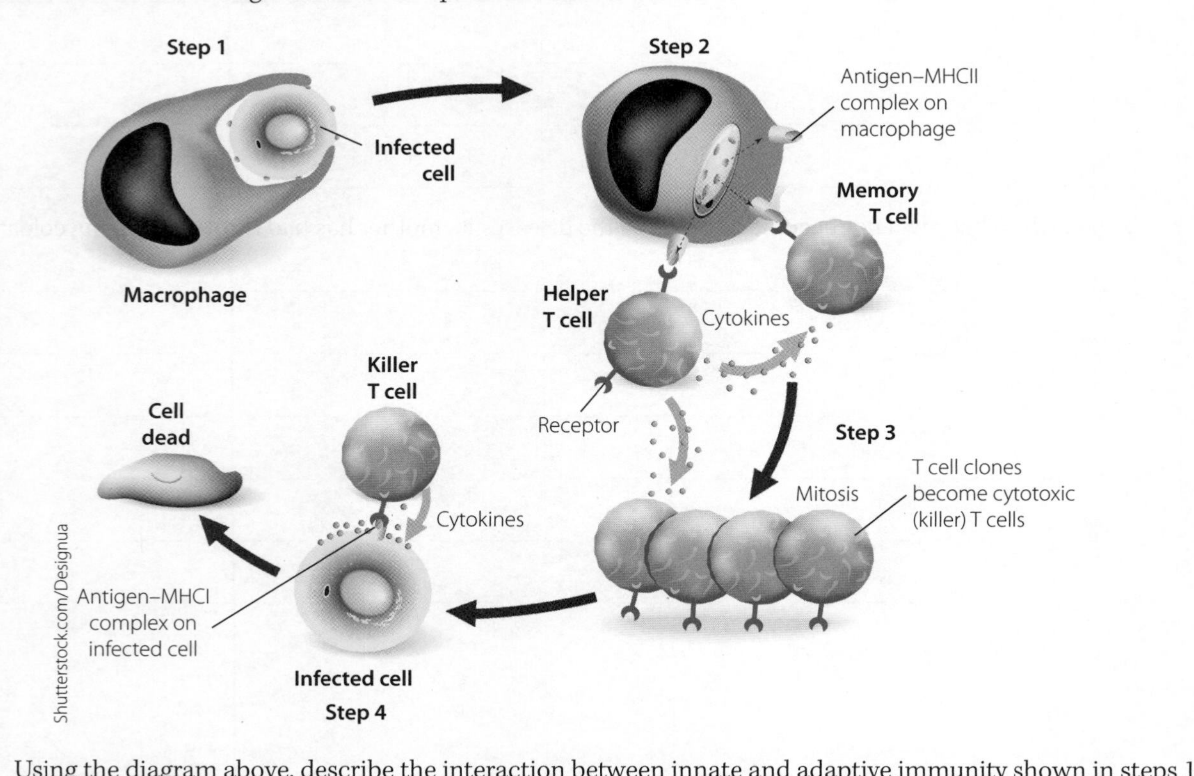

1 a Using the diagram above, describe the interaction between innate and adaptive immunity shown in steps 1 and 2.

____

____

____

____

____

____

b Explain the process being depicted in step 4.

____

____

____

____

c B cells and T cells do not work independently. Identify the step where B cells could be added to the diagram.

____

2 Explain the role of helper T cells in cytotoxic T cell and B cell activation.

3 Compare cytokines and foreign antigens.

12 Prevention, treatment and control

INQUIRY QUESTION: HOW CAN THE SPREAD OF INFECTIOUS DISEASES BE CONTROLLED?

WS 12.1 Local, regional and global spread of infectious disease

STUDENT BOOK
Pages 432–6

LEARNING GOALS

Describe a range of local, regional and global factors involved in the spread of infectious disease.

Interpret data related to local, regional and global factors.

In order to limit the spread of disease, health care workers and health management authorities must first understand what factors influence the spread of disease, and how these factors interact with each other.

1 Complete the table to identify the factors as having either a local, regional or global impact on the spread of disease.

Factor	Description of how factor influences the spread of disease	Scale of impact (local, regional, global)
Waste disposal		
Antibiotic misuse		
Regional geography		
Overcrowding		
Food trade		
Seasonal climate variations		
Population migration		
Local cultural practices		

9780170449625

2 The maps below present information relevant to the global spread of infectious disease. ('DDD/1000 Pop' = defined daily doses per 1000 individuals.) *Klebsiella pneumoniae* are bacteria that are commonly found in the human intestine, but can be associated with blood or respiratory infections. Carbapenems is an antibiotic used to treat severe bacterial infections.

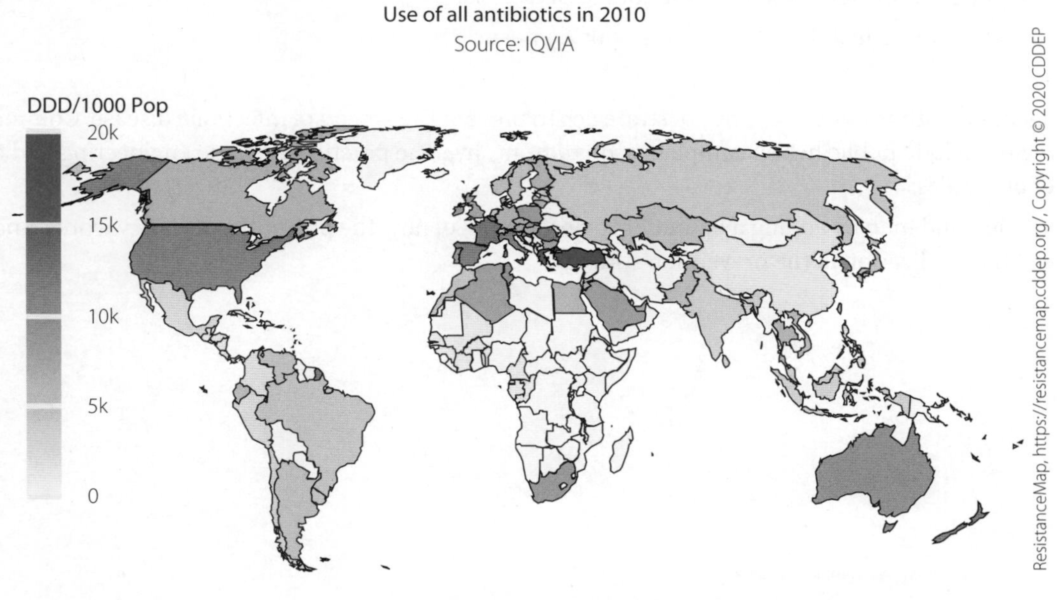

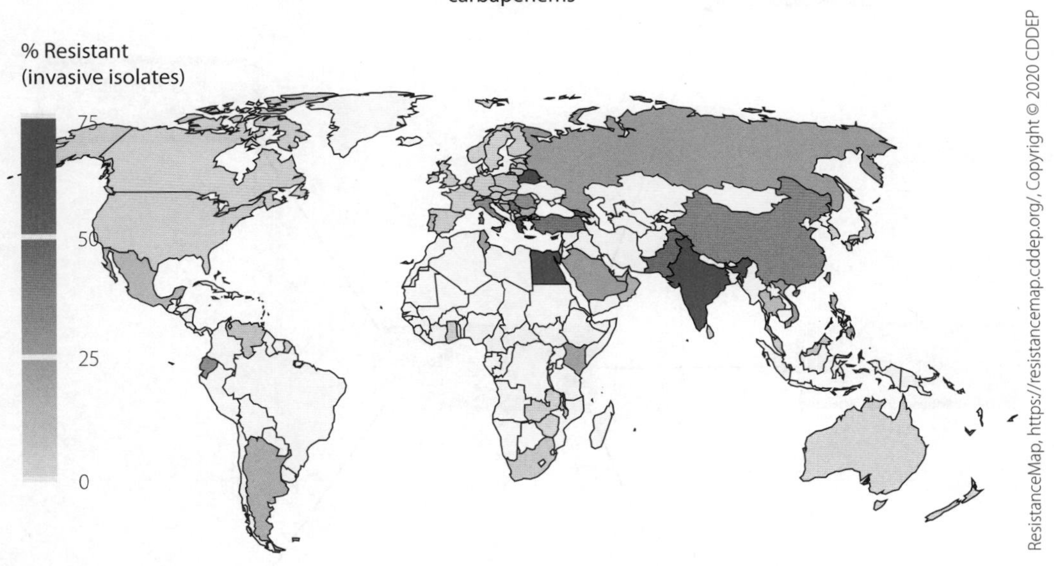

a What information is provided by the maps?

b Explain the relationship between the two figures. Include specific reference to the stimulus as part of your answer.

WS 12.2 Preventing the spread of disease

STUDENT BOOK
Pages 436–46

LEARNING GOALS

- Describe procedures employed to prevent the spread of disease.
- Design an investigation to test the effectiveness of using pesticides.

Health management authorities use a range of strategies to prevent the spread of infectious disease. Commonly applied strategies include public health campaigns, quarantine, hygiene practices, genetic engineering and the application of pesticides.

1 Complete the mind map by adding a definition, a description of how the prevention strategy works, and an example of the application of the prevention strategy.

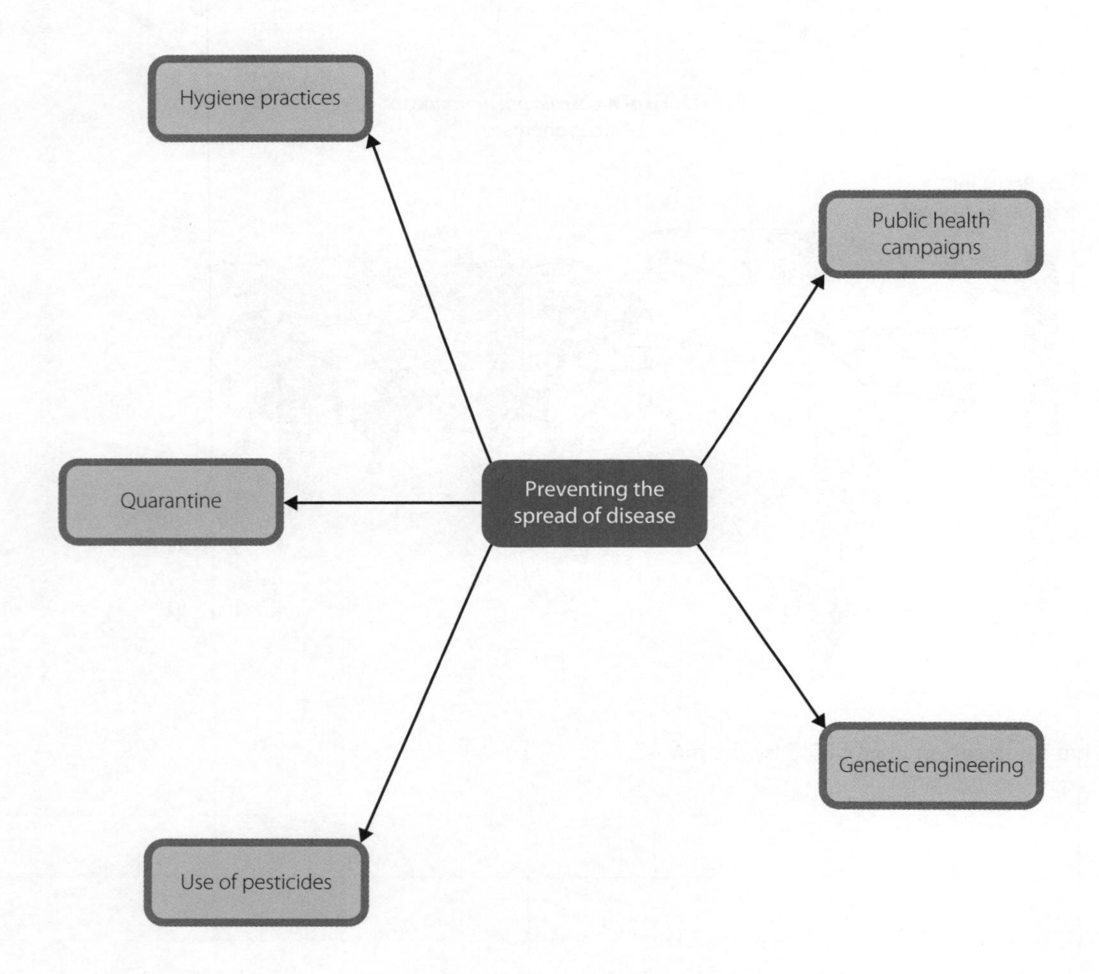

9780170449625

2 Esme is a winemaker who wanted to design an investigation to test the effectiveness of using the fungicide Flint 500 to reduce powdery mildew outbreaks on her grape vines.

a Design a method that Esme could use to test the effectiveness of using Flint 500 on her property. Include a risk assessment, and ensure your method is valid and reliable.

b Explain how you ensured the investigation you designed in part **a** was valid and reliable.

WS 12.3 Understanding immunity

STUDENT BOOK
Pages 442–3

LEARNING GOALS

- Define passive and active immunity.
- Interpret data relevant to vaccination and passive immunity.

1 Define the terms *vaccination* and *immunity*. Include the terms from the word bank in your answer.

immune system	attenuated	injection	memory cells	vaccine

____

____

____

____

____

2 Complete the table to compare passive and active immunity.

Passive immunity	Similarities	Active immunity

3 *Helicobacter pylori* is a bacterium that causes gastritis. Clyne et al. published a study in 1997 examining the role of passive immunity in preventing infection by *H. pylori*. In the study, bacteria were incubated in human milk taken from a woman infected with *H. pylori* and from a non-infected woman. Bacteria were then incubated with Kato III cells (cells from a gastric adenocarcinoma cell line), and adherence of bacteria to these cells was measured.

Shown below are some of the data collected by the study. The graph shows the percentage of Kato III cells that tested positive for *H. pylori* adherence, after *H. pylori* had been incubated in different types and concentrations of milk.

On the *x*-axis, 'PBS' stands for phosphate buffered saline and was a technique used in the study to harvest the bacteria from the agar plates. The numbers 2–6 refer to different treatments applied to human milk.

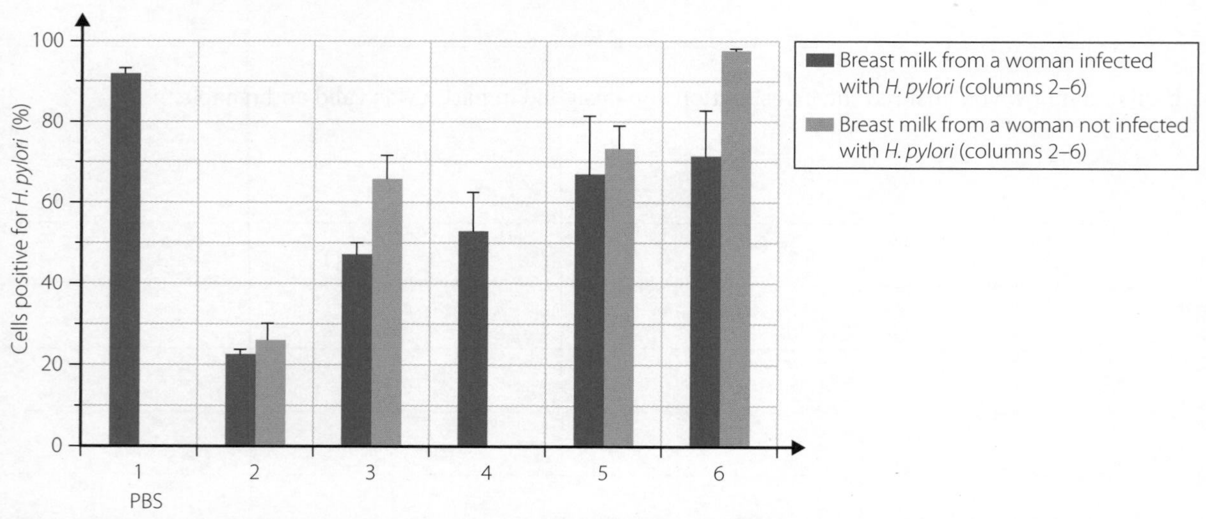

Source: Clyne M, Thomas J, Weaver L, et al (2018). *In vitro evaluation of the role of antibodies against Helicobacter pylori in inhibiting adherence of the organism to gastric cells.* Gut 1997;40:731-738. Figure 7, https://gut.bmj.com/content/gutjnl/40/6/731.full.pdf

a Identify one way the authors could have ensured their study was valid.

______

______

b Describe the trend shown in the graph on the previous page.

> **HINT**
> Include specific reference to data in your answer.

______

______

______

c What can we learn about passive immunity from the data?

______

______

______

______

4 Cheesy gland or caseous lymphadenitis (CLA) is a bacterial disease in sheep that causes abscesses in the lymph nodes and body organs, particularly the lungs. It is caused by a toxin released by ***Corynebacterium pseudotuberculosis*** and transmission between hosts can be via coughing or through contact with infected pus.

Prevalence of CLA associated with different lice control practices

Lice control method	Producers using this method (%)	Average CLA prevalence for this method (%)
None	6	14
Plunge dip	37	29
Shower dip	39	31
Blackline	22	27

Source: Cheesy Gland Caseous Lymphadenitis in Sheep Agfact A3.9.21, NSW Department of Primary Industries, Division of Animal Industries 2nd edition, 1996

The average CLA prevalence associated with using different CLA vaccination programs

Cheesy gland vaccination program used	Resulting incidence of cheesy gland (%)
Complete program 2 shots plus annual boosters	3
Incomplete program No vaccination	29
1 shot as lamb, no booster	33
1 shot as lamb plus booster	31
2 shots as lamb, no booster	22

Source: Cheesy Gland Caseous Lymphadenitis in Sheep Agfact A3.9.21, NSW Department of Primary Industries, Division of Animal Industries 2nd edition, 1996

a What information is provided by the tables?

______

______

b Using the information from both tables, what treatment and prevention plan would you recommend to minimise the risk of spreading cheesy gland to uninfected sheep?

______

______

______

WS 12.4

Effectiveness of pharmaceuticals for controlling infectious disease

LEARNING GOALS

Describe antivirals and antibiotics.

Interpret data related to antibiotic effectiveness.

When used correctly, pharmaceuticals can be very effective in controlling infectious disease outbreaks. Two major classes of pharmaceuticals used to treat infectious diseases are antivirals and antibiotics.

1 Describe the term *antibiotic* using the words in the word bank.

growth	bacterial	slowing	virus

2 Explain why antiviral drugs do not kill a virus, but rather slow the progression of the disease the virus causes.

3 Curtis et al. (2004) tested the antimicrobial properties of garlic at different temperatures, and over time, on *Escherichia coli* (*E. coli*). Some of the results of their investigation are shown in the graphs below.

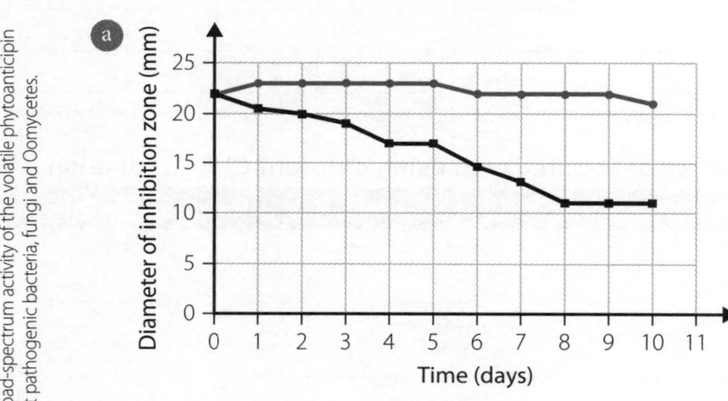

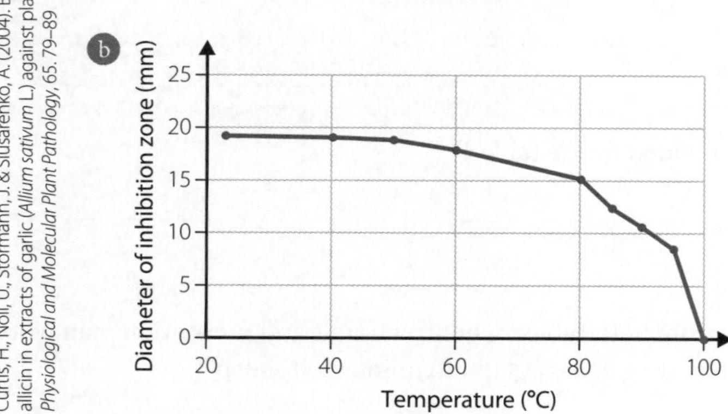

Curtis, H., Noll, U., Störmann, J. & Slusarenko, A. (2004). Broad-spectrum activity of the volatile phytoanticipin allicin in extracts of garlic (*Allium sativum* L.) against plant pathogenic bacteria, fungi and Oomycetes. *Physiological and Molecular Plant Pathology*, 65. 79–89

a Diameter of inhibition zone of *E. coli* colonies over time when grown with garlic extract stored in the dark at room temperature (black squares) or at 4°C (pink circles). **b** Diameter of inhibition zone of *E. coli* colonies when grown with garlic extract that was heated for 10 minutes at various temperatures.

9780170449625

a Use the graphs to complete the table below.

Variable	Graph A	Graph B
Independent		
Dependent		

HINT

Include units in your answers.

b Use a diagram to describe the phrase *inhibition zone.*

c Describe the trends shown in each graph.

d Using data from the graphs, recommend the best treatment of garlic to inhibit the growth of *E. coli.*

Environmental management and quarantine methods

LEARNING GOAL

Investigate and evaluate environmental management methods used to control the 2015 Zika virus epidemic in Brazil.

Brazil experienced a Zika virus epidemic in 2015. Using molecular studies, researchers suggested that Zika virus was introduced during the 2014 Sprint Championship canoe race that was held in Rio de Janeiro. Zika originated in French Polynesia, and participants from this region attended the competition with active Zika transmission.

1 What is the difference between an epidemic and a pandemic?

____

____

____

____

2 Use information from secondary sources to complete the table with regards to the Zika virus.

a	Pathogen	
b	Transmission	
c	Incubation period	
d	Symptoms	
e	Treatment	

9780170449625

3 Graph 1 below shows the suspected and confirmed cases of Zika virus in Brazil during 2015 and 2016. Epidemiological weeks start on a Sunday and end on a Saturday.

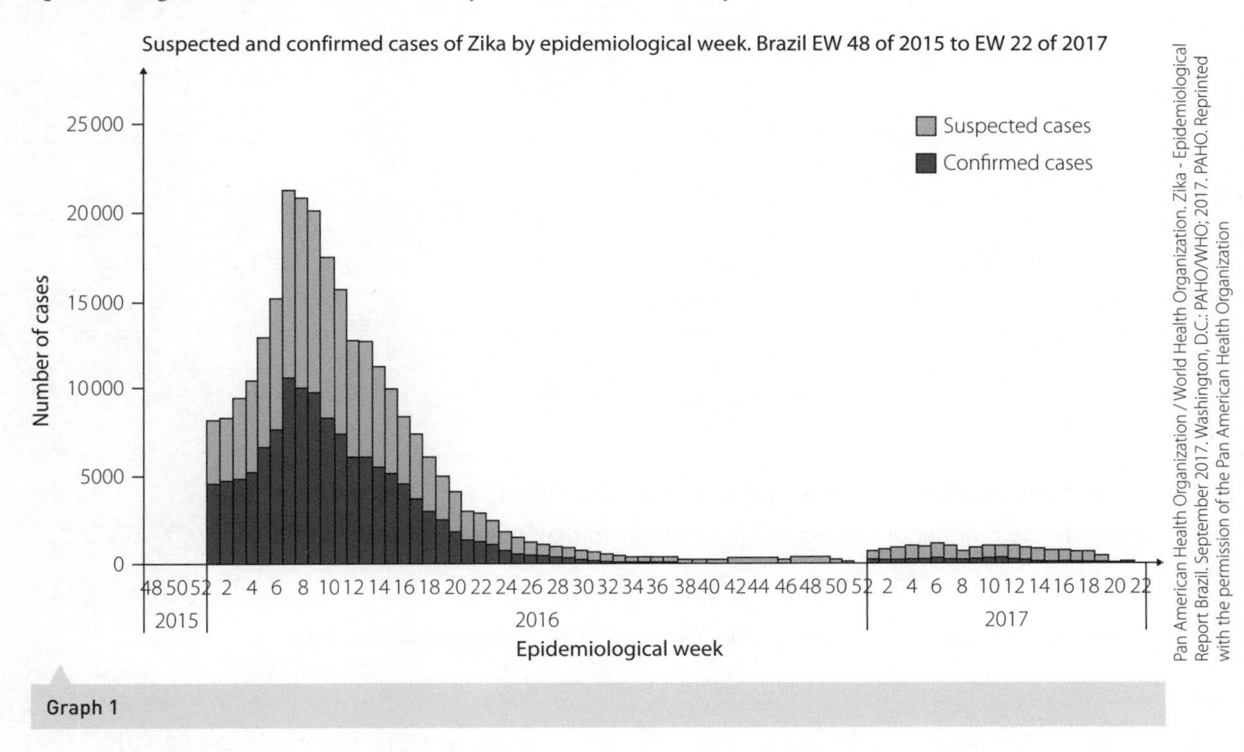

Graph 1

Pan American Health Organization / World Health Organization. Zika - Epidemiological Report Brazil. September 2017. Washington, D.C.: PAHO/WHO; 2017. PAHO. Reprinted with the permission of the Pan American Health Organization

Graph 2 below shows the number of Zika virus cases across the regions of Brazil during 2016.

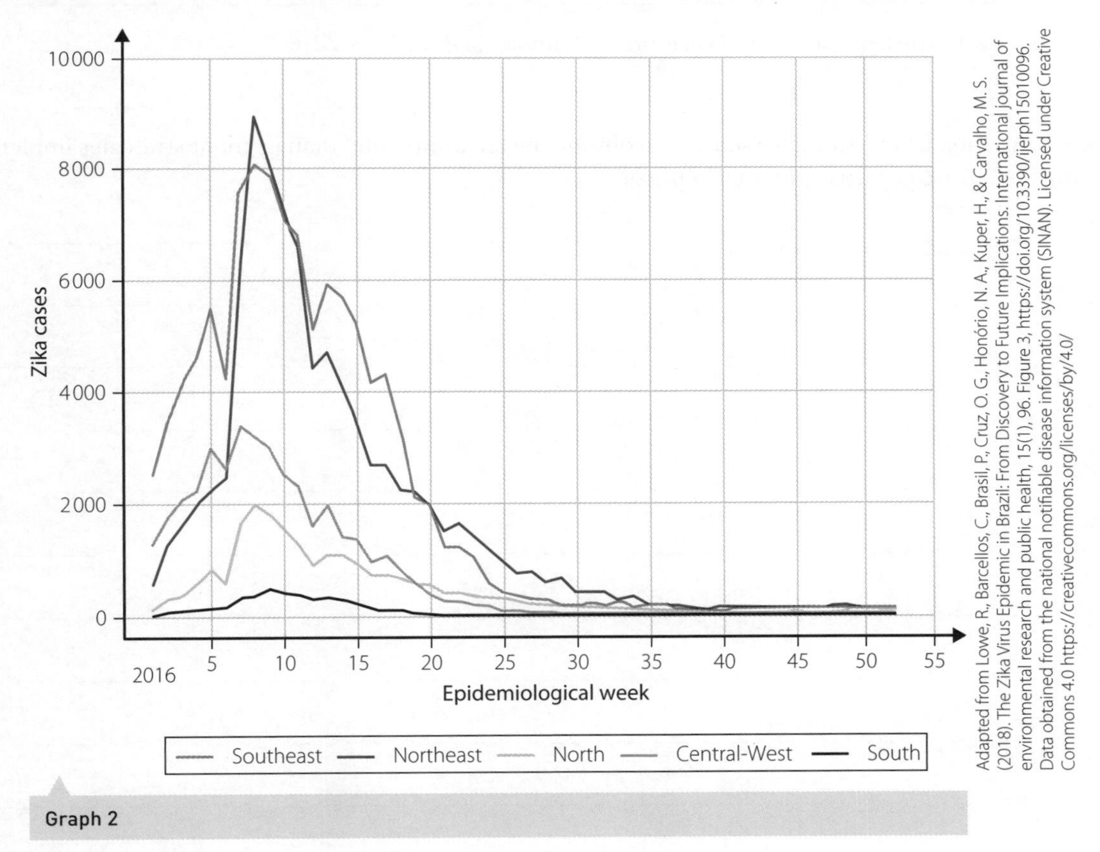

Graph 2

Adapted from Lowe, R., Barcellos, C., Brasil, P., Cruz, O. G., Honório, N. A., Kuper, H., & Carvalho, M. S. (2018). The Zika Virus Epidemic in Brazil: From Discovery to Future Implications. International journal of environmental research and public health, 15(1), 96. Figure 3, https://doi.org/10.3390/ijerph15010096. Data obtained from the national notifiable disease information system (SINAN). Licensed under Creative Commons 4.0 https://creativecommons.org/licenses/by/4.0/

a Describe the information provided in Graph 1 and Graph 2.

b Use Graph 1 and Graph 2 to complete the following information.

i Approximate number of suspected Zika cases in epidemiological week 8, 2016:

ii Approximate number of confirmed Zika cases in epidemiological week 8, 2016:

iii Approximate number of Zika cases in the Central-West region in epidemiological week 20, 2016:

iv Region with the least amount of cases in epidemiological week 15, 2016:

4 a Use information from secondary sources to outline the environmental management strategies implemented during the 2015 Zika virus epidemic in Brazil.

b Assess the validity of one source used in part **a**.

c Evaluate the effectiveness of environmental prevention strategies implemented in Brazil during the 2015 Zika virus epidemic.

HINT

The verb *evaluate* requires you to make a judgement based on criteria or to determine the value of something. Either approach would be possible for this question. Answers should be structured under subheadings and information given using dot points.

Incidence and prevalence of infectious disease

LEARNING GOAL

Interpret data relating to the incidence and prevalence of dengue fever in Thailand.

Dengue fever is an arboviral disease that affects humans. Vector transmission is via the *Aedes aegypti* mosquito. Globally, 2.5 million people in tropical and subtropical regions are at risk of the disease. Gathering incidence (number of new cases) and prevalence (number of all cases) data when outbreaks occur assists with the effective implementation of disease control and prevention measures.

1 Describe how data related to prevalence and incidence of disease can be used by health management authorities.

____

____

____

____

2 Graph 1 and Graph 2 below are taken from Limkittikul et al. (2014). Graph 1 shows the prevalence and incidence data of dengue fever in Thailand from 2000 to 2011, and Graph 2 provides dengue incidence by age group category from 2000–2010.

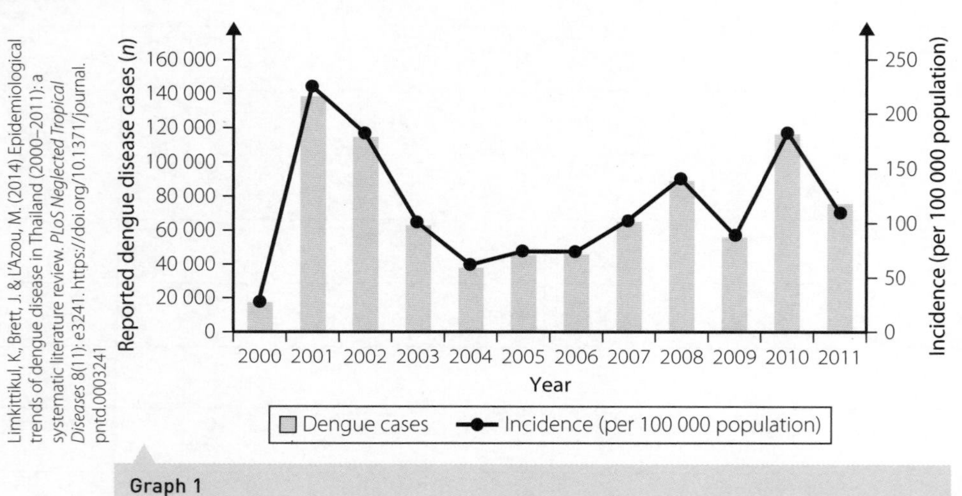

Graph 1

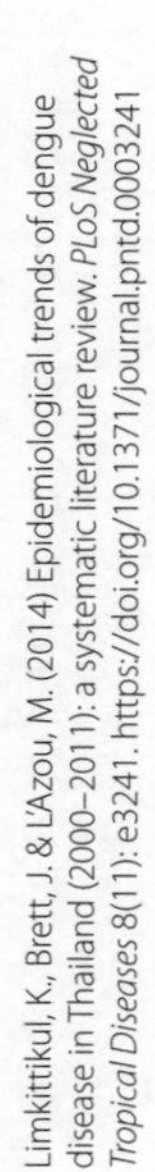

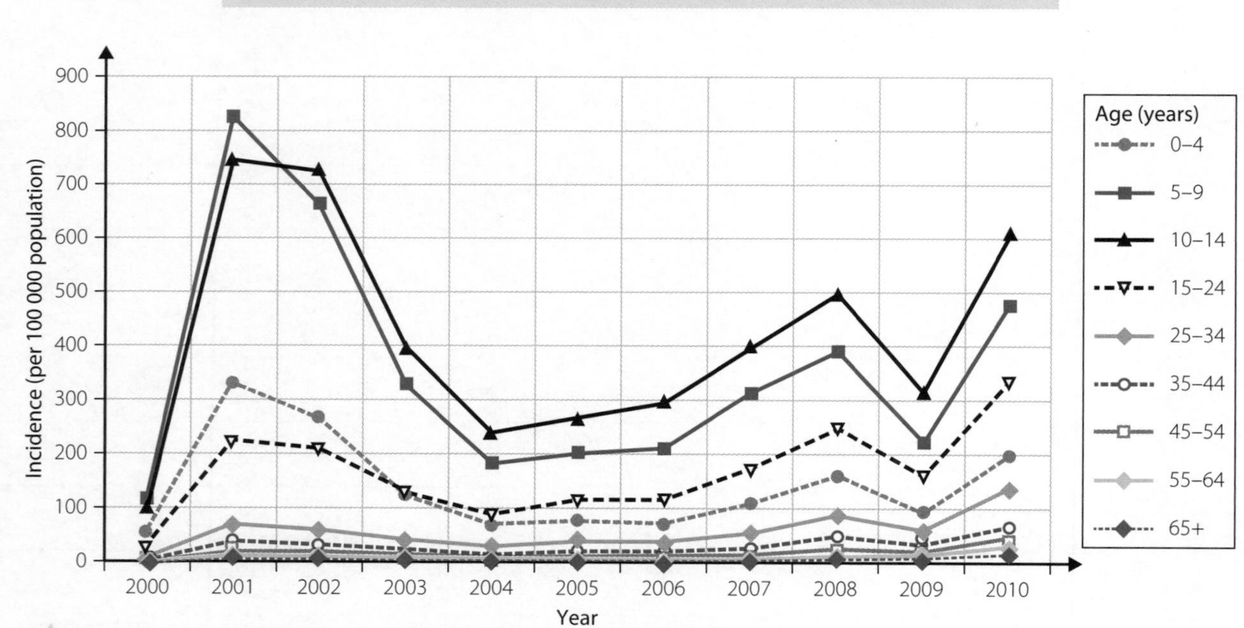

Graph 2

9780170449625

a Describe the trend in prevalence shown in Graph 1.

b Describe the relationship between prevalence and incidence in Graph 1.

c Use the data in Graph 1 to complete the table below.

Year	Incidence per 100 000	Calculated incidence (%)
2000		
2003		
2006		
2009		

HINT

Use the worked examples on page 455 of *Biology in Focus Year 12* Student Book to help you complete the calculations.

d Use Graph 2 to complete the information below.

i Two age groups with the highest dengue incidence between 2000 and 2010:

ii Two groups with the lowest dengue incidence between 2000 and 2010:

iii Age group with the largest range of incidence between 2000 and 2010:

iv Incidence of dengue in 15–24 year olds in 2005:

v Year when dengue incidence was lowest for all age groups:

3 Graph 3 below is from Vannavong et al. (2019) and shows the incidence of dengue fever in two provinces in South East Asia between 2010 and 2013. Lakhonpheng district is located in Lao People's Democratic Republic and the Manchakhiri district is in Thailand.

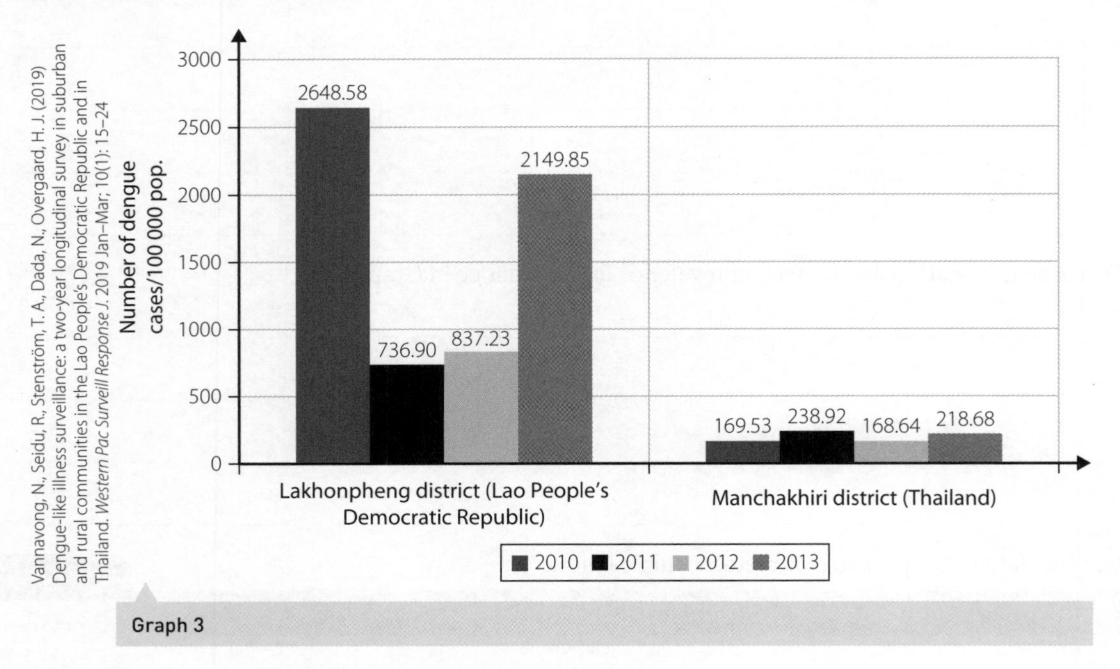

Graph 3

a Describe the trend in Graph 3.

b Socioeconomic factors can be used by researchers to determine risk factors for infectious disease. The table below is an extract of socioeconomic data also taken from Vannavong et al. (2019) (percentages in parentheses).

		Lao People's Democratic Republic		Thailand	
		Suburban	Rural	Suburban	Rural
Total households per village		215	130	272	139
No. of selected households		122	112	115	122
Room occupancy rate	>2.5 persons/room	61 (50.0)	43 (38.4)	21 (18.3)	32 (26.2)
	≤2.5 persons/room	61 (50.0)	69 (61.6)	94 (81.7)	90 (73.8)
Wealth status	Poor	38 (31.1)	91 (81.2)	5 (4.3)	19 (15.6)
	Intermediate	51 (41.8)	15 (13.4)	34 (29.6)	66 (54.1)
	Rich	33 (27.1)	6 (5.4)	76 (66.1)	37 (30.3)
Housing material	Cement	24 (19.7)	1 (0.9)	23 (20.0)	16 (13.1)
	Cement-wood	48 (39.3)	16 (14.3)	80 (69.6)	74 (60.7)
	Wood	50 (41.0)	95 (84.8)	12 (10.4)	32 (26.2)

Vannavong, N., Seidu, R., Stenström, T. A., Dada, N., Overgaard, H. J. (2019) Dengue-like illness surveillance: a two-year longitudinal survey in suburban and rural communities in the Lao People's Democratic Republic and in Thailand. *Western Pac Surveill Response J.* 2019 Jan–Mar; 10(1): 15–24

Are there any correlations between the socioeconomic factors shown in the table and the trends seen in Graph 3? Use data to support your conclusion.

Predicting and controlling the spread of disease: 1918 influenza pandemic

STUDENT BOOK
Page 457

LEARNING GOAL

Describe strategies used to control the spread of the 1918 influenza pandemic.

The 1918 influenza pandemic was caused by the H1N1 virus. It infected around one-third of the world's population at the time (around 500 million people), and resulted in around 50 million deaths.

Health management and prevention techniques applied today are the culmination of years of experience in managing infectious disease outbreaks globally. Lessons from the past and from other cultures can be used to improve the effectiveness of tools applied to current and future disease outbreaks.

Health prevention techniques utilised today can be summarised under eight headings, and are included in the table below.

1 Use information gathered from secondary sources to summarise health prevention and management measures employed during the 1918 influenza pandemic under the appropriate headings.

Health management category	Implemented during 1918 pandemic (Y/N)?	Details of strategies implemented	Comment on effectiveness of strategy/strategies
Hygiene			
Quarantine			
Vaccination			
Public health campaigns			
Pesticides			
Genetic engineering			
Pharmaceuticals			

Contemporary application of Aboriginal and Torres Strait Islander protocols

LEARNING GOALS

Describe uses of traditional Aboriginal and Torres Strait Islander medicines.

Assess the intellectual property value of a trademark application.

Aboriginal and Torres Strait Islander cultures in Australia approach health management with a holistic view that includes physical, social and mental wellbeing, and most do not separate mind and body health. Traditional forms of medicine include bush medicine, traditional healers and healing songs. Western medicine has a wonderful opportunity to learn from these traditional healers and currently there is a lot of western scientific attention on investigating traditional plant remedies.

1 Identify two ways that plant materials are prepared by Aboriginal and Torres Strait Islander peoples for medicinal purposes.

____

____

2 Many traditional medicinal plants have been found to contain chemical compounds known to western science as medicinally beneficial. Use information from secondary sources to complete the table below.

Compound	Medicinal properties
Tannin	
Oil	
Latex	
Mucilage	
Alkaloid	

3 Use an example to describe how Aboriginal and Torres Strait Islander peoples traditionally used natural resources for medicinal purposes.

____

____

____

____

____

____

9780170449625

4 Use an example to describe a contemporary application of traditional Aboriginal and Torres Strait Islander medicine.

5 *Pittosporum angustifolium* is an Australian tree that has been used in traditional bush medicine for a long time. In 2008, two non-Indigenous Australians were granted a patent for the exclusive rights to *Pittosporum angustifolium* leaf extracts, and in 2017 the same group filed a trademark application for the exclusive rights to the Indigenous name (Gumby Gumby) of this plant.

Many traditional land owners regularly use the plant for medicinal purposes and there are Indigenous-owned businesses that use this plant for commercial purposes. The Indigenous Elders were not consulted during the application process.

Current Australian intellectual property law allows any Australian individual or company to apply for the rights to any Indigenous knowledge. Australian Aboriginal customary law, developed over a long period of time and which connects people to the land, has a more benevolent view of who should be able to benefit from the medicinal properties of the plant.

Assess whether or not the trademark application for the exclusive use of the name 'Gumby Gumby' should have been filed.

HINT

The verb *assess* requires you to identify points for and against, and to make a judgement. Use these subheadings in your answer.

Module 7: Checking understanding

1 An example of direct transmission of a pathogen would be:

A drinking water containing the protozoan *Giardia*.

B being bitten by a malaria-infected female *Anopheles* mosquito.

C a pregnant woman passing hepatitis B to her unborn child.

D touching a door handle that was previously touched by someone carrying the influenza virus.

2 The image below depicts a pathogen.

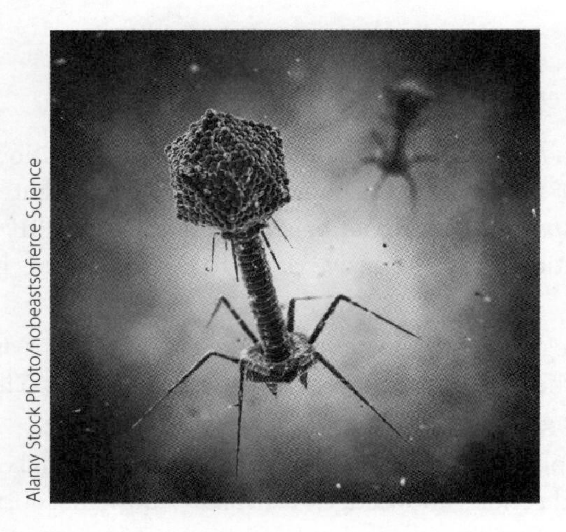
Alamy Stock Photo/nobeastsofierce Science

This pathogen can be best described as:

A cellular.

B unicellular.

C non-cellular.

D multicellular.

3 Importing animals from one country to another can lead to the spread of infectious diseases. Before allowing the animals past border control, the country receiving the animals should:

A give all the animals vaccinations.

B ensure all the animals are genetically modified.

C put the animals into a quarantine station for 14 days.

D wash the holding pens with antibacterial soap after the animals are removed.

4 The term *incidence of disease* can be best defined as:

A the number of new cases occurring during a specific time period.

B all of the cases of disease that occurred during the outbreak of a pandemic.

C all of the cases of disease that occurred during the outbreak of an epidemic.

D the proportion of the population that has the disease at a particular time.

5 Use an example to define the term *infectious disease*.

____

____

____

____

____

____

6 Explain why scientists would test for the presence of microbes growing in water.

7 What information can be gained from measuring the zone of inhibition?

8 Use an example to describe how local factors can influence the severity of an infectious disease outbreak.

9 Compare antivirals and antibiotics.

10 Assess the effectiveness of tylose formation in the xylem of plants to maintain the overall health of the plant.

11 People who suffer from heartburn take antacid tablets to increase the pH of their stomach. Explain how this affects the immune system.

12

Naive B cell is activated by corresponding antigen

↓

Process A

Plasma cell ↓ Process B

Cell X ↓ Cell X remains in the body for many years, offering continued immunity to pathogen

Refer to the diagram to answer the following questions.

a Describe process A.

b Identify cell X.

c Describe process B.

13 Tuberculosis is a bacterial disease spread by airborne droplets entering the lungs. Once in the lungs, the tuberculosis bacterium can reproduce rapidly. These bacteria are destroyed by phagocytes.

Describe the process by which a phagocyte destroys a bacterium.

NON-INFECTIOUS DISEASE AND DISORDERS

Reviewing prior knowledge

1 Below is a diagram of an enzyme being denatured.

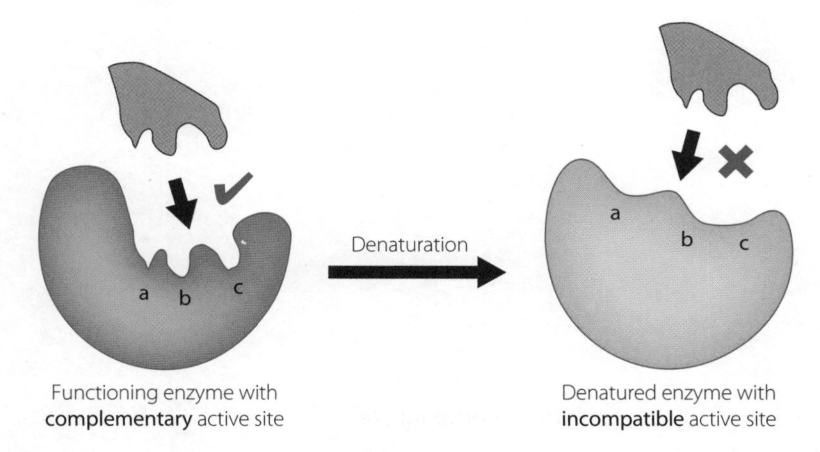

a Identify two environmental conditions that can cause denaturation in enzymes.

b Explain why denaturation affects the functioning of an enzyme.

2 Analyse the graph below to explain how increasing the temperature of an organism would affect metabolic processes.

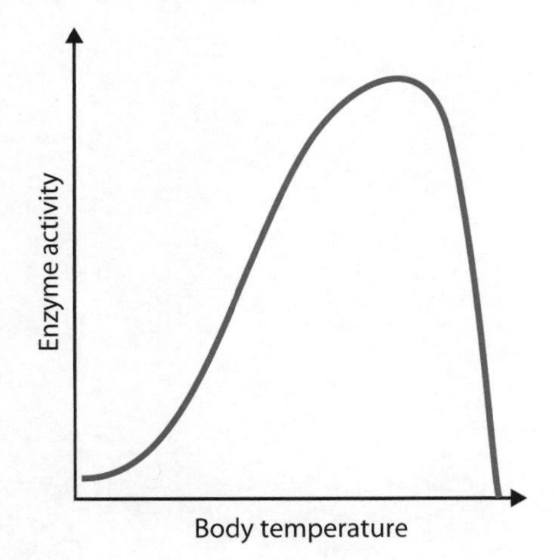

3 Define the term *mutation*.

4 What is genetic engineering?

5 Use a diagram to describe the structure of DNA.

6 The nervous system is made up of which three components?
- **A** Brain, heart and spinal cord
- **B** Brain, spinal cord and nerves
- **C** Nerves, arteries and veins
- **D** Nerves, liver and heart

7 What is the main purpose of the nervous system?
- **A** To fight off diseases
- **B** To distribute energy throughout the body
- **C** To provide communication between different parts of the body
- **D** To break down food

8 Nerve cells are called:
- **A** neutrons.
- **B** nuclei.
- **C** neurons.
- **D** nebulas.

9 The main function of the urinary system is to:
- **A** excrete waste and extra fluid.
- **B** create faeces.
- **C** keep waste in the body.
- **D** get rid of nutrients.

10 What is the function of the kidneys?
- **A** To balance the amount of water in the body
- **B** To filter waste out of the blood
- **C** Both A and B
- **D** Neither A nor B

11 The main components of urine are:
- **A** urea and water.
- **B** nitrogen and salt.
- **C** uric acid and sugar.
- **D** protein, carbohydrates and nucleic acids.

13 Homeostasis

INQUIRY QUESTION: HOW IS AN ORGANISM'S INTERNAL ENVIRONMENT MAINTAINED IN RESPONSE TO A CHANGING EXTERNAL ENVIRONMENT?

WS 13.1 Negative feedback

STUDENT BOOK
Pages 470–3

LEARNING GOAL

Describe the components of a negative feedback loop with named examples.

1 Number each action a–e to match each stage in the stimulus–response flow diagram below.

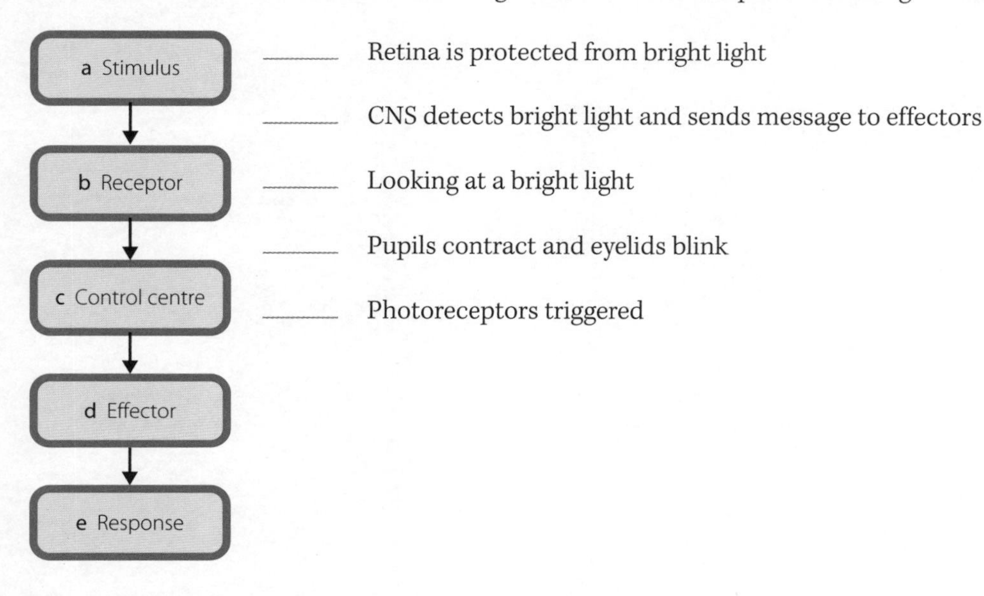

_____ Retina is protected from bright light

_____ CNS detects bright light and sends message to effectors

_____ Looking at a bright light

_____ Pupils contract and eyelids blink

_____ Photoreceptors triggered

2 On the right is a graph showing homeostasis.

a Explain the trends shown in the graph with reference to negative feedback, using a named example.

Normal range

Time

b Draw a negative feedback diagram to represent the example you used in part **a**.

9780170449625

WS 13.2 Temperature regulation in endotherms

STUDENT BOOK
Pages 473–5

LEARNING GOAL

Describe the mechanisms of temperature regulation in endotherms.

Endotherms respond to changes in environmental temperature in many different ways.

1 Complete the table by providing an animal example, outlining the purpose of the response and describing how it regulates temperature.

Response	Animal example	Response to hot or cold	Description of mechanism for cooling or heating
Sweating			
Shivering			
Contraction of pili muscles			
Vasodilation			
Licking forearms			
Curling up			

2 Explain why licking forearms could be classed as a behavioural and a physiological adaptation.

3 Draw an annotated diagram to describe how a penguin's 'shunt' blood vessels allow it to thermoregulate.

WS 13.3 Glucose regulation in humans

STUDENT BOOK
Pages 475–6

STUDENT BOOK Pages 475–6

LEARNING GOAL

Explain how blood glucose level is regulated in humans.

1 Explain why too much or too little glucose in the blood can cause problems to humans.

2 Aliyah woke up, had breakfast and then went for a run. Explain how her glucose levels would have fluctuated throughout the morning.

3 Describe the role of the following hormones.

a Insulin

b Glucagon

4 a Use the data in the table to draw a line graph.

Time of day	Blood glucose levels (mmol/L)
6.00	3.8
7.00	4.4
8.00	6.2
9.00	4.2
10.00	3.8
11.00	3.8
12.00	4.5
13.00	6.2
14.00	4.0
15.00	3.8
16.00	3.8
17.00	3.8
18.00	4.0
19.00	6.0
20.00	4.0

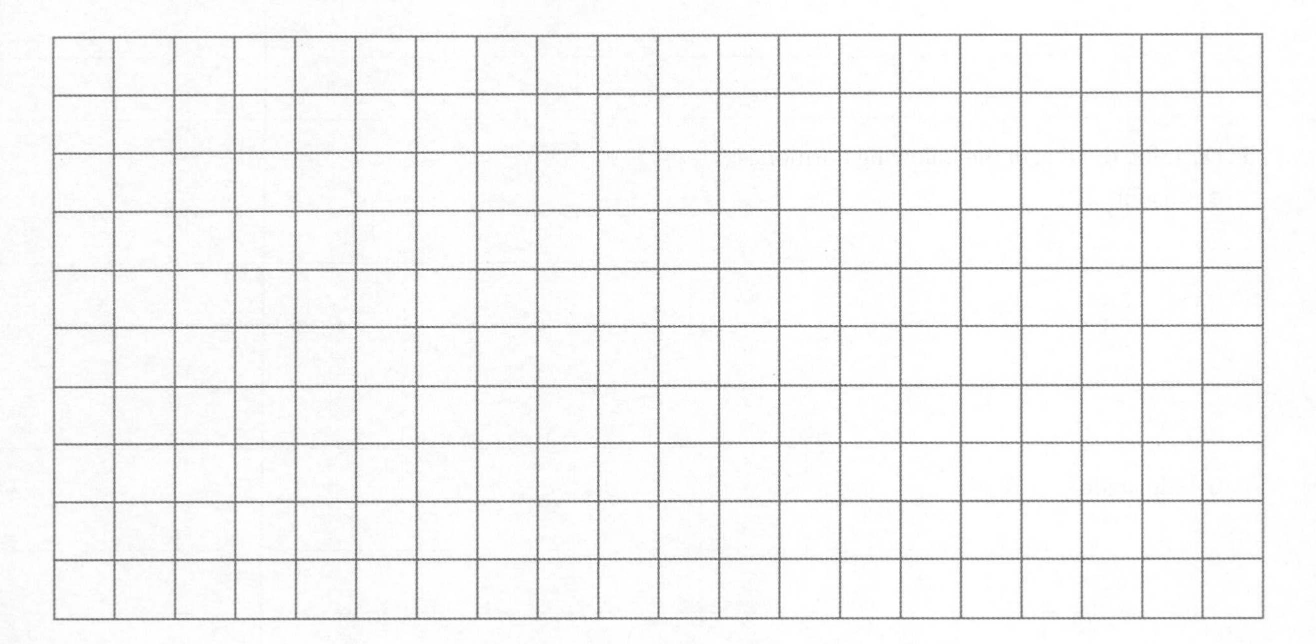

b Using your graph, suggest the times at which the person ate throughout the day.

c Draw lines on your graph to indicate where you would expect insulin levels in the blood to be throughout the day.

WS 13.4 Osmoregulation in animals

LEARNING GOAL

Explain the process of osmoregulation in animals.

1 Use the image below to outline a physiological and a behavioural adaptation that allow the kangaroo rat to conserve water.

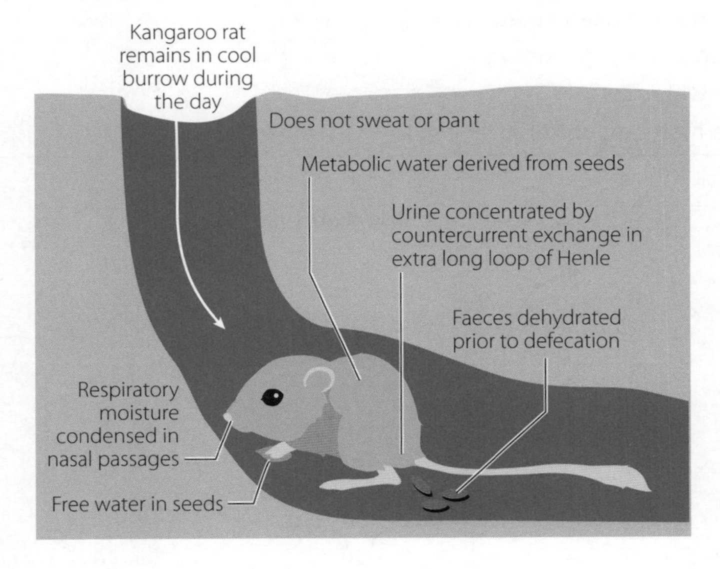

____

____

____

____

2 Below is a diagram indicating the length of the loop of Henle found in the kidneys of three mammals. Water is reabsorbed back into the body in the loop of Henle.

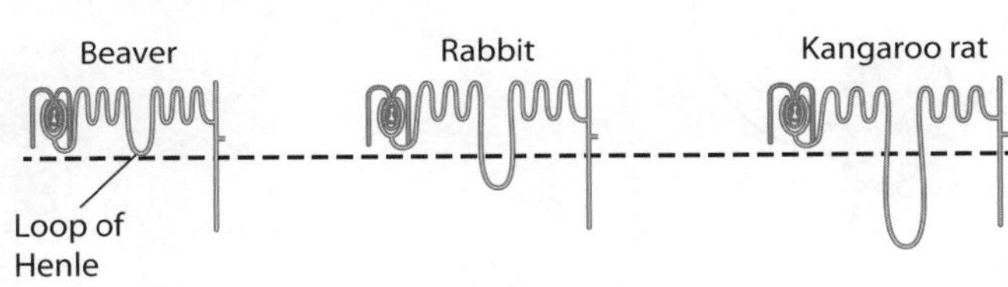

Using the information above and research about the natural environment of each of these animals, explain a link between the length of an animal's loop of Henle and its habitat.

____

____

____

____

____

____

____

____

3 Use the labels below to complete the negative feedback loop outlining the role of antidiuretic hormone (ADH) in osmoregulation.

Higher volume of less concentrated urine produced
Blood water content low
Higher amount of water reabsorbed from the kidney
High amount of water intake; increased water level in blood
Pituitary reduces amount of ADH produced
Lower volume and more concentrated urine
Less water is reabsorbed by the kidney
Excess sweating or water loss
Pituitary releases a high amount of ADH

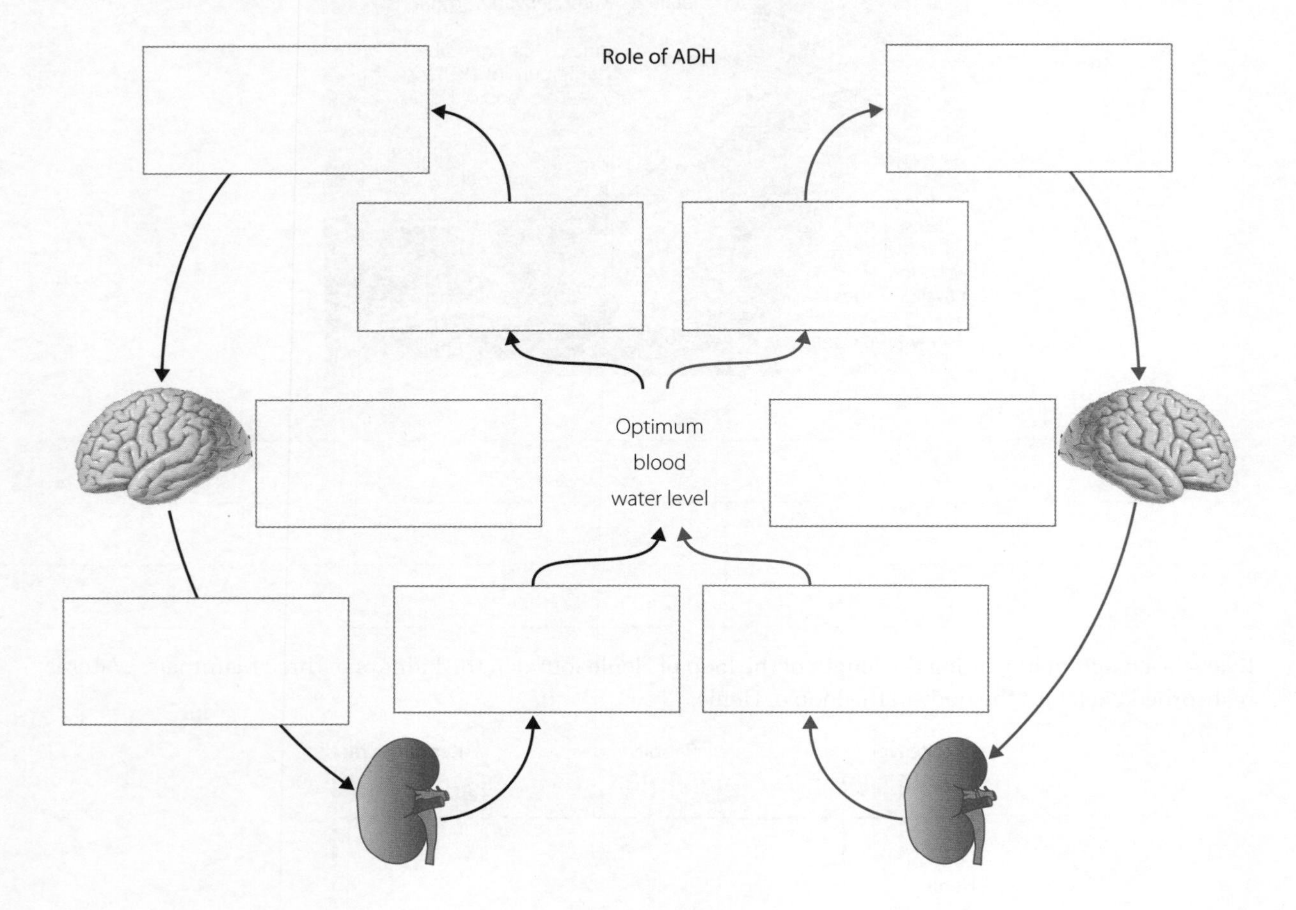

WS 13.5 Osmoregulation in plants

LEARNING GOAL

Explain the process of osmoregulation in plants.

Plants, just like animals, need to regulate their internal environment. Osmoregulation is especially important for plants that live in extremely dry and/or extremely salty environments.

Xerophytes are desert plants and are exposed to high levels of heat and evaporation.

Halophytes are plants that live in salty environments, so they lose water into their environment via osmosis if osmotic balance cannot be achieved.

1 Complete the tables to describe some of the adaptations of these types of plants that assist in osmoregulation. Use Australian examples. You may need to conduct research to complete the tables.

a Xerophyte:

Example chosen: ______________________

Adaptation	Description	How it limits water loss
Rolled leaves		
Thick waxy cuticle		
Low growing		
Opening stomata at night		

b Halophyte:

Example chosen: ______________________

Adaptation	Description	How it limits water loss
Tissue partitioning		
Salt excretion		
Root exclusion		

WS 13.6 Hormones and the endocrine system

STUDENT BOOK
Pages 489–92

LEARNING GOAL

Explain the roles of hormones and neural pathways in controlling internal environments.

1 The image below shows a neuron of someone suffering from multiple sclerosis (MS). MS is considered to be an autoimmune disease. In people with MS, the immune system attacks Schwann cells, which produce and maintain the myelin sheath surrounding the axon of neurons. People with MS suffer from loss of muscle control, which often leads to slow and jerky movements.

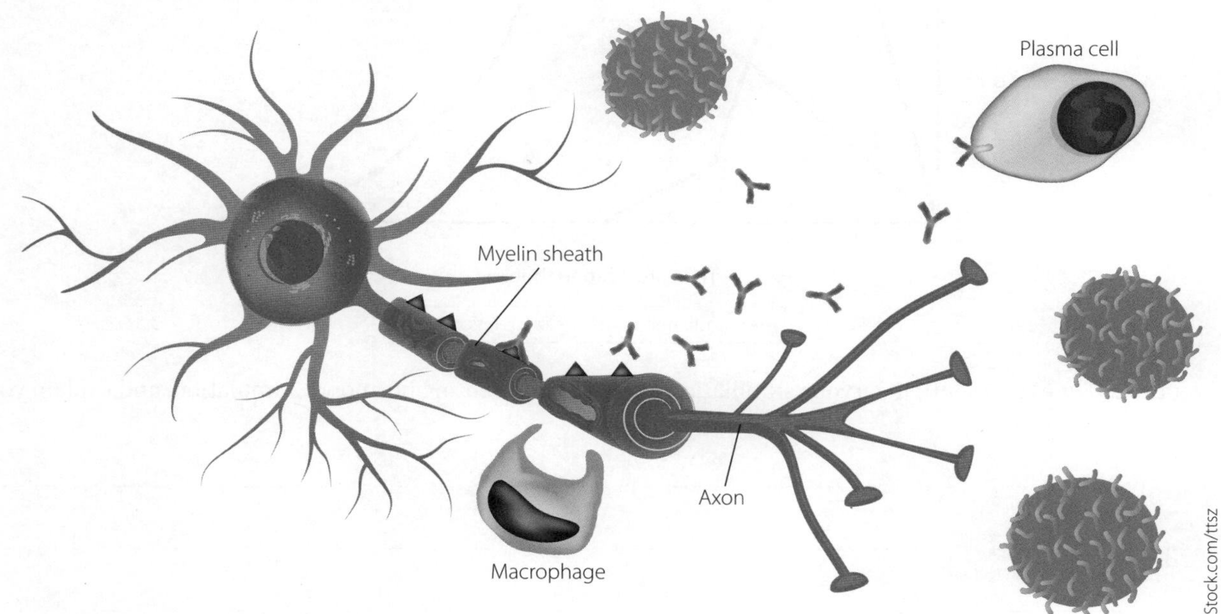

a Explain why damage to the myelin sheath can lead to the symptoms described above.

____

____

____

____

____

b Using your knowledge from Module 7, explain how the adaptive immune response is responsible for this condition.

____

____

____

____

2 The stomach produces more acid when stimulated by food than when dormant. The graph below shows a hormonal and a nervous stimulation.

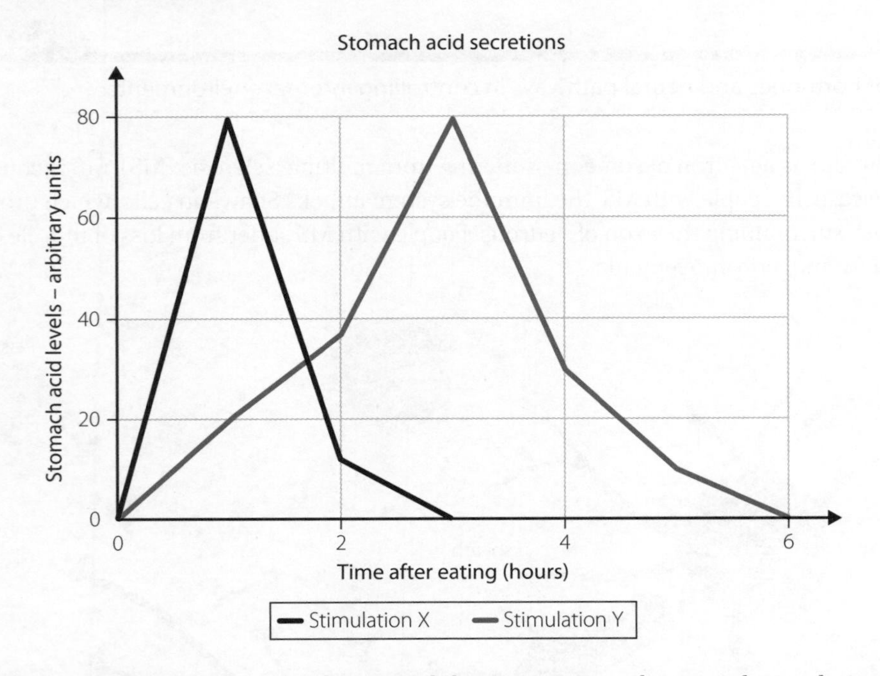

Identify the line representing nervous stimulation and that representing hormonal stimulation, and explain your answer.

14 Causes and effects

INQUIRY QUESTION: DO NON-INFECTIOUS DISEASES CAUSE MORE DEATHS THAN INFECTIOUS DISEASES?

WS 14.1 Causes and effects of non-infectious diseases

LEARNING GOAL

Describe causes and effects of non-infectious diseases in humans.

Non-infectious diseases are the leading cause of death and disability worldwide. These diseases are not caused by pathogens, but rather can be attributed to genetics, environmental exposure or nutritional imbalances.

1 Explain how genetic diseases are caused. Include the terms from the word bank in your answer.

mutation	chromosome	protein	cell division

2 Complete the table to summarise details regarding some non-infectious diseases.

Category of disease	Name of disease	Cause of disease	Symptoms
		Three copies of chromosome 21	
Environmental exposure			
			Extreme weight loss due to excessive strenuous exercise, self-induced vomiting, taking laxatives and appetite suppressants.
	Melanoma		

9780170449625

3 Use a table to compare undernutrition and overnutrition.

4 One way to measure the effect of non-infectious diseases is to quantify what is known as the burden of disease. Burden of disease measures the impact of living with illness or injury and dying prematurely. The measure is given as disability adjusted life years (DALY) and measures the years of healthy life lost due to illness, disability or early death.

The table below shows the DALY rates for all age groups by Australian states and territories in 2015.

Age-specific total burden (DALY) rates, by life stage and state and territory, 2015

State/ Territory	Age group (years)				
	Under 15	15–44	45–64	65–84	85+
ACT	59.8	96.9	213.7	456.2	1023.2
NSW	55.0	108.1	229.7	504.5	1053.6
NT	91.9	158.2	301.1	647.5	1453.1
Qld	62.3	118.9	224.9	503.8	1063.8
SA	54.3	118.8	242.6	504.7	1049.1
Tas	53.9	125.2	259.0	569.5	1160.3
Vic	51.2	110.3	217.3	487.9	1033.5
WA	50.9	115.7	217.5	477.7	1033.3

Note: DALY rate is expressed as DALY per 1000 population.

a Identify the age group with the highest DALY measures.

b Which age group is least impacted by the burden of disease?

c In which state or territory are 15–44 year olds least impacted by the burden of disease?

d Which state or territory is most impacted by the burden of disease across all ages?

e DALY measures provide a summary of population health against an ideal life expectancy. Outline one way that DALY measures might be useful for health care management authorities.

WS 14.2 Collecting and representing health data

STUDENT BOOK
Page 524

LEARNING GOAL

Graph prevalence and mortality data related to coronary heart disease in Australia.

Collecting data regarding the prevalence and mortality of non-infectious diseases enables health care authorities to plan for and allocate health care resources and infrastructure to maximise benefits for patients and their families.

Coronary heart disease (CHD) is the most common form of cardiovascular disease and includes heart attack and angina. The following prevalence and mortality data are extracts from data provided by the Australian Institute of Health and Welfare.

Table 1 Prevalence of self-reported CHD among persons 18 and over, by age group and gender, 2017–18

Age group	Males	Females
18–44	16 300	13 300
45–54	24 800	10 500
55–64	83 800	51 600
65–74	111 100	66 800
75+	137 900	64 200
Persons (number/ age-standardised rate*)	373 900	206 400

* Age-standardised to the 2001 Australian Standard Population

Table 2 CHD deaths per 100 000 population by age group and gender, 2018

Age group	Males	Females
<35	0	0
35–44	9	2
45–54	33	8
55–64	80	19
65–74	170	50
75–84	486	243
85+	2049	1506

1 Graph the prevalence of CHD by age group and gender as provided in Table 1.

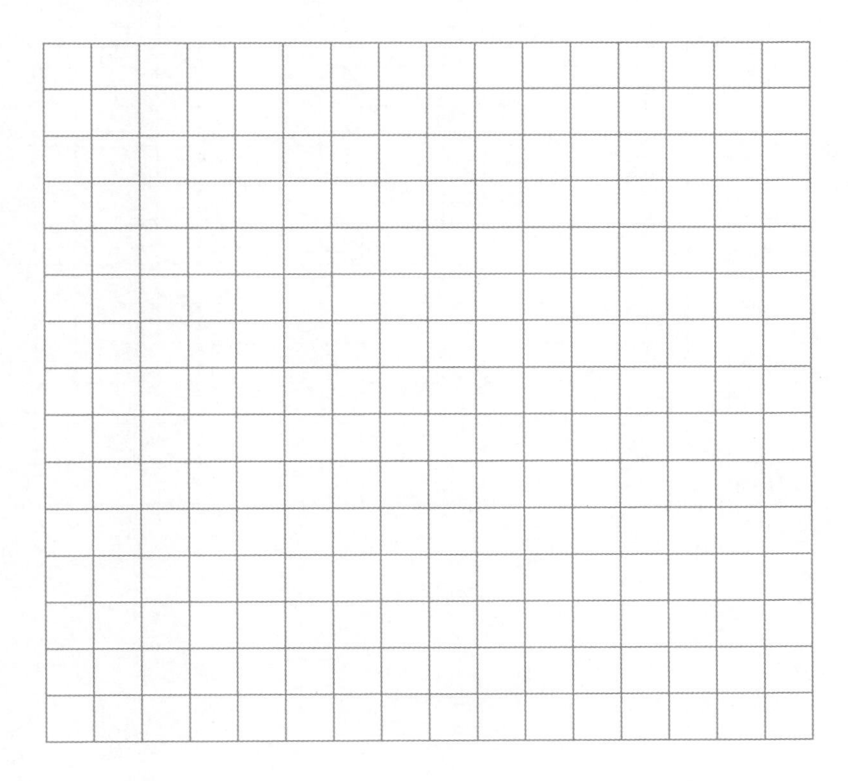

2 Ask a class member to use the rubric below to mark your graph in question **1**.

Criteria	Present	Missing
Draws a column graph		
Includes a key for male and female columns		
Title		
Axis labels		
Even scale		
More than 50% of grid utilised		
Data correctly plotted		

3 Graph the data in Table 2 in the space below, making sure you consider all feedback received in question **2**.

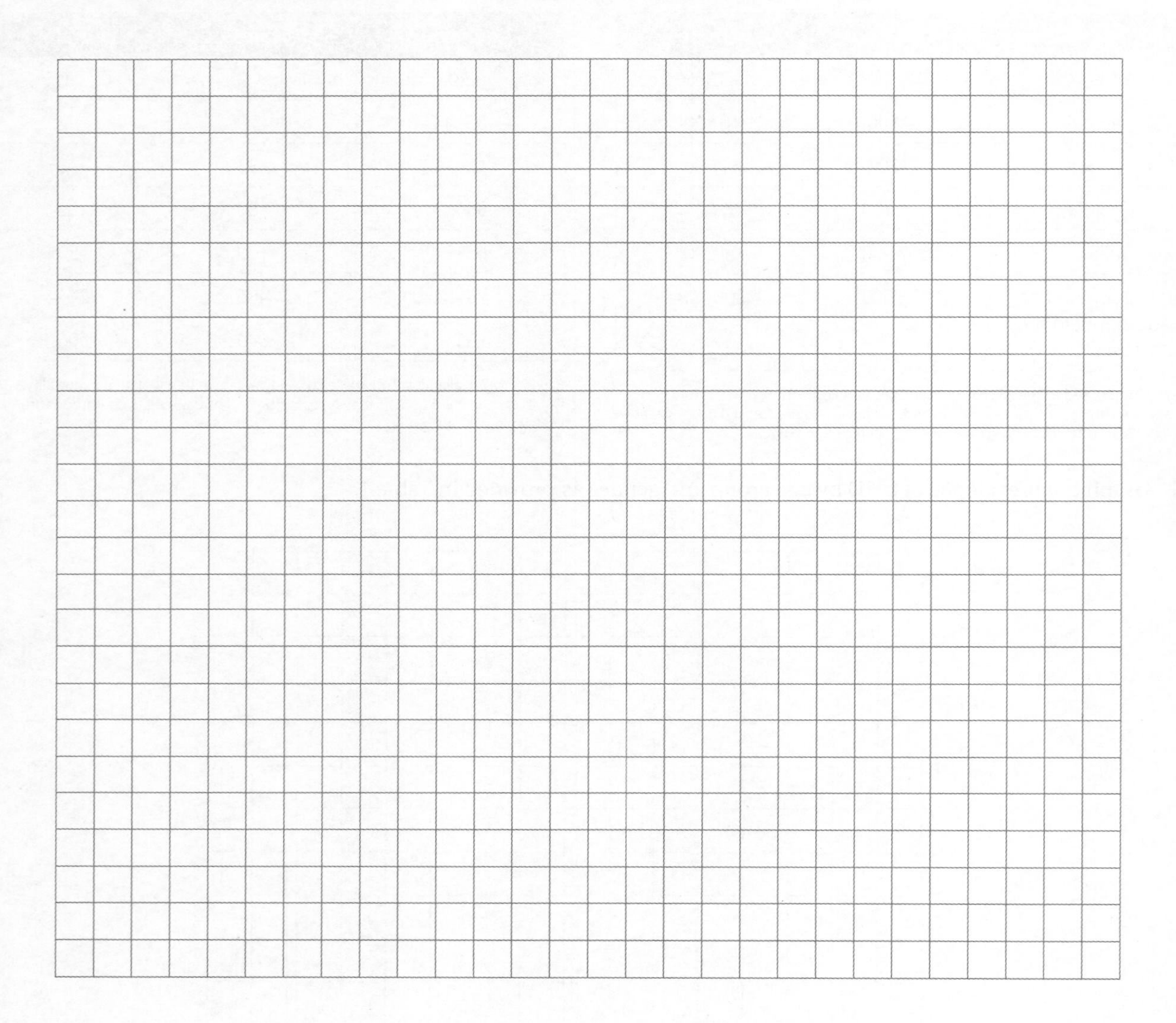

9780170449625

15 Epidemiology

INQUIRY QUESTION: WHY ARE EPIDEMIOLOGICAL STUDIES USED?

WS 15.1 Analysing patterns of incidence and prevalence in populations

STUDENT BOOK
Pages 542–3

LEARNING GOAL

Describe and interpret data related to incidence and prevalence of melanoma of the skin.

Melanoma is a skin cancer that forms in the cells that produce pigment. While not as common as other types of skin cancer, melanoma can be one of the most serious because the tumours can easily spread. The table below provides a summary of the incidence and mortality (2005–2014) and prevalence (31 December 2014) for skin melanoma in Australia.

State/territory	Incidence rate (per 100 000)	Median age at diagnosis	Prevalence as at 31 December 2014	Median age	Mortality rate (per 100 000)	Median age at death
New South Wales	57.1	64.3	4222	38.5	7.4	72.5
Victoria	46.9	61.9	2433	38.9	6.3	71.8
Queensland	69.0	64.4	3597	35.5	7.2	69.0
South Australia	45.3	63.2	801	39.8	6.2	72.0
Western Australia	49.8	62.8	1274	36.0	6.1	71.0
Tasmania	54.3	56.2	322	40.0	6.8	73.0
Northern Territory	25.9	61.5	77	31.3	2.2	61.0
Australian Capital Territory	38.6	63.7	156	34.4	5.3	73.0

1 Use the data in the table above to complete the information below.

a State/territory with highest incidence rate:

____

b Lowest median age at diagnosis:

____

c Average mortality rate across all states and territories:

____

d Lowest incidence rate:

____

e Average prevalence of melanoma in Australia:

____

2 Graph the prevalence of melanoma in each state using the grid below.

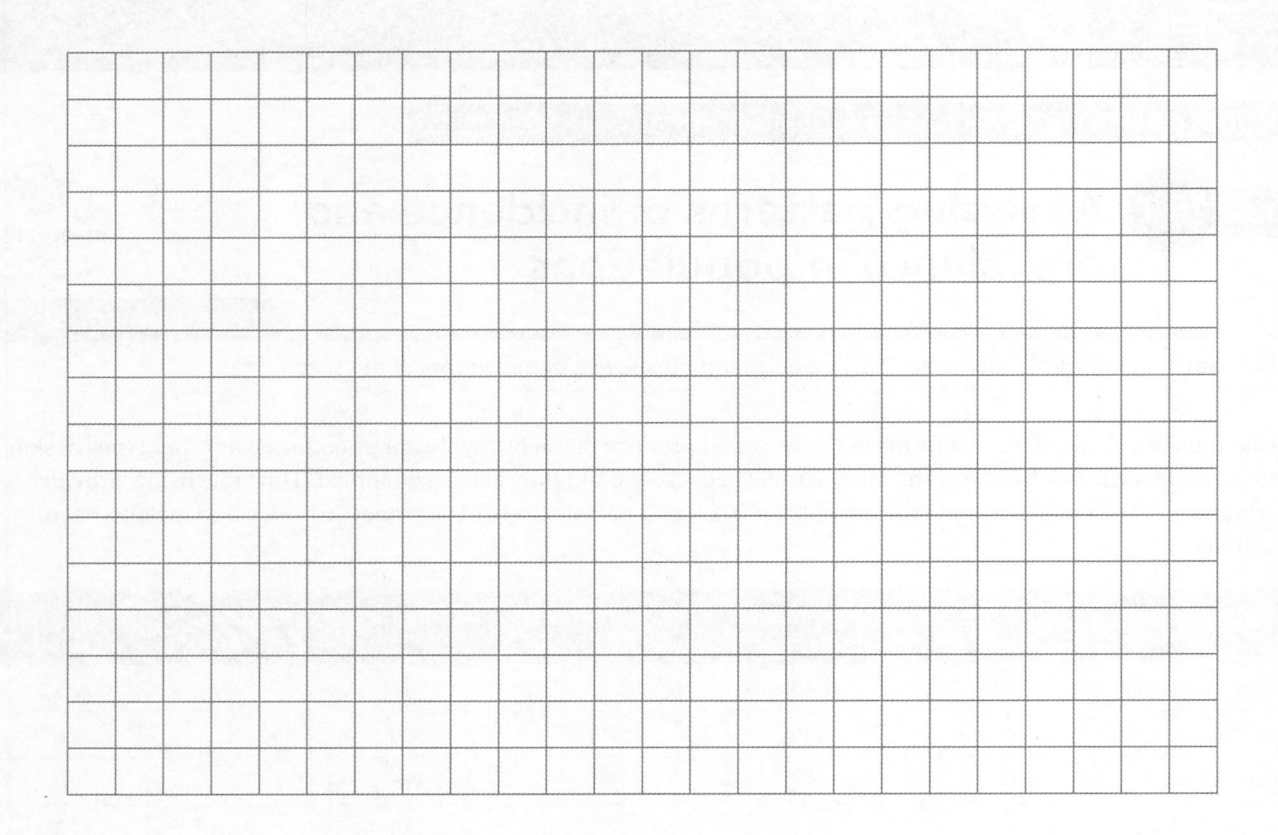

3 Describe the trend in the graph.

Treatment, management and future directions for research

LEARNING GOAL

Investigate possible treatment options and future research directions for the management of obesity.

Obesity is a disease caused by the consumption of more kilojoules than energy expended. This disease can result in negative health impacts such as type 2 diabetes and reduced life expectancy.

1 a Using information gathered from secondary sources, describe one surgical method used to treat obesity.

b Assess the reliability of one source you used in part **a**.

2 The human gastrointestinal tract is home to a large collection of microorganisms that constitute the gut microbiota. A lot of current research has focused on establishing the relationship between gut microbiota and obesity, with most of these studies analysing microbes found in stool samples. While most bacteria in the colon are anaerobic, it is only possible to culture bacteria aerobically. Some studies have suggested that the use of antibiotics can impact gut microbiota and contribute to obesity. Directions for future research include the application of prebiotics/probiotics and faecal transplants as ways to treat obesity.

a Use the information provided to identify one possible investigation question related to current research on the management of obesity.

b Identify the independent and dependent variables for the investigation you identified in part **a**.

c Identify one possible limitation that may impact the validity of your investigation.

d Explain why the limitation you identified in part **c** will impact the validity of your investigation.

e Use information from secondary sources to outline how one of the possible future directions of research aims to provide effective treatment for obesity.

Evaluating epidemiological studies

STUDENT BOOK
Pages 554–63

STUDENT BOOK Pages 554–63

LEARNING GOAL

Evaluate the method and benefits of an epidemiological study.

Epidemiological studies are used to determine the distribution, risk factors and preventative measures associated with diseases. Epidemiology utilises information from other branches of science to provide meaningful data to health care authorities. There are three types of epidemiological studies: descriptive, analytical (including case-control and cohort studies) and intervention studies.

1 Fay is a student who is interested in the link between cosmic ionising radiation and cancer. The World Health Organization has declared that this form of radiation causes cancer, along with reproductive problems. Aircrew (pilots and flight attendants) and passengers are exposed to this radiation when they fly.

a Outline the steps Fay would take in an early epidemiological study to investigate a link between cosmic ionising radiation and cancer in people who work as aircrew. Include the name of the type of study in your answer.

____

____

____

b After her initial investigations, Fay decided to conduct a case-control study. A summary of her investigation is provided below.

> An Australian airline was contacted and agreed to contact employees, inviting them to take part in the study.
>
> An email was sent to 10 000 employees who were actively working as aircrew at the time of receiving the email. A total of 683 people responded and agreed to take part in the study. Of these, 49 respondents disclosed a cancer diagnosis received at some point during their working life.
>
> All respondents were surveyed regarding the number of flight hours per week, their age, smoking status, exercise regime, diet and whether or not they had or have cancer.
>
> A follow-up data collection was conducted each year for four years. Over the course of the study, 215 people changed employers and were not surveyed each year.
>
> Data were analysed for statistical significance.

Construct a table to summarise the possible sources of error in Fay's study and describe how these errors may impact the validity of the study. Use the column headings 'Source of error', 'Type of error (random, systematic: selection bias, information bias, confounding factor)' and 'Impact on validity'.

9780170449625

c Evaluate the validity of Fay's method.

HINT

The verb *evaluate* requires you to make a judgement based on criteria or to determine the value of something.

2 Many researchers employed around the world have found that working as aircrew can increase the risk of breast cancer in those the role for more than five years. Use information gathered from secondary sources to outline two benefits of conducting an epidemiological study of breast cancer prevalence in aircrew.

16 Prevention

INQUIRY QUESTION: HOW CAN NON-INFECTIOUS DISEASES BE PREVENTED?

WS 16.1 The effectiveness of public health programs

STUDENT BOOK
Pages 570–9

LEARNING GOAL

Evaluate the effectiveness of the National Bowel Cancer Screening Program (NBCSP).

A proactive approach to disease prevention can result in significant reductions in health care costs to governments and health care providers, along with improved quality of life for individuals and the wider population. One strategy currently used is the application of educational programs and campaigns that aim to inform the public about the risks and effects of a disease.

Bowel cancer is one of the most commonly diagnosed cancers in Australia. It has one of the highest mortality rates of all cancers and affects around one in 12 Australians during their lives. The National Bowel Cancer Screening Program (NBCSP) was introduced in Australia in 2006. The program aimed to actively recruit people for bowel cancer screening to allow for early detection and prevention of the disease.

1 Use information from secondary sources to complete the table below. This table summarises how the implementation of the NBCSP aligned with the six success criteria for public health campaigns.

HINT

The Australian Department of Health provides a lot of information about its programs, including the NBCSP, online.

HINT

See page 571 of *Biology in Focus Year 12* Student Book for more detail on these success criteria.

Criterion	How it was implemented in the NBCSP
Evidence base for action	
Package of evidence-based interventions	

(*Continued*)

Criterion	How it was implemented in the NBCSP
Effective performance management including monitoring, evaluation and program improvement	
Public and private sector partnerships	
Communication of accurate information	
Political commitment	

2 The graph below shows the survival rate for bowel cancer patients. Bowel cancer data were taken from the appropriate cancer registry. From these data, individuals were placed into four groups (see table below).

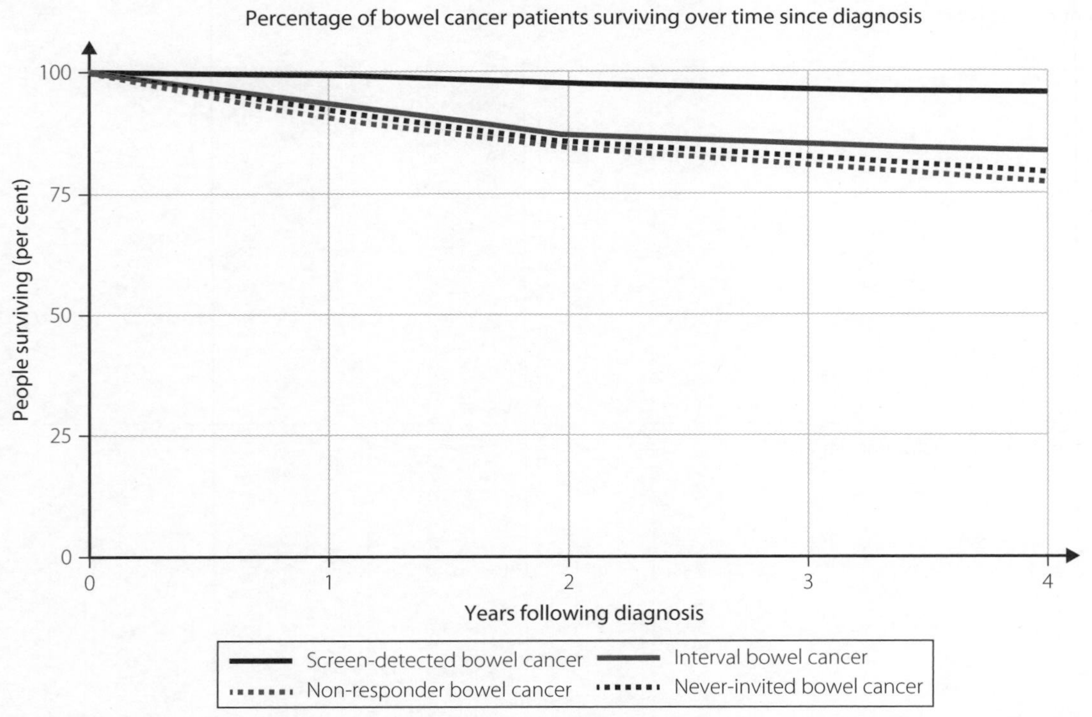

Study group	Definition
Screen-detected	Individuals invited to participate in the screening program who subsequently received a positive diagnosis for bowel cancer after the screening
Interval	Individuals invited to participate in the screening program who subsequently received a negative or inconclusive diagnosis for bowel cancer; these participants were later diagnosed during a 2-year follow-up period
Non-responder	Individuals who declined the invitation to participate in the screening program
Never invited	Individuals with a positive bowel cancer diagnosis who were not invited to participate in the screening program

Use the information provided in the graph to assess whether the NBCSP should be considered successful.

WS 16.2 The benefits of genetic engineering

STUDENT BOOK
Pages 580–3

LEARNING GOAL

Evaluate the effectiveness of genetic engineering to prevent the immune disorder SCID-X1.

Genetic engineering can be utilised by scientists to prevent inherited diseases. While currently being used to treat only a small number of diseases, the technological applications in this area are rapidly expanding.

The severe immune disorder known as X-linked severe combined immunodeficiency (SCID-X1) causes babies to be born with very poor immune systems. Also known as 'bubble boy disease', this inherited condition can be life threatening.

Use information from secondary sources to complete the questions.

1 Outline the cause of the disease SCID-X1.

2 Describe the symptoms of the disease.

3 Why is SCID-X1 also known as 'bubble boy disease'?

4 Explain how genetic engineering technologies can help treat SCID-X1.

5 Identify any limitations in the use of genetic technologies to treat this disease.

6 Outline the benefits of using genetic engineering to treat this disease, compared with other available treatment options.

7 Evaluate the effectiveness of genetic engineering to treat SCID-X1.

8 a Provide the reference of one source you used to answer questions **1–7**. Write the reference in Harvard style.

b Assess the accuracy of the source you identified in part **a**.

17 Technologies and disorders

INQUIRY QUESTION: HOW CAN TECHNOLOGIES BE USED TO ASSIST PEOPLE WHO EXPERIENCE DISORDERS?

WS 17.1 Hearing loss and technology

STUDENT BOOK
Pages 594–8, 622

LEARNING GOAL

Investigate a range of causes of hearing loss and the technologies that are used to assist with hearing loss.

Types of hearing loss

There are two main categories of hearing loss: congenital or progressive. Some people have a mixture of both types, or only have hearing loss in one ear. To use technology to assist with the effects of hearing loss, the type must be understood.

1 Distinguish between congenital and progressive hearing loss.

2 Use the graphs below to complete the questions that follow.

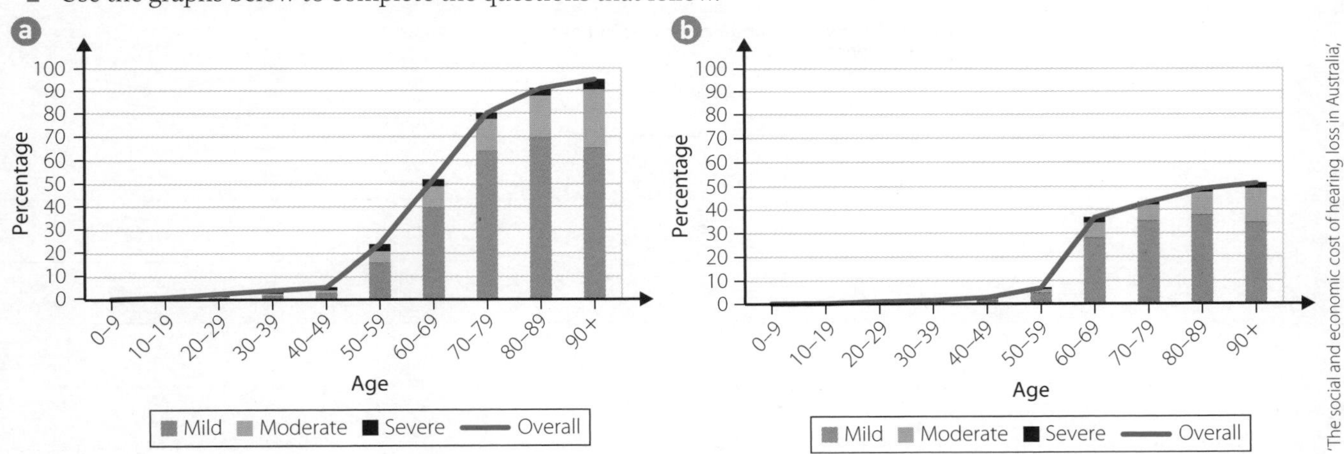

Prevalence rates of hearing loss by severity and age for **a** males and **b** females in Australia in 2017

a Estimate the percentage of 80–89-year-old males and females in each category of hearing loss in 2017.

b Describe the difference in the data for males and females for the 80–89-year-old age group in 2017.

c Identify the trends in overall hearing loss for males and females in 2017.

3 Describe the projected growth of hearing loss shown in the graph below. Use data from the graph in your answer.

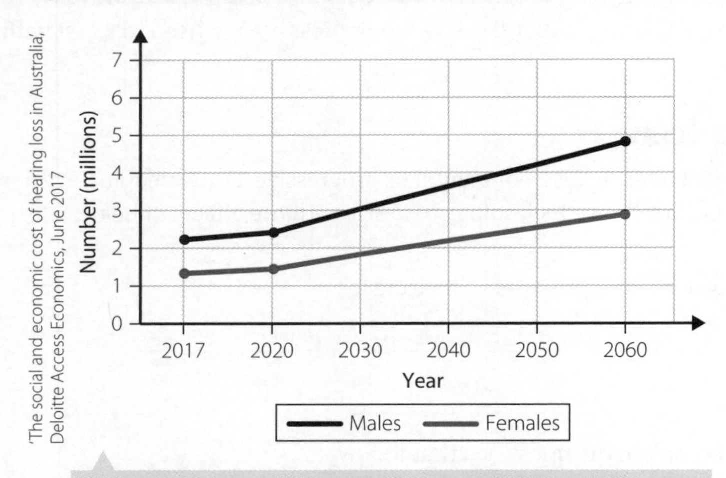

Projected growth of number of cases of hearing loss in Australia over time, 2017–2060

Conductive hearing loss

Conductive hearing loss is the result of sound vibrations being unable to pass freely from the outer to the inner ear. Anatomical or functional defects in the ear canal, ear drum or ossicles prevent conduction even though the cochlea and auditory nerve may be functional.

Some conductive deafness is genetic or congenital; for example, deformed or missing outer ear canals or abnormal bone growth that prevents the ossicles moving freely. Trauma or severe infection can also permanently damage the eardrum. Sometimes the cause of conductive hearing loss can be temporary: a build-up of wax in the outer ear canal, or fluid in the middle ear preventing normal eardrum and/or ossicle function (such as when a person has a cold virus or middle ear infection). Hearing returns once these temporary issues are resolved.

Sensorineural hearing loss

Sensorineural hearing loss is the result of defective hair cells in the cochlea or a defective auditory nerve. The outer and middle ear structures may function normally but no signal travels to the brain. As for conductive deafness, sensorineural hearing loss can be congenital or can be the result of trauma or severe infection. A point of difference is that sensorineural hearing loss is rarely temporary; it does not resolve because neurons do not regenerate after damage.

4 **a** Use the text to create a mind map about types of hearing loss. Brainstorm other ideas to add to the diagram.

b Complete the table using ticks where applicable to classify the following hearing loss scenarios.

	Hearing loss scenario	Sensorineural	Conductive	Permanent	Temporary	Progressive
i	A child says that he cannot hear properly. A toy bead is found deep in his outer ear canal.					
ii	A musician has damaged hair cells from exposure to loud music.					
iii	A woman has cochleas that are inflexible at the oval window, in both ears.					
iv	A middle-aged man has found that his auditory nerves are defective.					
v	A baby is born with defective hair cells.					
vi	The ossicles in the middle ear of a teenager have grown and are beginning to join into one bone.					
vii	The cartilage in a woman's outer ear canal grows and eventually closes the canal completely.					

Technologies to assist with hearing loss

Hearing aids

A hearing aid is a device designed to amplify sounds and direct them into the ear canal. All hearing aids contain the following components: a battery, one or more microphones to pick up external sound, an amplifier for each microphone, a speaker and a casing enclosing the components. Hearing aids are only effective for people with some residual hearing (the ability to detect some sound). The user has choices over the size and placement of the device. The opening of the ear canal must be sealed with a mould attached to the casing or by the device itself. Hearing aids can be programmed to amplify frequencies specific to individual users.

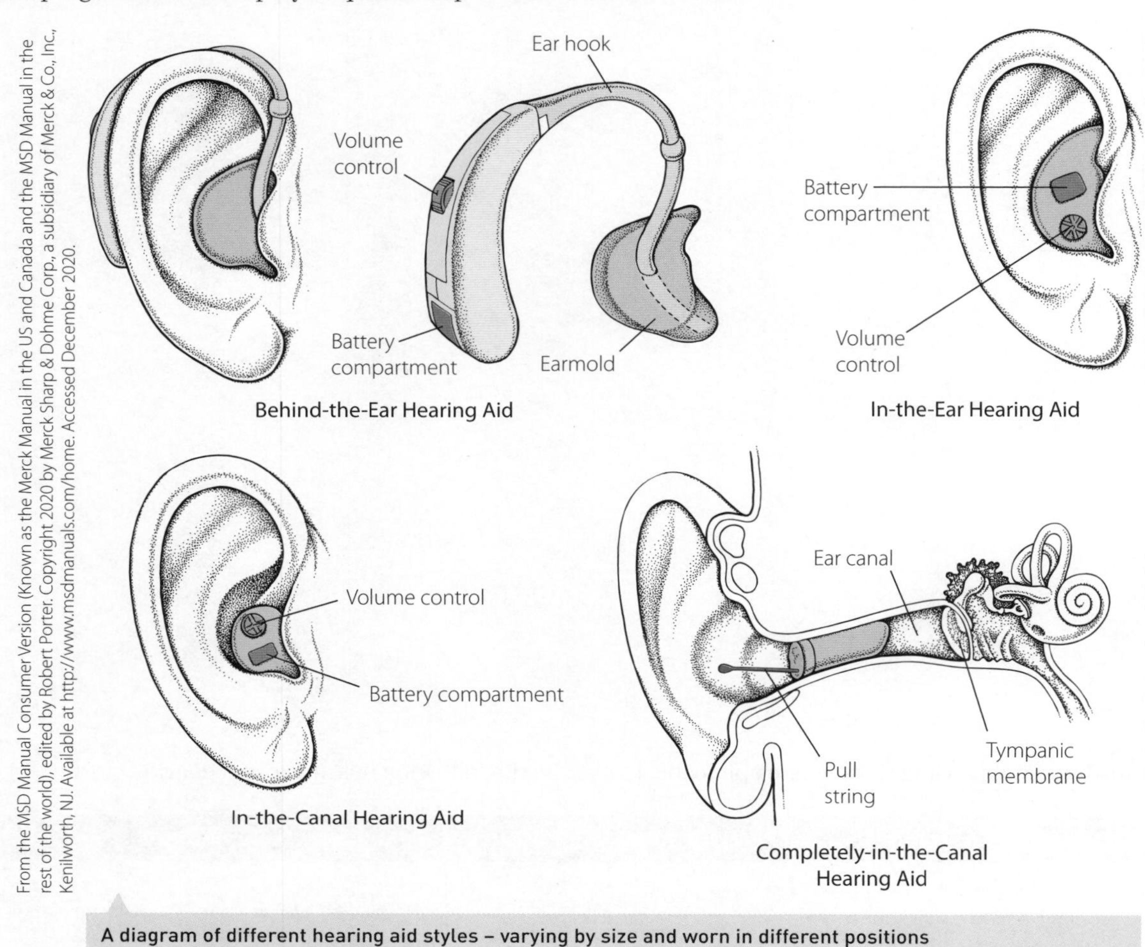

A diagram of different hearing aid styles – varying by size and worn in different positions

5 a Suggest a reason for sealing the ear canal with an ear mould.

b Predict problems for the user of these types of hearing aids.

> **HINT**
> Structure your answer in a list of dot points.

c Assess specific problems with hearing aids that may be encountered by a user with deformed or no ear canals, and a user who produces excessive wax in the ear canal.

Bone conduction implants

HINT

Bone conduction implants are also known as bone-anchored hearing implants or bone-anchored hearing aids. Regardless of the name, all refer to a component surgically inserted into the skull.

Bone conduction implants utilise the fact that sound waves travel better through solids than through air. The system consists of three parts:

- Sound processor: The battery-powered external sound processor captures sounds via a microphone and converts them into vibrations.
- Abutment: A connection that transfers the vibrations from the sound processor to the implant.
- Titanium implant: The surgically implanted titanium structure in the bone behind the ear. It transfers the sound vibration through the skull bone directly to the cochlea, bypassing the outer and middle ear.

6 a Use the text to label the parts of the bone conduction implant on the images below. Draw in the path of the sound/vibration if someone rang a bell near this person. Add any other notes to the pictures.

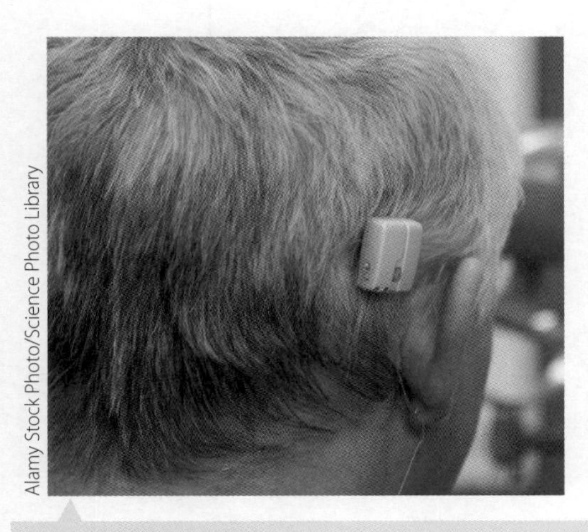

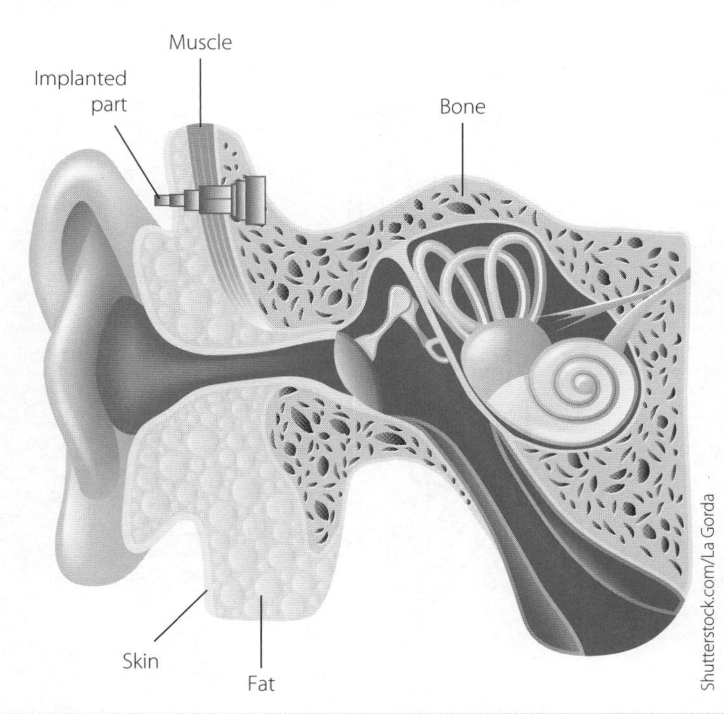

Location of external part of bone conduction hearing system (left) and skull cross-section showing position of abutment and implant (right)

b Assess the suitability of a bone conduction implant for the scenarios outlined in the table. Justify your answers.

	Scenario	Suitable (S)/ Unsuitable (U)	Reason
i	A person has no ear canal, but has normal anatomy and function of other structures.		
ii	A child has a bead stuck in their ear canal, which is affecting their hearing.		
iii	A baby is born with a non-functional auditory nerve.		
iv	A person has mild permanent conductive hearing loss.		

Cochlear implants

A cochlear implant is suitable for someone with no residual hearing. It is designed to bypass absent or damaged hair cells. The implant is surgically placed under the skin behind the ear. A wire with 22 stimulating electrodes is laid into a channel carved from the bone of the skull and inserted through the wall of the cochlea. This is a significant surgery.

7 a Trace the path of information from position 1 to 8 on the diagram below with a highlighter and add arrows.

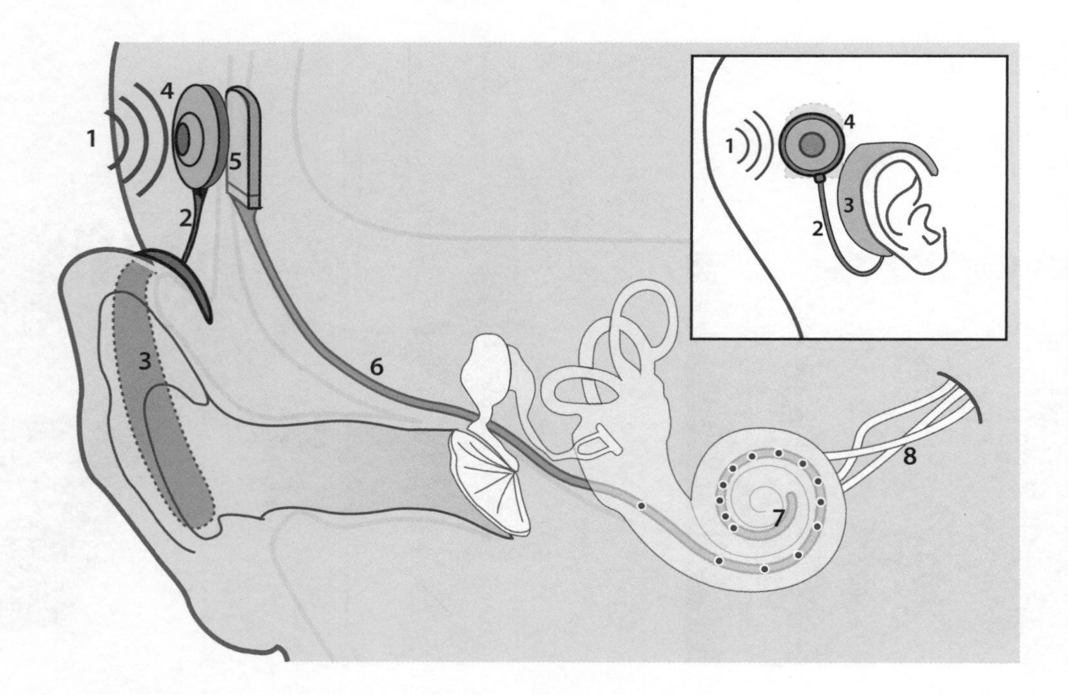

b The table below contains statements about the transmission of sound information through a cochlear implant. The statements are in the wrong order. Allocate the correct number from the diagram to each statement.

Number on diagram	Process resulting in the brain hearing sound
	The stimulated neurons send an electrochemical message to the brain along the **auditory nerve**. The brain can then try to interpret the stimulation as a meaningful sound.
	The coded signal travels via a cable to the **transmitting coil** in the headset.
	The **speech processor** analyses the information and converts it into an electrical code.
	Radio waves from the transmitter coil carry the coded signal through the skin to the **implant** inside.
	The information from the **microphone** is sent to the speech processor via a cable.
	Sounds and speech are detected by the microphone on the **headset**. The headset is held in place by a magnet.
	The implant package decodes the signal. The signal contains information that determines how much electrical current will be sent by a wire to the different electrodes.
	Electrical current passes down the lead wires to the selected **electrodes**. The frequency or pitch of the original sound determines which electrodes stimulate the neurons in the **cochlea**.

c Use the bold terms in part **b** to add labels to the diagram in part **a**.

HINT

Confused about *cochlea* and *cochlear?* We're not just spelling the same thing two ways! *Cochlea* is the *noun* referring the structure in the inner ear, whereas *cochlear* is an *adjective*, a describing word, so an implant that goes into the cochlea (the structure) is a cochlear implant (describing something about where the implant goes).

8 A cochlear implant does not restore or create normal hearing. The electrodes are only designed to stimulate the nerve cells lined up with the frequencies of sound associated with speech, approximately between 250 and 8000 Hz. Hearing through a cochlear implant sounds different from normal hearing, but it allows people with severe hearing problems to participate fully in oral communication.

a Use a highlighter on the diagram to trace the frequencies associated with the electrodes on a cochlear implant.

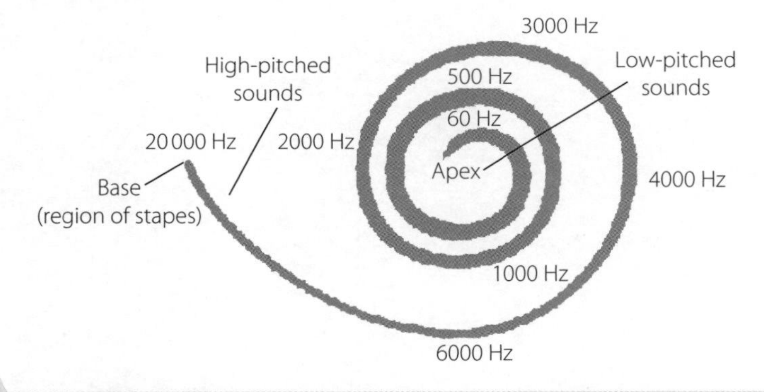

Vibrations of different frequencies stimulate hair cells at different positions along the cochlea.

b Describe the types of sounds that will not be heard by someone with a cochlear implant compared to a person with no hearing loss.

______

______

c Give examples of everyday sounds that a cochlear implant may not deliver effectively.

______

______

WS 17.2 Correcting vision

STUDENT BOOK
Pages 607–13

LEARNING GOAL

Explain a range of causes of visual disorders and investigate technologies that are used to assist with the effects of a disorder.

Hyperopia and myopia

Hyperopia and myopia are two common vision problems, often referred to as being 'long sighted' and 'short sighted', respectively. Most people become more hyperopic with age, as a result of the lens of the eye becoming less flexible over time.

1 Construct two annotated sketches that explain hyperopia and myopia. Only include relevant structures in your drawings. Include reference to the distance of the object being viewed.

HINT

Remember these disorders with this little trick:

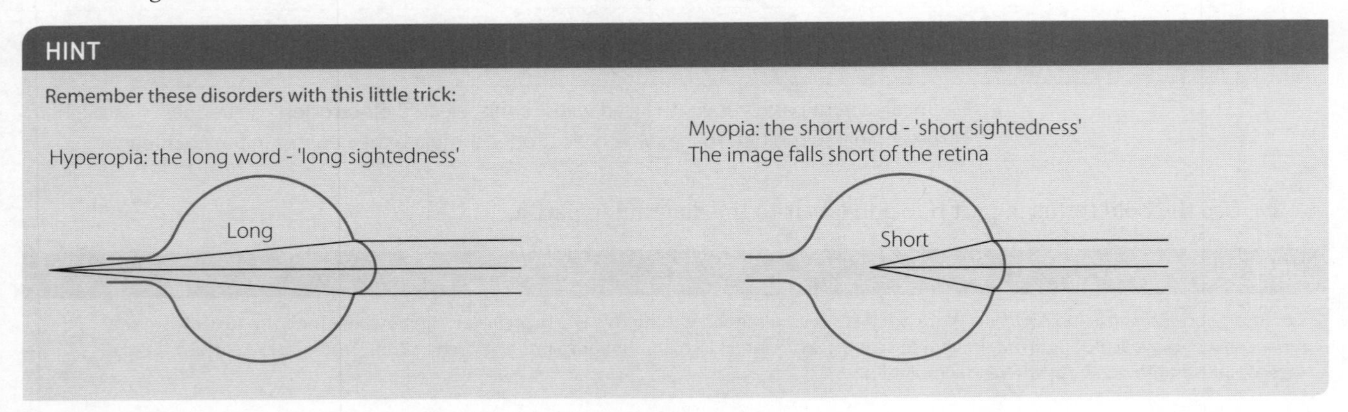

2 Use the words in the word bank to complete the passage. Some words may be used more than once.

converged	blurry	toward the back	behind
diverged	focused	forward	in front of

For someone with myopia, the ____________ image of a distant object falls ____________ the retina. As a result, the image that lands on the retina and sends a message to the brain is ____________. The light rays need to be more ____________ so that the focused image is moved ____________ of the eye, onto the retina. For someone with hyperopia, the ____________ image of a close object falls ____________ the retina. As a result, the image that lands on the retina and sends a message to the brain is ____________. The light rays need to be more ____________ so that the ____________ image is moved ____________, onto the retina.

3 Spectacles or contact lenses can change the way light enters the eye, compensating for the defects causing hyperopia and myopia. Use the words in the word bank to complete the passage. Some words may be used more than once.

converged	focused	divergence	concave	hyperopia
diverged	convergence	convex	myopia	

____________ requires ____________ lenses in front of the eye. As a result the light rays are ____________ slightly before the light enters the eye. This extra ____________ is enough to compensate for the insufficient ____________ of the defective lens. A ____________ image will then form on the retina. ____________ requires ____________ lenses in front of the eye. As a result the light rays are ____________ slightly before the light enters the eye. This extra ____________ is enough to compensate for the insufficient ____________ of the defective lens. A ____________ image will then form on the retina.

4 Brainstorm the benefits and limitations of contact lenses and spectacles. Design a table to summarise your findings.

5 The LASIK (laser in situ keratomileusis) procedure is a surgical treatment, using a laser, that permanently changes the shape of the cornea. It is most commonly used for correcting myopia.

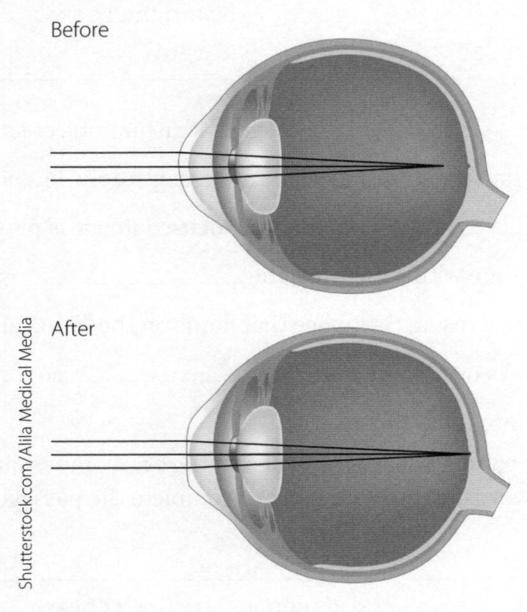

Use the diagram to help explain why flattening the cornea corrects myopia permanently.

____

____

____

____

____

____

6 Research the LASIK procedure. Use a flow chart to summarise the steps it involves.

WS 17.3 Kidney function and dialysis

LEARNING GOAL

Explain a range of causes of loss of kidney function and kidney replacement therapy.

Normal kidney function

1 Complete this passage using the words in the word bank. Some words may be used more than once.

water	nitrogenous	correct	urinary	urine
nephrons	excess	constant	ureters	urea

The kidneys are part of the ________________ system and they perform two roles: excretion of the ________________ wastes made by our cells, and osmoregulation, keeping the amount of ________________ and salts relatively ________________ in the blood. The kidneys contain microscopic structures called nephrons that filter the blood. The ________________ remove wastes that would reach toxic levels if they stayed in the blood and also ensure the blood retains the ________________ levels of water, salts, amino acids, glucose and other nutrients. The liquid removed by the nephrons will become ________________ and leaves the kidneys via the ________________. It consists of mostly of ________________, the principal ________________ waste, mixed with any water considered to be ________________ to the body's needs.

2 Blood travels from the heart toward the kidneys via the aorta. A large artery, the renal artery, branches off and takes blood into the kidney. Internally, the blood vessels divide and become smaller until they are capillaries. Capillaries surround the nephrons and their large surface area facilitates the passive and active transport between the nephrons and the blood. A dissection of the kidney reveals a bright red area where these capillaries are concentrated. This is called the medulla. The outer region of the kidney is called the cortex. After blood has been filtered, the capillaries combine to make larger vessels and eventually the blood leaves the kidney via the renal vein. The renal vein re-joins the aorta and the filtered blood continues around the body.

Use the text to label the diagram of the urinary system and the kidney.

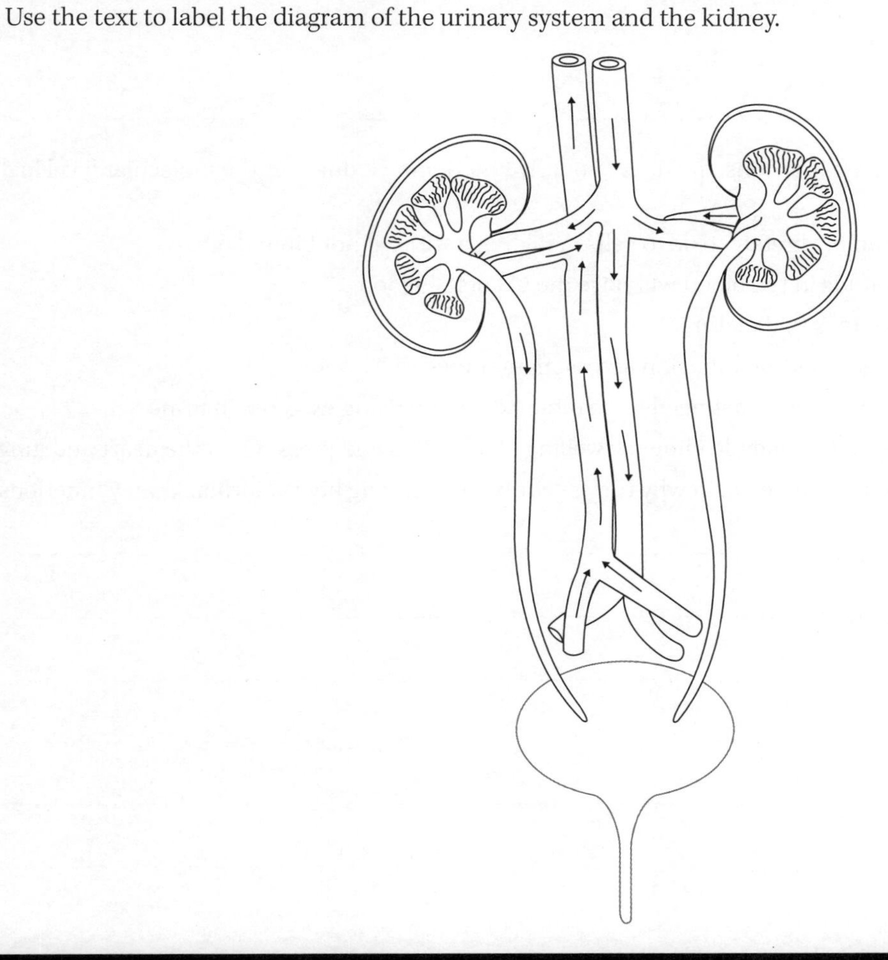

9780170449625

Loss of kidney function

Some kidney failure is sudden, or acute. Acute loss of function can be caused by trauma, infection, clots or severe or sudden dehydration. Kidney function may return when the underlying cause is resolved, but it is possible for permanent loss of function to occur. More commonly, permanent kidney damage occurs because of chronic kidney disease (CKD), where progressive loss of kidney function occurs over time.

3 Explain the differences between acute and chronic kidney disease.

4 Use secondary sources to gather information about these diseases that can lead to chronic kidney disease. In dot points, list your findings about how the disease leads to loss of kidney function.

Common cause of CKD	What occurs in the kidney
Polycystic kidneys	
Hypertension	
Diabetes	
Glomerular diseases	

5 Regardless of the underlying cause, the consequences of poorly functioning kidneys at the molecular level include the following.

- Nitrogenous compounds can be directly toxic to cells in high quantities if not filtered out.
- Excess nitrogenous compounds in the blood will raise the pH of the blood.
- Incorrect pH can cause enzymes to denature.
- There is interference with osmosis and diffusion across membranes.
- Nutrients such as glucose or proteins that need to be retained end up being excreted in urine.
- Too much water is retained in the body, leading to swelling of the limbs and pressure on the heart and lungs.

Choose two of these consequences and explain why they occur by considering how a normal kidney functions.

9780170449625

Haemodialysis

A person is diagnosed with end-stage kidney disease; their kidneys are failing to support their body's needs. Without treatment, they have only a few weeks to live. Available treatments are to replace kidney function either by dialysis or kidney transplantation. There are waiting lists for donated organs, so transplant candidates usually start dialysis first, and can continue indefinitely until a suitable donor kidney is available. All forms of dialysis remove wastes and excess water from the blood, but dialysis is not as efficient as a living kidney. Haemodialysis involves using a machine to filter the blood, whereas peritoneal dialysis works without a machine by involving the abdominal area called the peritoneal cavity. Patients on dialysis devote considerable time to their dialysis but can be still quite unwell between treatments. It is difficult to maintain full-time work and quality of life when on a dialysis program.

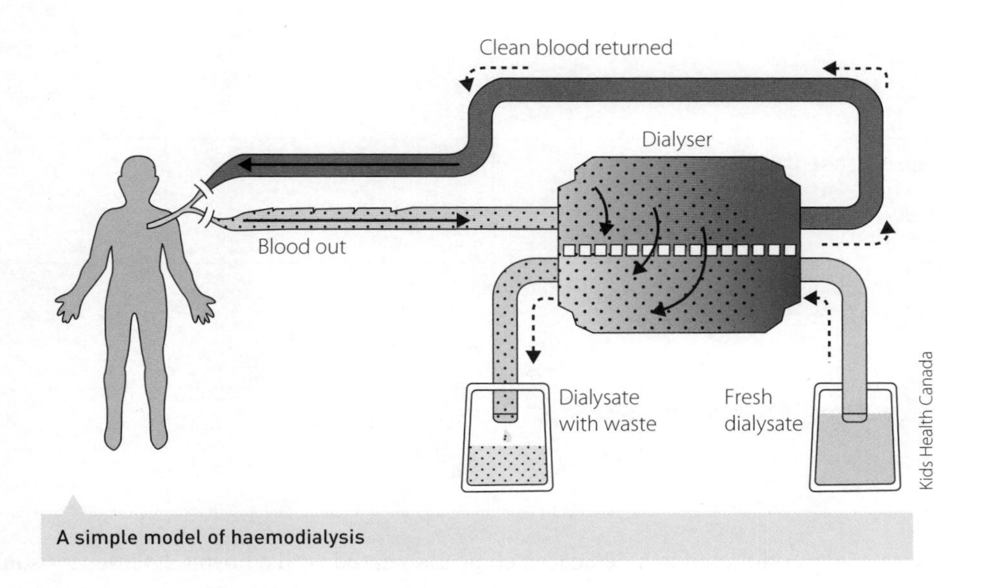

A simple model of haemodialysis

6 a Use the diagram to number these statements in chronological order. Note that the last two processes effectively occur at the same time, so there will be two processes labelled '5'.

_____ Used dialysate containing waste is discarded. This is the equivalent of urine.

_____ The substances that the body needs to retain do not pass through the semi-permeable membrane into the dialysate.

_____ Blood flows along one side of a semi-permeable membrane and dialysis solution (dialysate) flows on the other side.

_____ Blood flows out of the body into a tube.

_____ Blood is returned to the body.

_____ The waste products diffuse from the blood (an area with lots of waste products) into the dialysate (an area with no waste products).

b The diagram above is a very simple model of dialysis. Add components and notes to the diagram to make the model more accurate, using the information below.

> Blood exits the body under high pressure via an artery and enters a pump that ensures blood keeps going around the circuit. Before blood enters the dialyser, a blood thinning medication is added. In the dialyser, blood flows through thousands of thin tubes with semi-permeable membranes, which are all surrounded by dialysate. The fresh dialysate is warmed to body temperature before it enters the dialyser. Blood flows through a bubble trap before it enters the body through a vein.

c Suggest reasons for the following processes in haemodialysis.

	Process	Reason for process
i	Blood goes through a pump.	
ii	Blood thinner is added.	
iii	Blood goes through a bubble trap.	
iv	Dialysate is warmed.	
v	There are thousands of tiny tubes in the dialyser.	
vi	Dialysate and blood flow in different directions.	

7 Haemodialysis is usually conducted in a dedicated clinic associated with a hospital. Patients usually visit three times a week and are connected for about six hours a day for the rest of their life, or until they have a kidney transplant. Some patients opt to have a haemodialysis machine at home, but the patient or someone else needs to be trained for about six weeks to insert the needles into the artery and vein and to run all aspects of the process.

Brainstorm the positives and negatives of undertaking haemodialysis at a clinic and at home.

HINT

Your answer could be listed, drawn in a mind map or presented in a table.

WS 17.4 How can technology help?

LEARNING GOAL

Evaluate the effectiveness of technology in assisting with the effects of a disorder.

1 Prepare an extended response to evaluate the effectiveness of a named technology in assisting with the effects of a disorder.

The disorders listed in the syllabus are:	The technologies listed in the syllabus include:
• hearing loss • vision problems • loss of kidney function.	• hearing aids, bone conduction and cochlear implants • spectacles, laser surgery • dialysis.

You may have also investigated other technologies such as brain stem transplant, bionic eye or cataract surgery. If you get a question in an exam that specifies a particular technology from the syllabus, you must write your answer in response to that technology.

HINT

For this evaluation you must choose a *disorder* and a *technology* used to assist that disorder. *Evaluate* means to make a judgement about the effectiveness of the technology and support your answer with evidence. This requires you to know *details* about the disorder and the technology in order to weigh up the benefits and limitations before you make a judgement. When you have decided what you are interested in and know the most about, re-write the question for yourself. For example: *Evaluate the effectiveness of the bone conduction implant on assisting with the effects of hearing loss.*

Grab some scrap paper and make a plan. It is OK to use subheadings in a science extended response. You can even include tables or labelled diagrams to save space and to control your word count.

2 Assess one of your peers' responses to question **1** using the criteria in the table below.

	Thorough	Adequate	Limited	Missing
Summary of the disorder that describes what causes the disorder				
The effect of the disorder on the person				
Describe how the technology works				
Who can use the technology? Are some people unsuitable? Long-term suitability?				
Advantages/Benefits and Disadvantages/ Limitations				
Judgement about the effectiveness of the technology in addressing the disorder				

Module 8: Checking understanding

1 Explain why the internal environment of organisms needs to be maintained.

2 The prevalence of a disease can be best described as:

A the total number of new cases of a disease within a year.

B the number of deaths due to a disease within a time period.

C the number of new cases of a disease within a time period.

D the total number of cases of a disease within a time period.

3 Phenylketonuria (PKU) is a non-infectious disease with symptoms that include intellectual disability, seizures, low birth rates and potential heart problems. Mutations in the PAH gene result in low level production of the enzyme phenylalanine hydroxylase. The importance of diet in treating this disease was recognised in 1953. Patients are required to avoid foods that contain phenylalanine.

Based on the information provided, PKU can be described as a:

A nutritional disease.

B single gene abnormality.

C chromosomal non-disjunction.

D disease caused by chemical exposure.

4 Outline the role of epidemiological studies in the management of non-infectious diseases.

5 Distinguish between descriptive epidemiological studies and intervention epidemiological studies.

6 What impact do systematic errors have on an epidemiological study?

7 Which of the following options correctly identifies the functions of the ossicles, cochlea and auditory nerve, respectively?

	Ossicles	Cochlea	Auditory nerve
A	Receive sound waves and convert them into vibrations	Converts vibrations into electrical impulses	Equalises air pressure between the inside and the outside of ear
B	Receive sound waves and convert them into vibrations	Maintains correct balance in the body	Transports nervous impulses to the brain
C	Transfer vibrations to the oval window	Converts vibrations into electrical impulses	Transports nervous impulses to the brain
D	Receive sound waves and convert them into vibrations	Converts vibrations into electrical impulses	Transports nervous impulses to the brain

8 What are the receptor cells in the ear?

A Hair cells

B Organ of Corti

C Tectorial membrane

D Oval window

9 Which of the following statements about renal dialysis is correct?

A Dialysate has the same concentration as blood plasma without the waste and flows in the same direction to the blood in the dialyser.

B Dialysate has the same concentration as blood plasma without the waste and flows in the opposite direction to the blood in the dialyser.

C Dialysate has the same concentration as blood plasma and flows in the opposite direction to the blood in the dialyser.

D Dialysate has the same concentration as blood plasma and flows in the same direction to the blood in the dialyser.

10 An individual can see near objects clearly, but distant objects appear blurred. What is the name given to this visual disorder?

A Hyperopia

B Myopia

C Astigmatism

D Cataract

11 Assess the accuracy of this statement: *Hearing aids are only suitable for conductive hearing problems.*

Section I

20 marks

1 Why is accurate replication of DNA important?

A It leads to cell differentiation.

B It maintains genetic information.

C It allows for evolution of the species.

D It enables cells to modify their proteins.

2 In an RNA molecule, which nitrogenous base is found in place of thymine?

A Guanine

B Uracil

C Adenine

D Cytosine

3 Which option correctly describes the following pathogens?

	Prions	Viruses	Bacteria	Fungi	Macroparasites
A	Eukaryotes with cell walls; unicellular or multicellular	Unicellular prokaryotes with a cell wall	Non-cellular; do not contain genetic material	Multicellular eukaryotic organisms; can be endo- or ectoparasites	Consist of a capsid that encloses DNA or RNA
B	Non-cellular; do not contain genetic material	Consist of a capsid that encloses DNA or RNA	Unicellular prokaryotes with a cell wall	Eukaryotes with cell walls; unicellular or multicellular	Multicellular eukaryotic organisms; can be endo- or ectoparasites
C	Consist of a capsid that encloses DNA or RNA	Eukaryotes with cell walls; unicellular or multicellular	Multicellular eukaryotic organisms; can be endo- or ectoparasites	Unicellular prokaryotes with a cell wall	Non-cellular; do not contain genetic material
D	Multicellular eukaryotic organisms; can be endo- or ectoparasites	Unicellular prokaryotes with a cell wall	Consist of a capsid that encloses DNA or RNA	Non-cellular; do not contain genetic material	Eukaryotes with cell walls; unicellular or multicellular

4 When designing an investigation to test for microbes in different sources of water, the independent variable would be the:

A source of the water.

B amount of water used in each sample.

C number of species of bacteria that are cultured.

D temperature at which the samples are incubated.

5 Two parents had a child with a genetic disorder but neither of the parents expressed the disorder. Which of the following is most likely regarding the allele?

A The parents are homozygous recessive.

B The parents are homozygous dominant.

C The child is heterozygous.

D The parents are heterozygous.

6 A recombinant vaccine is a vaccine produced using recombinant DNA technology. This involves inserting the gene coding for an antigen (such as a viral protein) that stimulates an immune response into mammalian cells, expressing the antigen in these cells and then purifying it from them.

The following steps are in the wrong order.

1 Plasmid DNA from a bacterium is cut open with a restriction enzyme.
2 The HB antigen-producing gene is isolated from the HB virus.
3 The recombinant DNA, containing the target gene, is introduced into a yeast cell, and the recombinant yeast cell forms.
4 The isolated HB antigen-producing gene is inserted into the bacterial plasmid, forming the recombinant DNA.
5 The HB antigens are extracted, purified and bottled. The vaccine is ready for vaccination in humans.
6 The recombinant yeast cell multiplies in the fermentation tank and produces the HB antigens.

Choose the option below that indicates the correct procedure.

A 3, 4, 6, 5, 2, 6
B 1, 4, 2, 6, 3, 5
C 2, 1, 4, 3, 6, 5
D 1, 2, 3, 4, 6, 5

7 Which of the following changes would be expected if a CAUUUG sequence of bases mutated to CACUUG?

Second base

First base	U	C	A	G	Third base
U	UUU, UUC Phe; UUA, UUG Leu	UCU, UCC, UCA, UCG Ser	UAU, UAC Tyr; UAA Stop; UAG Stop	UGU, UGC Cys; UGA Stop; UGG Trp	U C A G
C	CUU, CUC, CUA, CUG Leu	CCU, CCC, CCA, CCG Pro	CAU, CAC His; CAA, CAG Gln	CGU, CGC, CGA, CGG Arg	U C A G
A	AUU, AUC, AUA Ile; AUG Met/Start	ACU, ACC, ACA, ACG Thr	AAU, AAC Asn; AAA, AAG Lys	AGU, AGC Ser; AGA, AGG Arg	U C A G
G	GUU, GUC, GUA, GUG Val	GCU, GCC, GCA, GCG Ala	GAU, GAC Asp; GAA, GAG Glu	GGU, GGC, GGA, GGG Gly	U C A G

A The amino acid sequence would be shorter than expected.
B The identity of one amino acid would change.
C The identity of more than one amino acid would change.
D The amino acid sequence would remain unchanged.

8 Which statement best describes the relationship between proteins and polypeptides?

A Proteins are composed of polypeptides.

B Polypeptides are composed of proteins.

C Proteins, unlike polypeptides, are composed of amino acids.

D Polypeptides, unlike proteins, are composed of amino acids.

9 Identify the role of tRNA during translation.

A Bond to the DNA strand and carry the code for polypeptide synthesis out of the nucleus

B Carry ribosomes to the site of protein synthesis

C Break apart mRNA and send it back to the nucleus so that it can be reused

D Carry amino acids to the mRNA for correct placement into the polypeptide chain

10 The hormone that induces labour and controls labour, including the delivery of the placenta, via a positive feedback mechanism is:

A ADH.

B oxytocin.

C HCG.

D insulin.

11 Zika virus is a disease spread by *Aedes* mosquitoes. The best description of regional factors capable of influencing the severity of a Zika virus epidemic would be:

A the disruption of the sewerage system during a hurricane.

B the migration of people carrying Zika virus between countries.

C the length of the incubation period of the Zika virus within the mosquito vector.

D precipitation patterns that influence the availability of mosquito breeding habitat.

12 The schematic section below shows the reproductive structures of a flower.

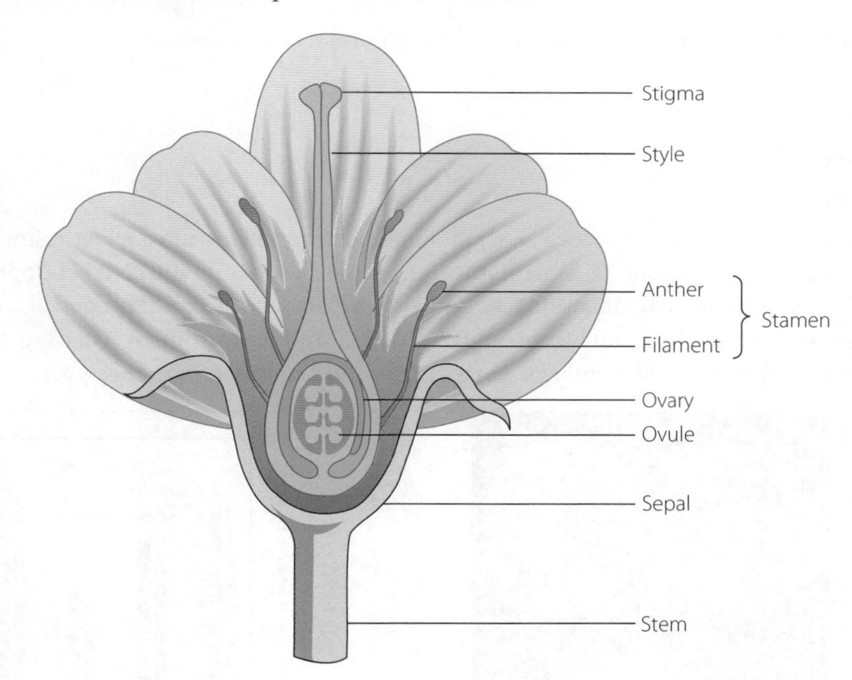

Roger is a kiwifruit farmer who wishes to use artificial pollination on his farm. The kiwifruit plant is dioecious; that is, flowers of each sex are found on separate plants. Which steps should he take?

A Remove the anther and brush directly onto the stigma of the same plant.

B Use a brush to remove the pollen from the stamen and then transfer the pollen to the stigma of a flower from a different plant.

C Use a brush to remove the pollen from the stigma and then transfer the pollen to the stamen of a flower from a different plant.

D Brush the pollen from the filament from one plant and place onto the style from a separate plant, after first removing the stigma.

13 The results of a study investigating the safety of childbirth in Kenya are shown below.

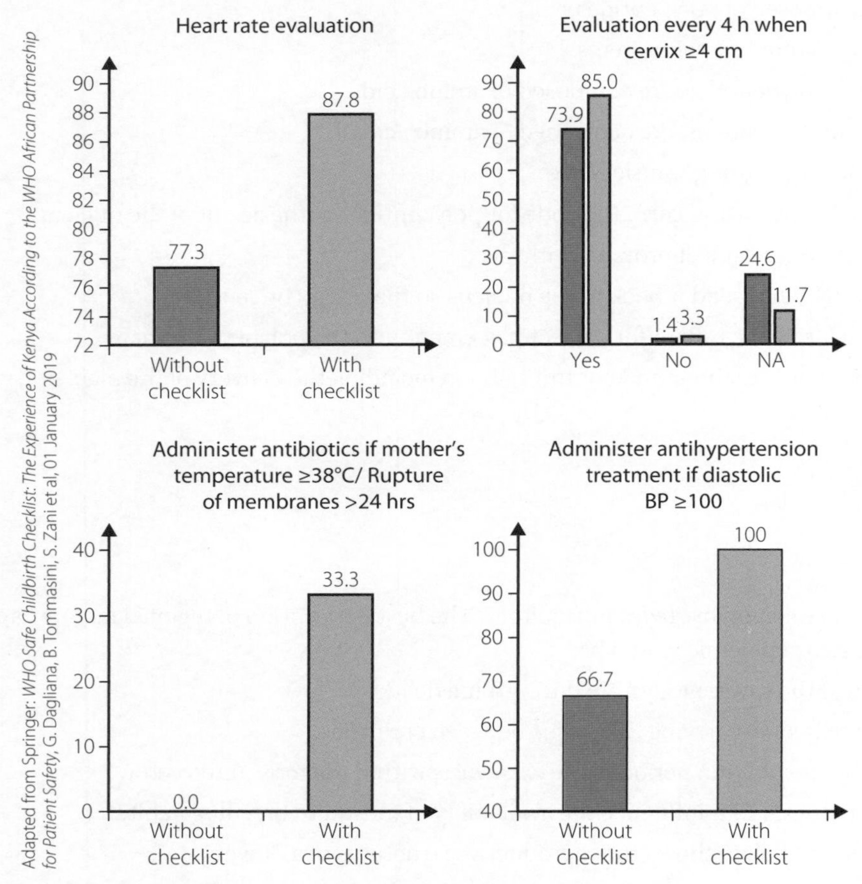

Based on these data, it can be assumed that the type of study conducted was a:

A cohort study.

B descriptive study.

C intervention study.

D case-control study.

14 The figure below shows some of the results of a study aiming to determine the most effective method of detoxifying water using semiconductor photocatalysts. The photocatalysts used were pure ZnO NP, Nd-doped ZnO NP and Er-doped ZnO NP. Products of the detoxification reaction also display antimicrobial properties. The figure shows the antibacterial effectiveness of the photocatalysts, compared with a standard antimicrobial drug, gentamicin, against three bacteria species: ***Escherichia coli*, *Staphylococcus aureus*** and ***Listeria monocytogenes***.

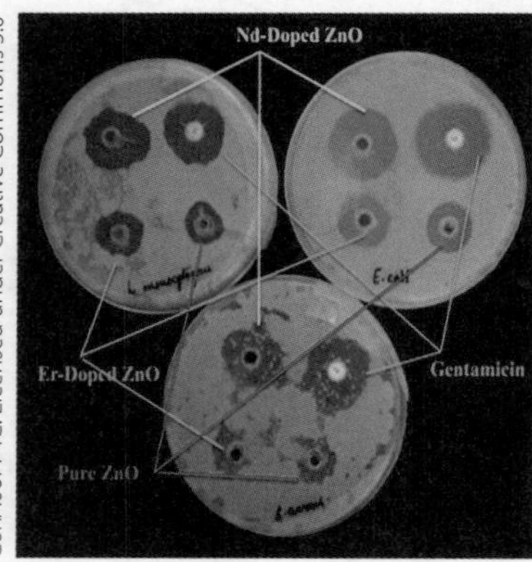

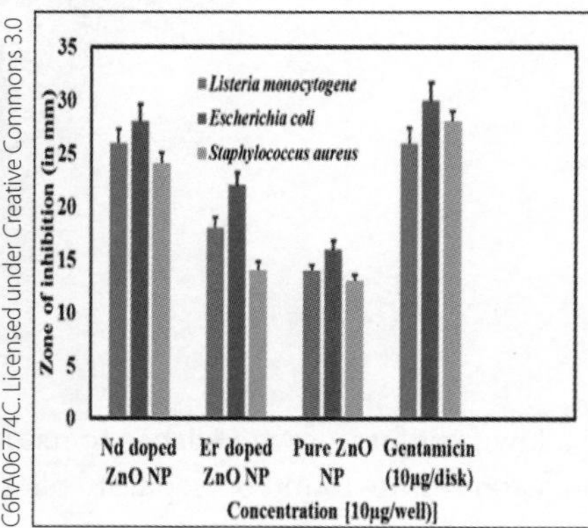

Which of the following is the best interpretation of the figure?

A Of the three photocatalysts, pure ZnO NP is the most effective antibacterial.

B Of the three photocatalysts, Nd-doped ZnO NP is the most effective antibacterial.

C All of the photocatalysts are more effective at preventing microbial growth, when compared to the drug gentamicin.

D Pure ZnO NP is a stronger antimicrobial than gentamicin.

15 The diagram below shows chromosomes in a cell during meiosis I.

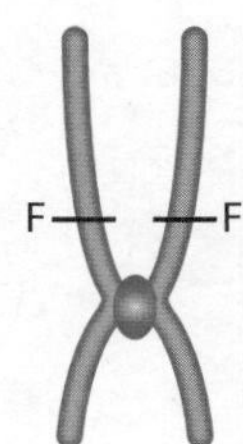

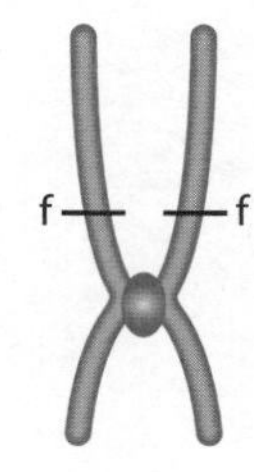

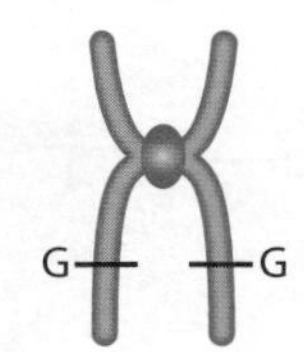

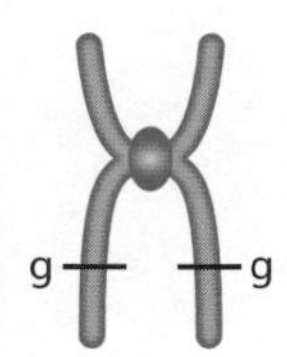

What is the genotype of the organism?

A FF, GG

B Ff, Gg

C ff, gg

D FF, ff, GG, gg

16 The following Punnett square shows a cross between a white chicken (W^W) and a black chicken (B^B), producing black offspring with white patches.

	C^W	C^W
C^B	C^WC^B	C^WC^B
C^B	C^WC^B	C^WC^B

What can be concluded from the Punnett square?

A White is dominant over black.

B Black is dominant over white.

C The inheritance of colour is incomplete dominance.

D The inheritance of colour is codominant.

17 The image below represents the human body's initial and secondary responses to a pathogen.

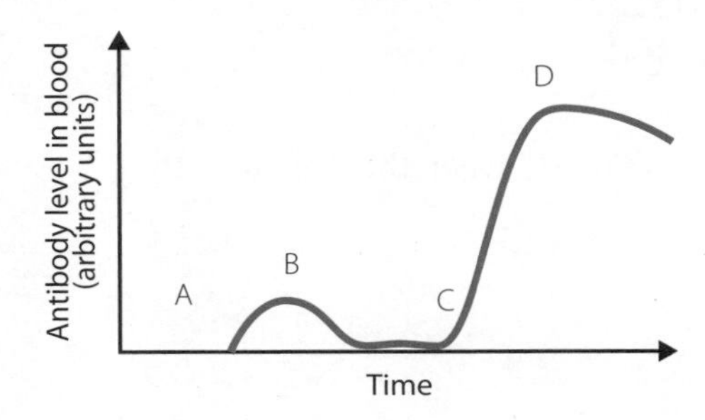

At what point does secondary exposure to the pathogen occur?

A A

B B

C C

D D

18 Below is a graph describing a change in a population.

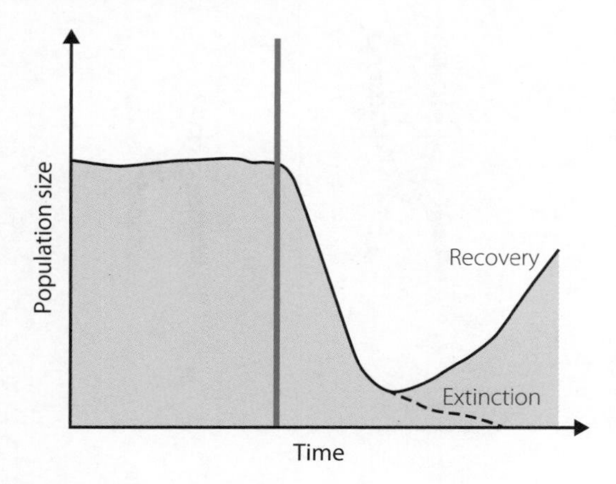

What does the pink line on the graph represent?

A Evolution by natural selection
B A genetic bottleneck event
C The founder effect
D Gene flow

19 The diagram below represents a type of chromosomal mutation.

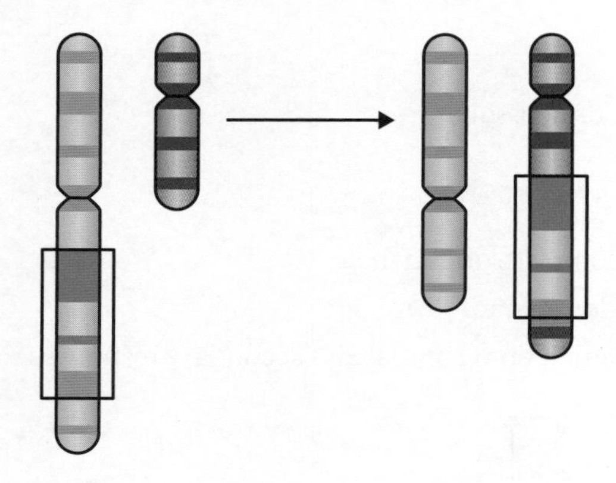

Identify the type of chromosomal mutation shown in the diagram.

A Insertion
B Deletion
C Inversion
D Duplication

20 The diagram below details the human body's response to an increase in temperature.

Which one of the following would be an appropriate response?

A Vasoconstriction
B Vasodilation
C Shivering
D Increasing metabolic rate

9780170449625

Section II

80 marks

Question 21 (5 marks)

Microbiologists can grow bacterial cultures in the laboratory using sterile media and optimum temperature, pH and gaseous conditions. Under these conditions the cells reproduce rapidly. No fresh medium is provided during incubation and the growth is plotted versus the incubation time. Note that the term 'growth' applies to the number of individuals in the population, not the size of the bacteria.

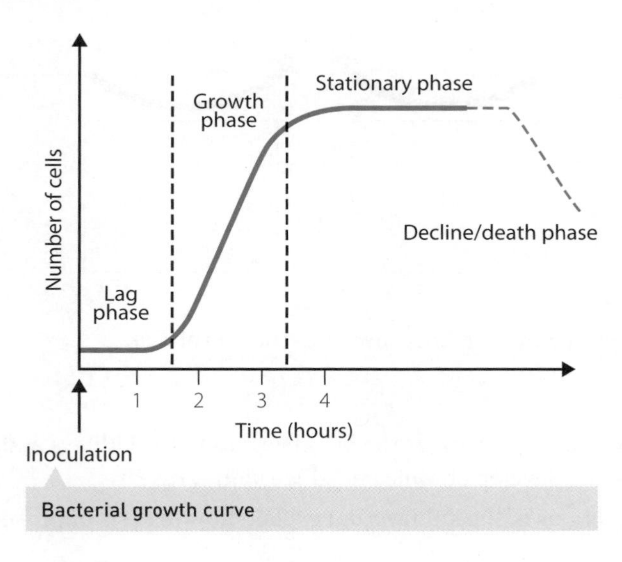

Bacterial growth curve

a What is being measured on the y-axis of the graph? (1 mark)

____

b Identify three controlled variables when conducting a replicated investigation to collect data for a bacterial growth curve. (3 marks)

____

____

____

c How long does the growth phase last for this species of bacteria (to the nearest half hour)? (1 mark)

____

Question 22 (15 marks)

a Read the text below.

> **Controlling blood glucose**
>
> Chemoreceptors in the pancreas detect levels of glucose in the blood.
>
> If blood glucose is high, the pancreas produces insulin. Insulin causes the liver to convert glucose into glycogen. Storing glycogen removes glucose from the blood. Once the levels of glucose in the blood decrease, the production of insulin decreases.
>
> If levels of glucose are too low, glucagon is produced by the pancreas. Glucagon causes the levels of glucose in the blood to increase because the liver converts stored glycogen into glucose.
>
> When the level of glucose increases back to normal, the production of glucagon is reduced.

i Use the text to correctly identify the hormones and their effect in each of the text boxes in the diagram below. (4 marks)

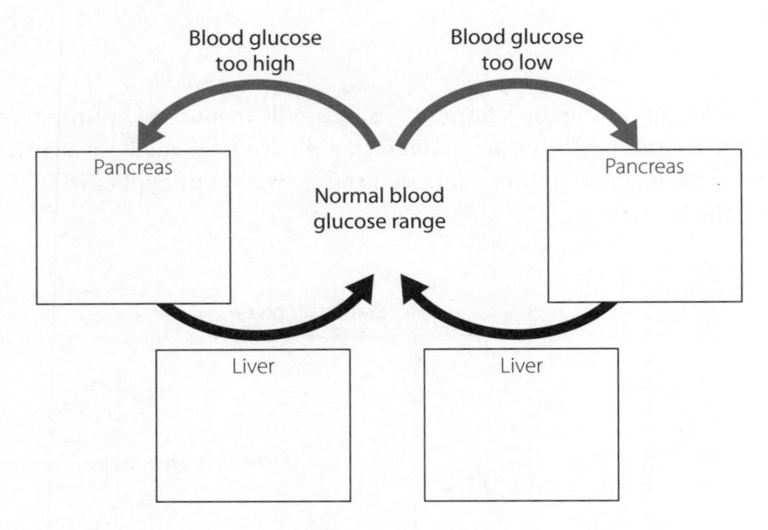

ii Name the process described by diagrams similar to the one above. (1 mark)

____

b Fluctuating blood glucose over time can damage the blood vessels in the kidneys and, especially in conjunction with high blood pressure, diabetics can develop chronic end-stage kidney disease.

Identify the two important functions of the kidneys and explain how dialysis improves the health of a person with loss of kidney function. (5 marks)

____

____

____

____

____

____

____

____

____

____

____

____

c Modern examples of biotechnology include the manufacture of useful products using genetically modified organisms. For example, insulin has been manufactured since the 1980s using recombinant DNA technology.

Outline an example of a genetically modified organism used to make a product useful to humans now or in the future. Outline the need for the product in society and the benefits of making the product using recombinant DNA technology. (5 marks)

Question 23 (3 marks)

Spectacles can help correct several vision defects that can occur in combination with each other. Consider just myopia (the inability to focus on distant objects) and hyperopia (the inability to focus on close objects).

Complete the table. Refer to the anatomy of the eye and the effect on the image created in people with these vision defects. Identify the shape of the corrective lenses in spectacles prescribed for myopia and hyperopia. (3 marks)

Vision disorder	Shape of eyeball and lens	Position of image formation	Shape of corrective lens
Hyperopia			
Myopia			

Question 24 (6 marks)

a Use an example to describe the indirect transmission of an infectious disease. (3 marks)

b Outline how the adaptation of a pathogen can facilitate the indirect transfer of a named infectious disease. (3 marks)

Question 25 (6 marks)

Mesothelioma is a cancer that usually attacks the lungs after exposure to asbestos. While Australia banned the use of all asbestos in 2003, the cancer can form decades after exposure. The table below shows the number and rate of mesothelioma in males and females from 2011–2018.

Number and rate (per 100 000 population) of people diagnosed with mesothelioma, by year and sex, 2011 to 2018

Year of diagnosis	Males		Females	
	No.	Rate	No.	Rate
2011	598	5.3	106	0.8
2012	610	5.2	129	0.9
2013	579	4.9	136	1
2014	624	5.1	145	1
2015	583	4.6	162	1.1
2016	615	4.7	158	1
2017	631	4.7	128	0.8
2018	520	3.7	142	0.9

Based on Australian Institute of Health and Welfare; AMR data. Licensed under Creative Commons 3.0 https://creativecommons.org/licenses/by/3.0/au/

9780170449625

a Use the grid below to graph the rate of mesothelioma diagnosis in males and females between 2011 and 2018. (4 marks)

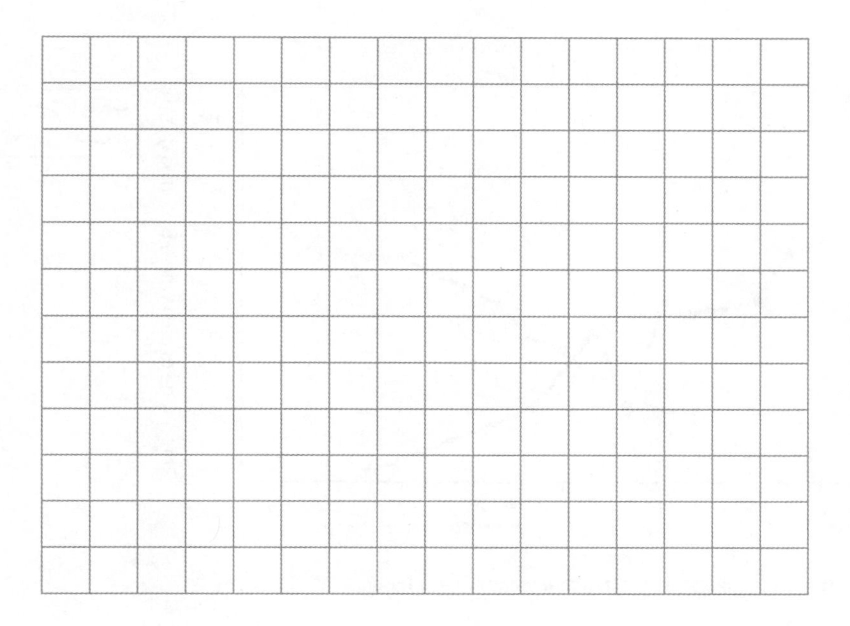

b Describe the trend shown in the graph you drew in part **a**. (2 marks)

Question 26 (5 marks)

Use named examples to distinguish between a lifestyle non-infectious disease and a chromosomal non-infectious disease. In your answer include reference to the cause(s) and effects of each disease. (5 marks)

Question 27 (7 marks)

Many scientists advocate the use of reproductive biotechnology, not only for economic benefits, but as a way to solve many environmental problems. Figures 1 and 2 below show the results of some studies investigating the application of biotechnologies.

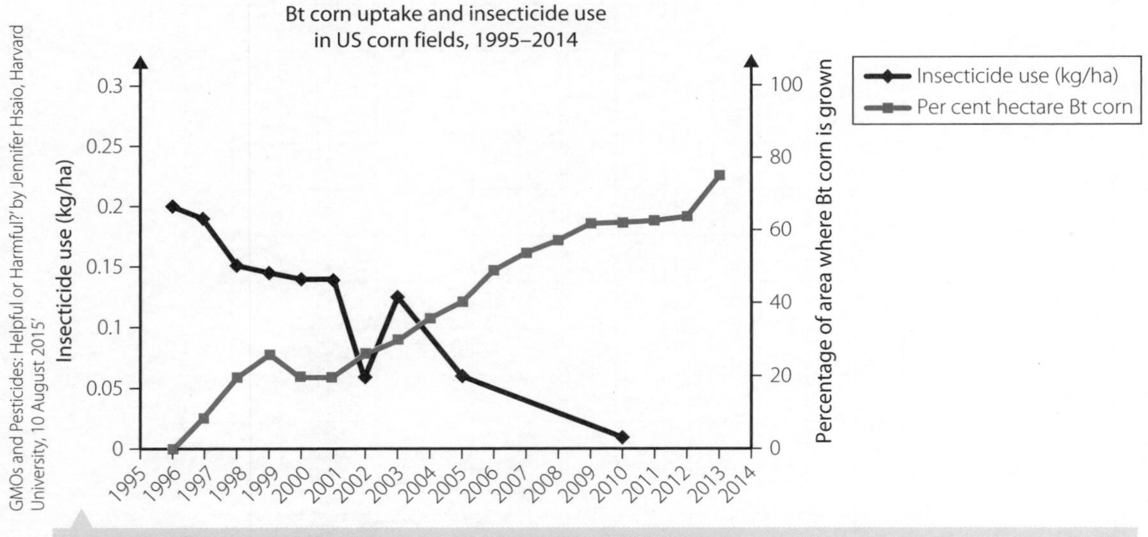

Figure 1 Bt corn uptake and insecticide use in US corn fields, 1995–2014

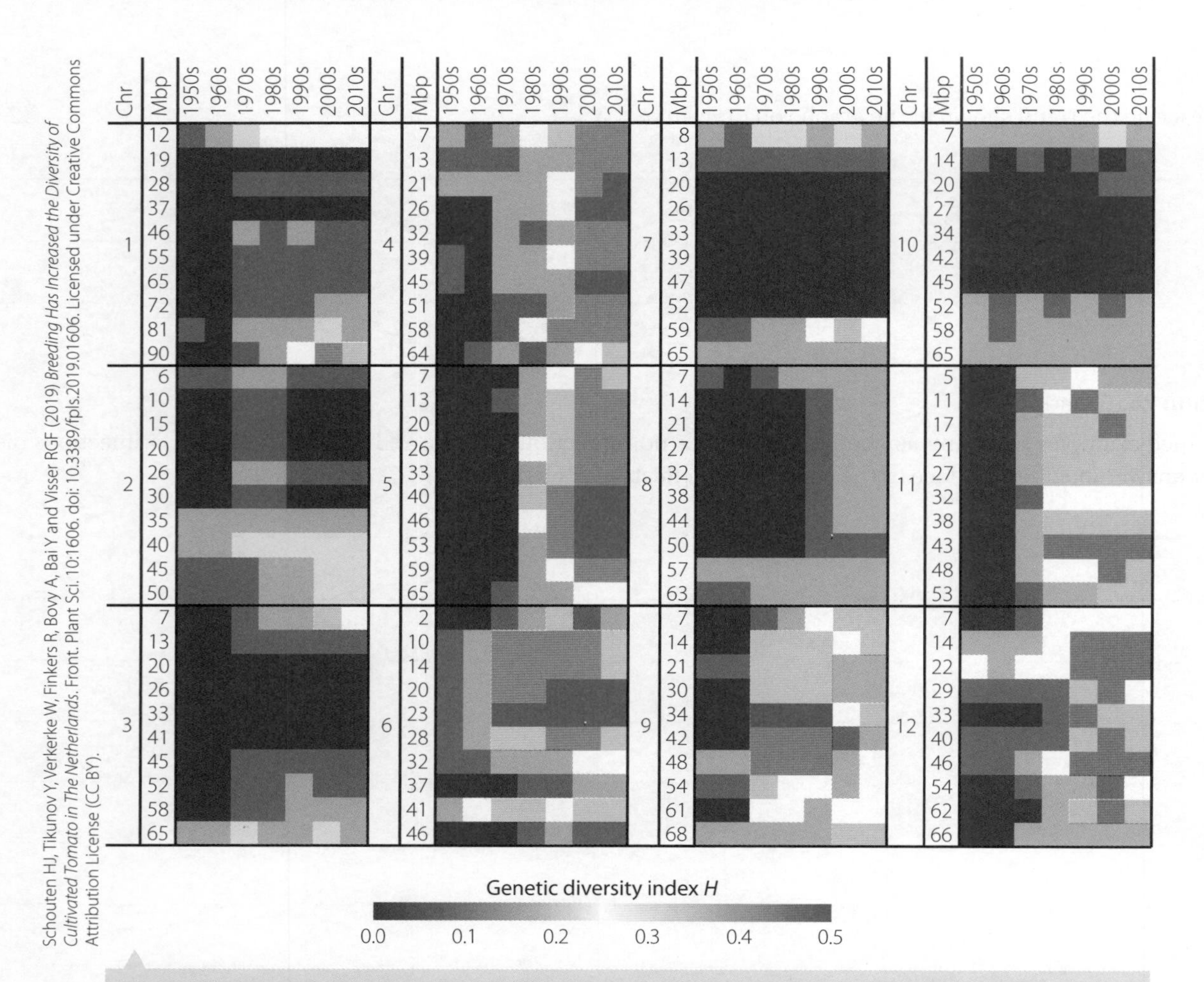

Figure 2 Heat map of the genetic diversity in the 12 chromosomes of tomatoes grown commercially from the 1950s to the 2010s. Dark grey colours represent low genetic diversity and pink colours indicate a high genetic diversity. Part of this study, conducted in the Netherlands, investigated the impact of introgression (or interbreeding) to create hybrids, and then backcrossing of hybrids, on the genetic diversity of tomatoes.

9780170449625

Use your understanding of reproductive biotechnologies, along with reference to the stimuli provided, to evaluate the following statement: 'Reproductive biotechnology will benefit biodiversity.' (7 marks)

Question 28 (2 marks)

Below is a diagram of a cell undergoing cell division.

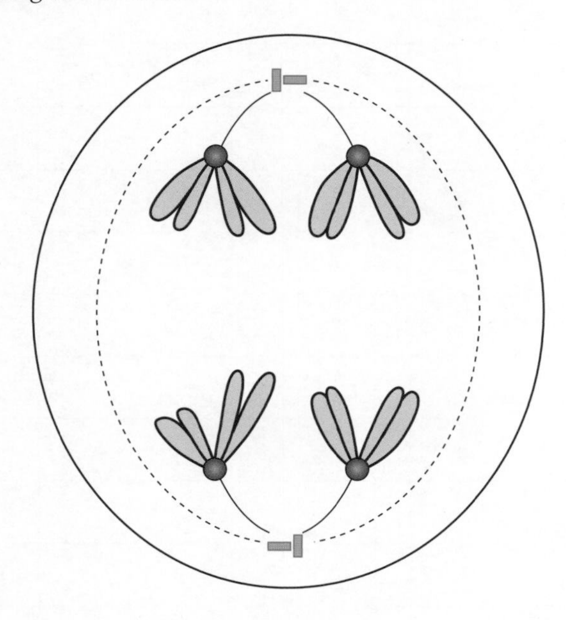

Describe the process taking place. (2 marks)

____

____

____

____

Question 29 (4 marks)

The Amish are an isolated religious group who live in America. This group was founded by 30 Swiss people, who moved to America many years ago. One of the 30 Swiss founders had a genetic disorder called Ellis-Van Creveld syndrome. The syndrome is due to a mutated allele. People with this disorder have extra fingers and toes as well as heart conditions. About 1 in 200 Amish are born with Ellis-Van Creveld syndrome; however, it is extremely rare in the global population.

Use your knowledge of population genetics to explain why this genetic syndrome is more common in the Amish community. (4 marks)

____

____

____

____

____

____

____

9780170449625

Question 30 (6 marks)

The diagram below shows how one type of white blood cell can protect the body from infection.

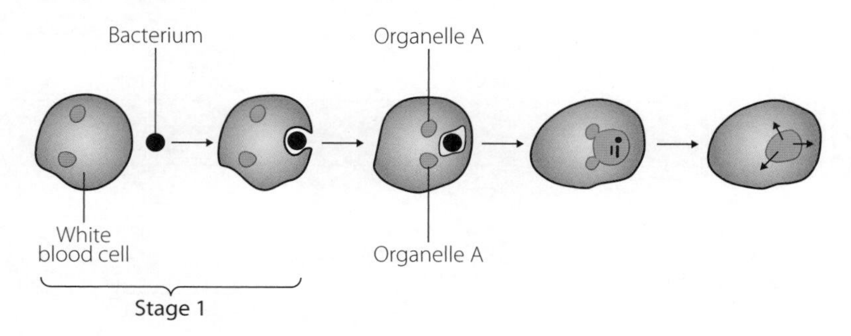

a Describe the process occurring in the diagram. (2 marks)

b Explain how the process above is part of both the innate and the adaptive immune systems. (4 marks)

Question 31 (4 marks)

Below is a diagram outlining a type of point mutation.

Original sequence

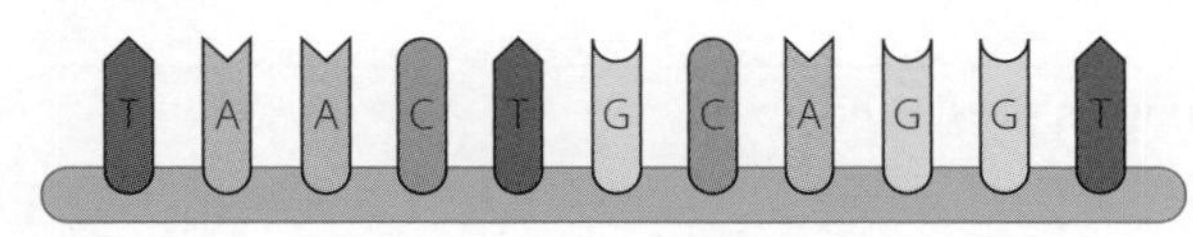

Point mutation

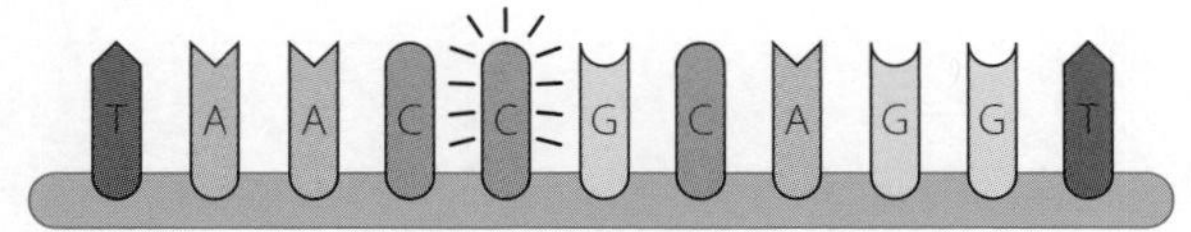

Evaluate the validity the following statement: 'This type of mutation may have no effect on the phenotype of the individual.' (4 marks)

Question 32 (8 marks)

The following information is from the Australian Government Department of Health website.

Influenza vaccines in pregnant women

In pregnant women, standard trivalent inactivated vaccine (TIV) is:

- approximately 50% effective in reducing laboratory-confirmed influenza
- 65% effective against inpatient hospital admissions for acute respiratory illness.

Vaccinating pregnant women against influenza also protects their infants against laboratory-confirmed influenza. This is due to transplacental transfer of high-titre influenza-specific antibodies. A recent systematic review concluded that maternal influenza vaccination reduces laboratory-confirmed influenza in infants <6 months of age by about 48%. The vaccine protects infants for up to 6 months after birth.

Source: Australian Technical Advisory Group on Immunisation (ATAGI). *Australian Immunisation Handbook*, Australian Government Department of Health, Canberra, 2018, immunisationhandbook.health.gov.au

a Assess the validity of the information presented above. (3 marks)

b Explain why giving a pregnant woman a flu vaccine protects the child for only up to 6 months after birth. (5 marks)

Question 33 (9 marks)

The African penguin is currently classified as endangered and is the focus of many research projects to try to protect the species. A recent survey has detected a previously unknown disease in several colonies on the southwest coast of South Africa.

All of the populations highlighted in Figure 3 regularly interbreed but all have their own feeding grounds.

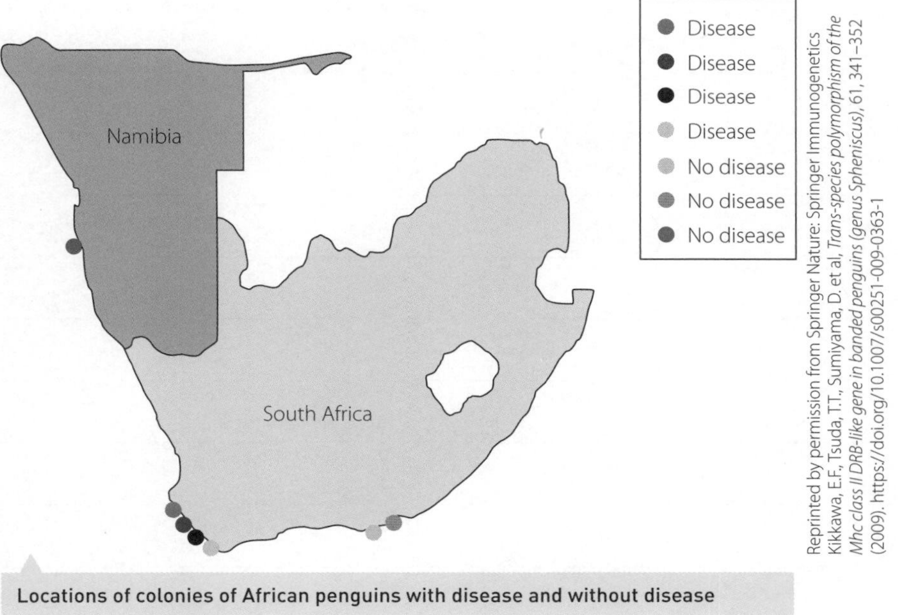

Locations of colonies of African penguins with disease and without disease

Reprinted by permission from Springer Nature: Springer Immunogenetics Kikkawa, E.F., Tsuda, T.T., Sumiyama, D. et al, *Trans-species polymorphism of the Mhc class II DRB-like gene in banded penguins (genus Spheniscus)*, 61, 341–352 (2009). https://doi.org/10.1007/s00251-009-0363-1

Design an appropriate epidemiological study to analyse the circumstances that would have led to the occurrence of this new disease in the bird population, and deduce whether the non-affected populations are at risk. (9 marks)

9780170449625

MODULE FIVE: HEREDITY

REVIEWING PRIOR KNOWLEDGE PAGE 1

1 Prokaryotic cells are more primitive and are simpler than eukaryotic cells. Prokaryotic cells have no membrane-bound organelles and a single chromosome. Bacteria and archaea are prokaryotic cells. Eukaryotic cells have a variety of membrane-bound organelles, such as nuclei, mitochondria and Golgi bodies. Animal and plant cells are examples of eukaryotic cells.

2 a i DNA
 ii Golgi body
 iii Endoplasmic reticulum
 iv Ribosomes

~continue in right column

 b i Found in prokaryotic cells; found in eukaryotic cells
 ii Found in eukaryotic cells
 iii Found in eukaryotic cells
 iv Found in prokaryotic cells; found in eukaryotic cells

3 Proteins are important structural components in cells, cell membranes and tissues. Some proteins function as enzymes. Proteins are made from chains of amino acids and they enter the body via the diet. Nucleic acids include two types: DNA and RNA. DNA stores the genetic information and is found mostly in the nucleus with small amounts in the mitochondria and chloroplasts. RNA is found in small amounts in the nucleus and larger amounts in the cytoplasm, and assists in the production of proteins.

4

Sexual reproduction	Similarities	Asexual reproduction
Offspring genetically unique from parent	Produces offspring	Offspring identical to parent (clone)
Requires two parents		Requires one organism

5 Sexual reproduction randomly combines the genetic makeup of two parents to produce a genetically unique offspring. This increases the genetic diversity of a population with each new offspring that is produced.

6 Adenine (A), thymine (T), guanine (G) and cytosine (C)

7 In coding DNA, a sequence of three nucleotides is called a codon. These codons instruct ribosomes to add a specific amino acid onto a polypeptide chain. The specific order of the amino acids results in a vast array of proteins being produced. These proteins ultimately control the phenotypic expression of the genome and so determine the features of an organism.

8 A population with a wide range of genetic variation will display many traits that could be selected for if there is a change to the environment or a specific selection pressure. This will aid in the long-term survival of the species because a favourable trait will lead to an adaptive advantage and therefore an adaptation to the new environment.

~continue in right column

Chapter 1: Reproduction

WS 1.1 PAGE 3

1 a Gamete
 b Haploid number
 c Meiosis
 d Chromosomes/DNA
 e Fertilisation
 f Zygote
 g Diploid number
 h Somatic cells
 i Mitosis
 j Asexual

2

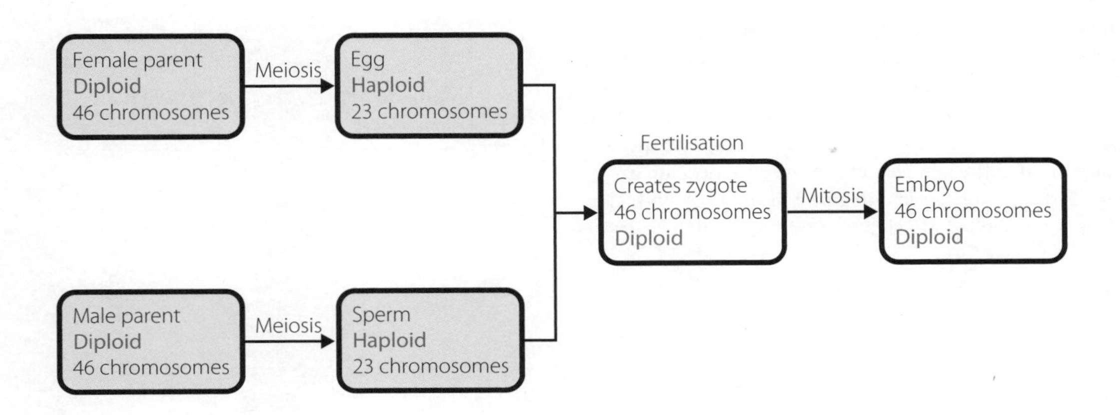

3 Organisms that reproduce asexually can produce large numbers of identical offspring because only one parent is involved. This means that the numbers of individuals can increase quickly and help ensure the continuity of the species if most survive and go on to reproduce for themselves. Organisms that reproduce sexually need two parents, which involves more risk and effort than using only one parent. The offspring produced are not identical to each other or to the parents. This helps ensure continuity of the species by introducing variety into the species. This results in a population more likely to survive changes in the environment or specific selection pressures.

4

	External fertilisation	Internal fertilisation
Differences in environment	• Always occurs in water • May occur in a large body of water (e.g. ocean or river) or a puddle • May involve copulation but never intercourse	• Only occurs inside the body of the female • Always involves copulation or intercourse
Similarities in environment	• The place where fertilisation occurs is wet/moist, so gametes do not desiccate • Sperm must meet egg and the nuclei fuse	

5 Broadcast spawning, if timed to environmental events such as lunar cycles, results in the sperm and egg being released at the same time, therefore increasing the chances of fertilisation. Courtship and copulation bring the male and female close to each other, so when gametes are released there is a high chance of fertilisation.

6 Positives:

- Water movements will carry the gametes away from the individuals that released them and after fertilisation there should be a large variety of gene combinations, leading to variety in the species.
- After fertilisation, the zygotes and embryos will be carried by water movement away from the parents and this will help colonisation of new substrates and will introduce genetic diversity to an area.

Negatives:

- Survival of gametes, zygotes and embryos is low because of predation, destruction by wave action or weather, or embryos not landing on a suitable substrate for survival. Gametes must therefore be produced in large quantities to ensure survival of the species. This will cost the parents a lot of energy.

7 Reptiles are more recently evolved than frogs. Frogs can survive in the terrestrial environment, but they rely on water for reproduction because they use external fertilisation and the young develop in water. Frogs are not adapted to living exclusively on land. In contrast, reptiles reproduce entirely outside of the water environment; they use internal fertilisation and have a waterproof shelled egg. The sperm are deposited in the female and the embryos develop in the shelled egg, which is laid in a nest or underground. The evolution of a shelled egg allowed reptiles to colonise drier areas of Earth because their reproduction does not rely on water.

WS 1.2 PAGE 7

1 *A table should have a border all the way around and include rows and columns.*

Risks	Strategies to minimise risks
Thorns or biting insects on the flowers may scratch or bite.	Closely check the plant before starting the dissection. Thorns can be cut off. Safely remove spiders or other insects.
Blades may cut or stab skin.	Use a safety blade with one non-sharp side and cover the blade when not in use. Cut away from the fingers. Wear covered shoes.
Flower secretions may irritate the skin or eyes.	Wear safety glasses, wash hands after dissection and wear gloves if hands are sensitive.
The microscope may be damaged/cause damage if dropped.	Carry the microscope with two hands. Wear covered shoes.

2

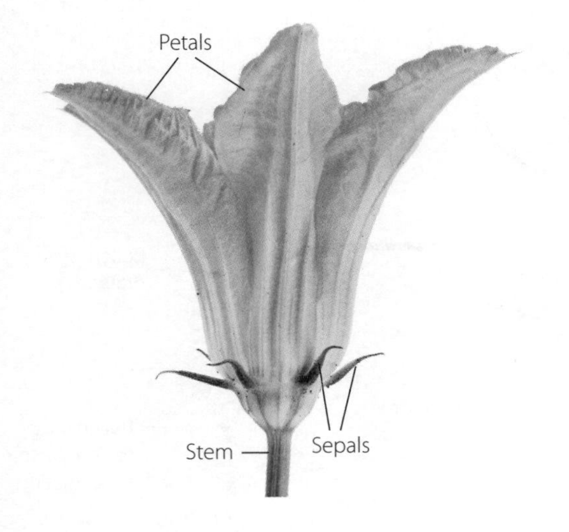

Shutterstock.com/ppl

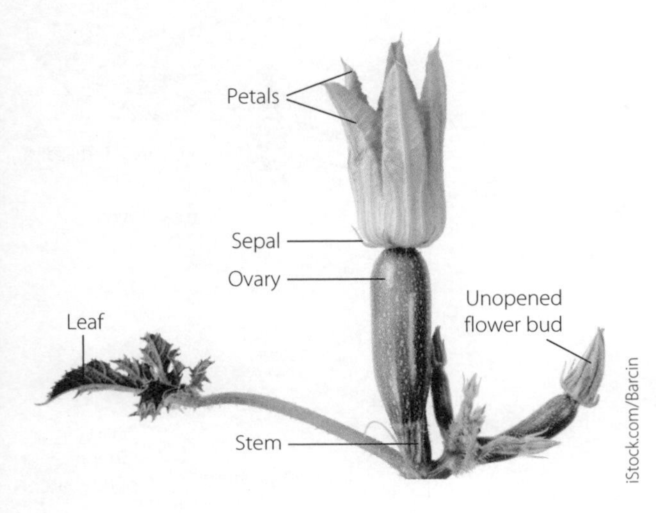

iStock.com/Barcin

3

Alamy Stock Photo/Custom Life Science Images

4

Shutterstock.com/Tish1

Shutterstock.com/Huy Thoai

5

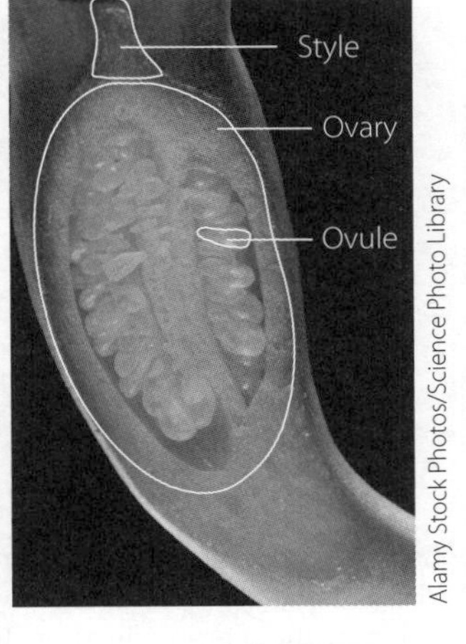

Alamy Stock Photos/Science Photo Library

6

Pollination	Both	Fertilisation
• Only involves the external parts of the flower • May involve a pollinator • Only involves the male gamete • Pollen moves from male to female parts of a flower	• Both are processes that are involved in sexual reproduction • Both processes must happen in order to create offspring	• Only occurs inside the female part of the flower • Nuclei of female and male gametes join • Results in an embryo developing

7

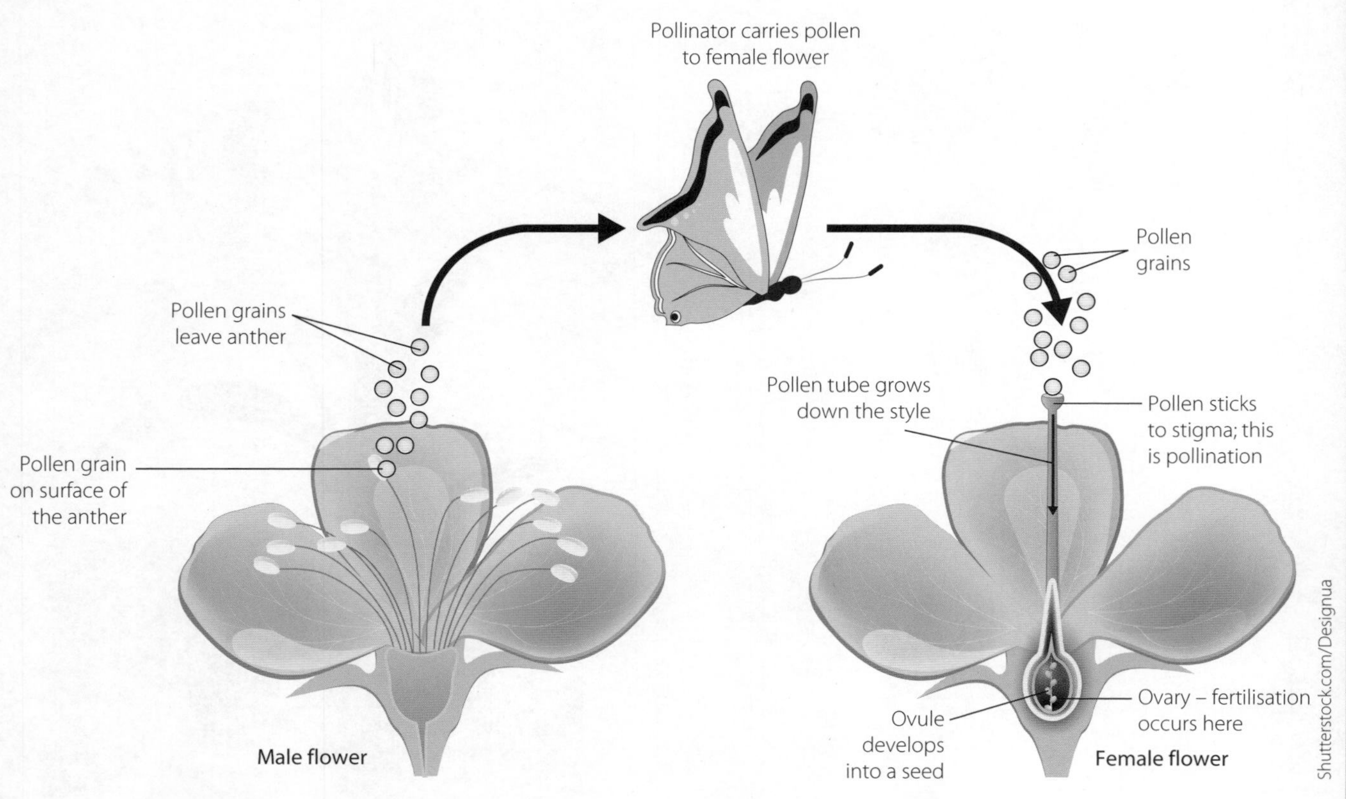

8 a

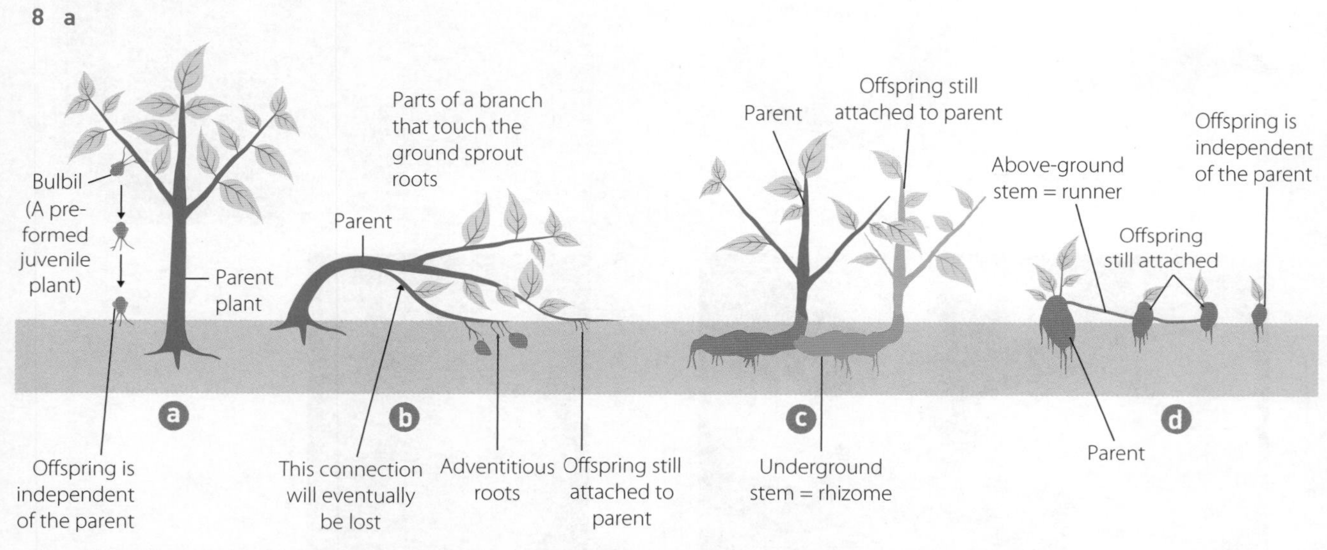

Adapted from Springer: *Evolutionary Ecology: The evolutionary ecology (evo-eco) of plant asexual reproduction*, Niklas, K.J., Cobb, E.D., 2017

b In asexual reproduction, offspring are produced from only one parent. In all the diagrams, a new plant has grown from an adult plant without any pollination. For example, in some pictures juvenile plants grow on the parent before they fall to the ground or they grow from runners or rhizomes. Another example shows a branch that touches the soil forming roots, and eventually the offspring separates from the parent. All the offspring have arisen from the single parent without any sexual reproduction.

WS 1.3 PAGE 12

1

Term	Definition
Ovulation	The release of ovum/egg from the ovary
Fertilisation	The fusion of sperm and egg nuclei
Oestrus	An increase in sexual behaviour at the time of ovulation
Menstrual cycle	The cycle of ovulation and shedding of the lining of the uterus
Endometrium	The lining of the uterus

9780170449625

2

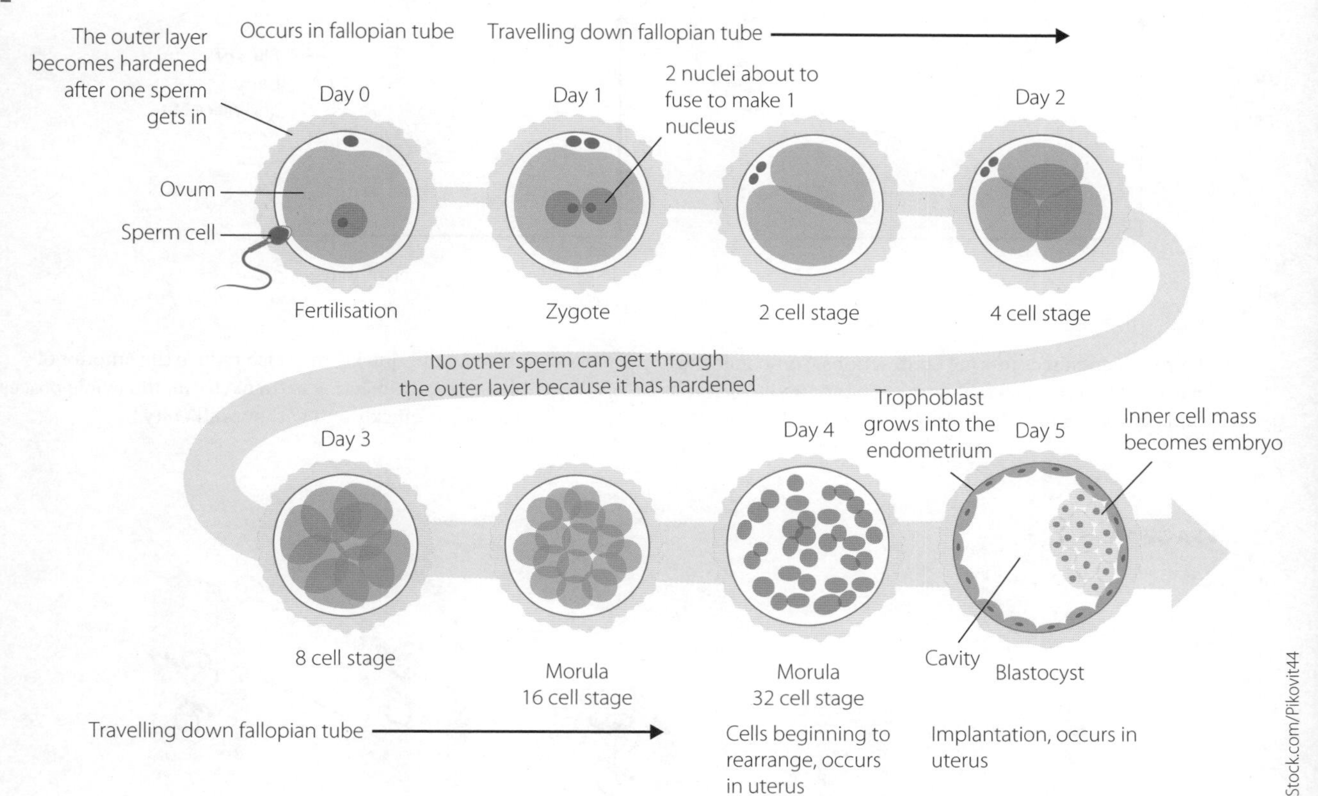

3 a

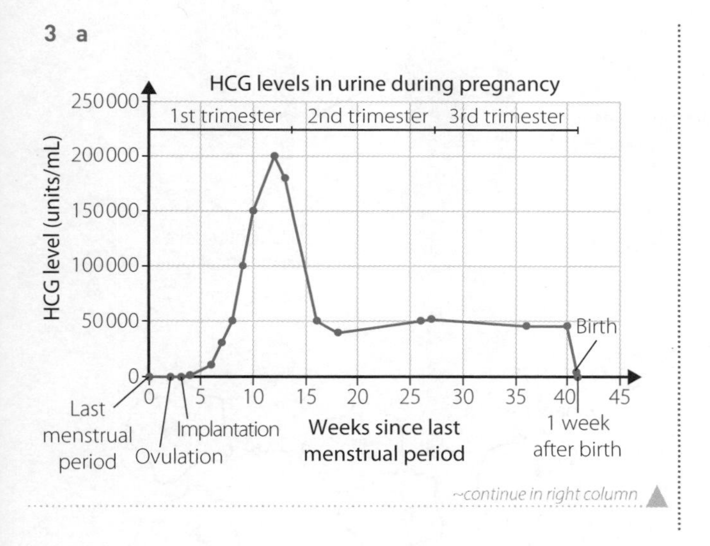

~continue in right column

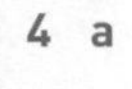

b HCG level is close to zero before implantation. It rises sharply, reaching a peak at 12 weeks' gestation, dropping until 16 weeks. After this time, levels remain approximately at the same level as at week 8. Once the baby is born, HCG drops to pre-pregnancy levels.

c From the table, at implantation the HCG levels are at most 50 units/mL. One week later the level has jumped to 500 units/mL and it is even higher a week later again. Implantation occurs about three weeks after the last menstruation. The table says these numbers are maximum levels of HCG (not average), so some people may have lower numbers. If the test is used too early, HCG levels may be too low for detection and may give a negative result even when the person is pregnant. By waiting until two weeks after ovulation, or four weeks since last menstruation, the result is likely to be more accurate. Waiting one further week would mean HCG levels would be even more detectable.

4 a

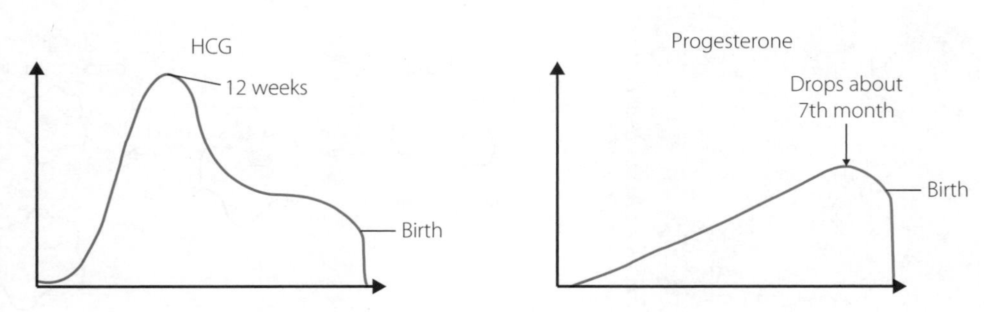

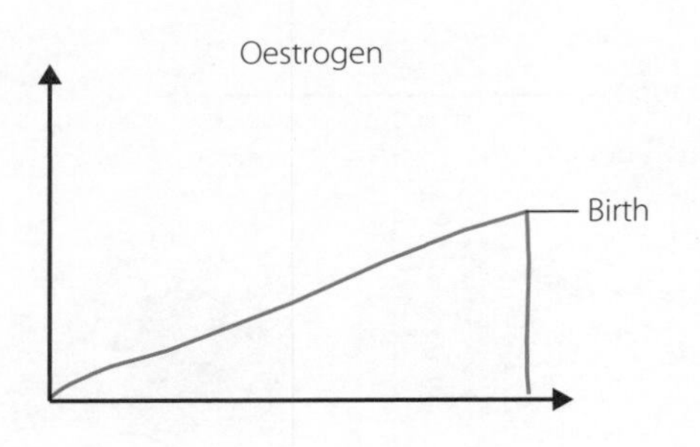

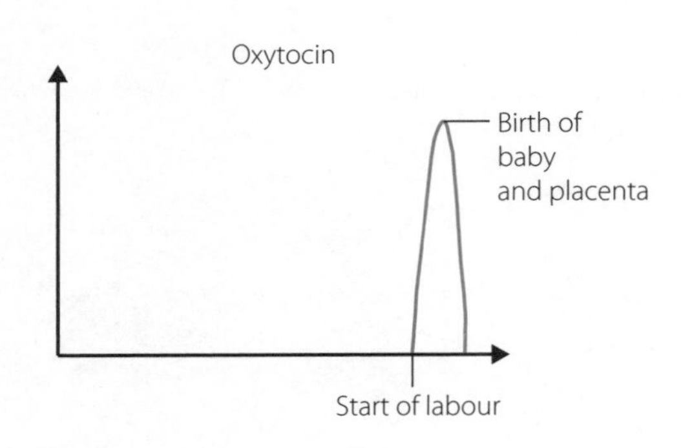

b Oxytocin enhances uterine contractions. Someone who has just been through labour may be too tired to go on and deliver the placenta over a long period. Getting the placenta out quickly may also reduce the amount of blood lost. By injecting extra oxytocin, the whole process may occur quickly and without difficulty.

~continue in right column

WS 1.4 PAGE 16

1

Parent cell

Bacterial chromosome

Extra copy of bacterial chromosome has been made

2 chromosomes move to opposite ends of cell

Cells start to split apart

Cytokinesis has happened

2 daughter cells with identical DNA

2 a 2 μm

b

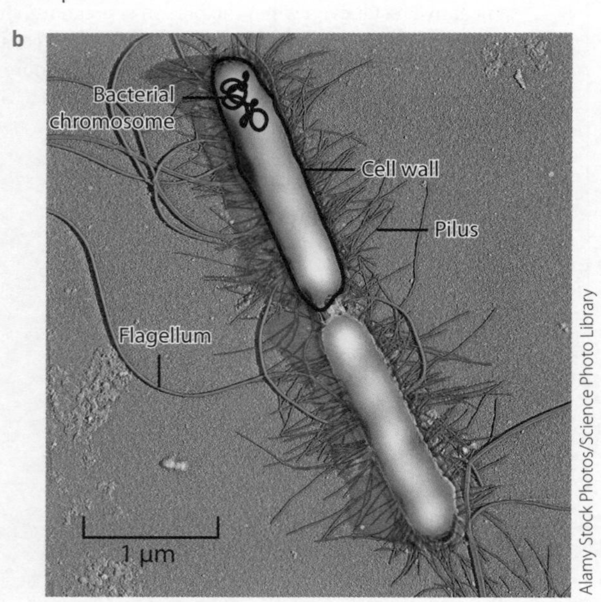

Alamy Stock Photos/Science Photo Library

3

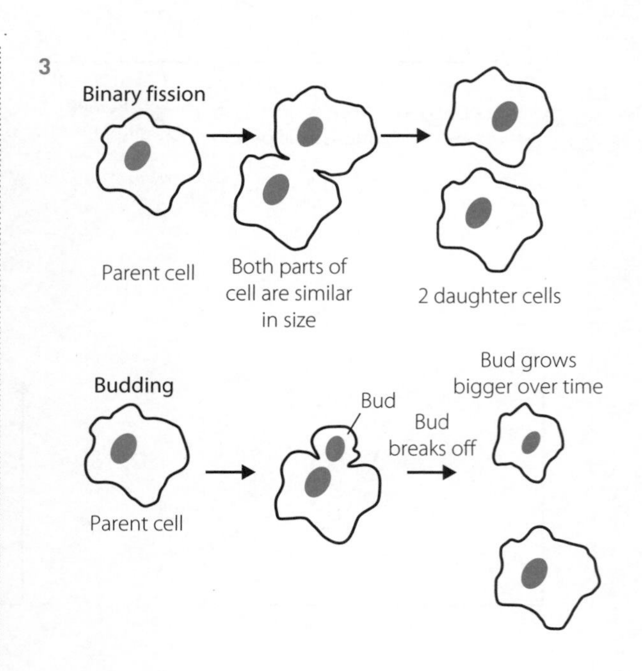

9780170449625

4 a&b

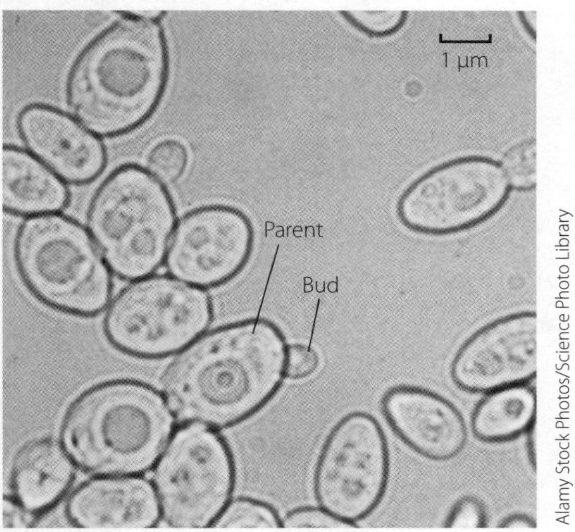

~continue in right column

5 Fungal spores are a unit for asexual reproduction and result in offspring, whereas bacterial spores do not make offspring. A single bacterium shuts down functions and is called a spore while it is surrounded by a protective layer. It resumes functions when conditions are more favourable.

6

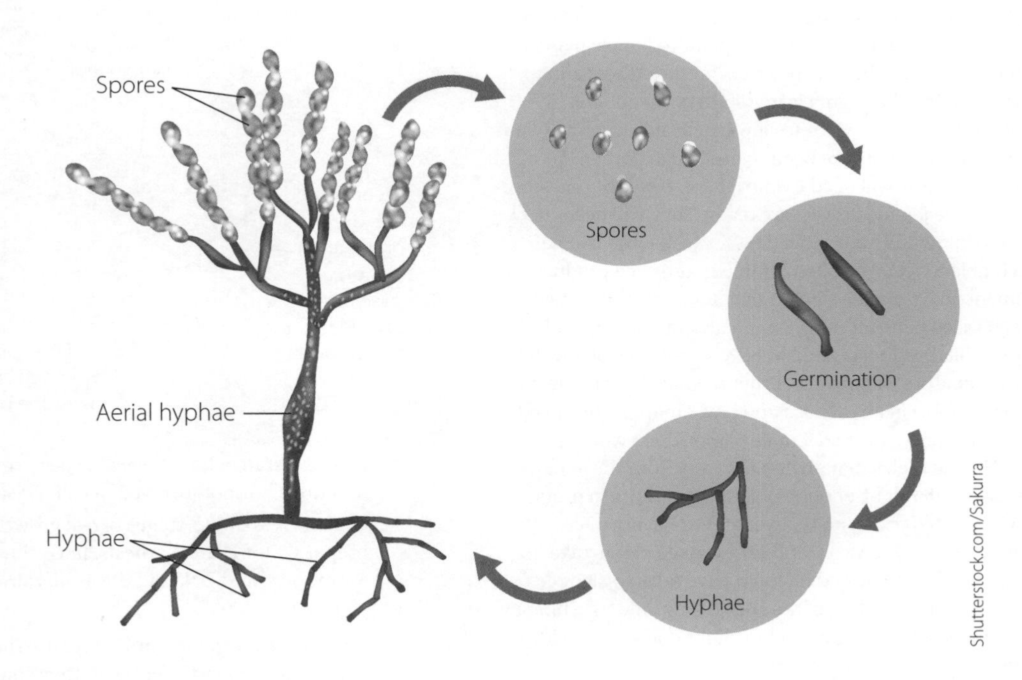

7 Some fungi reproduce by budding and others use spores. The yeast *Saccharomyces cerevisiae* uses budding. This involves a small part of the parent cell growing out of one side of the cell. The bud develops into an identical cell to the parent and eventually breaks off and grows into an adult cell. Spores are used by multicellular fungi such as *Penicillium*. The spores contain the same genetic material as the parent and are easily dispersed by wind. Each spore can give rise to a new *Penicillium* organism. Both examples allow the number of individuals to increase quickly when conditions are favourable. In addition, spores can be dispersed vast distances and allow the organism to spread in the environment. Asexual reproduction by budding or production of spores helps these fungi species continue to exist because numbers can build up quickly even if only one individual survives a change in the environment.

WS 1.5 PAGE 20

1 a Pollen is transferred from the male anther to the female stigma by bees or the wind. Pollen grains that land on the stigma grow a pollen tube down the style to the ovary. The sperm fertilises an ovum, resulting in a seed. A suitable sized fruit has more than 1000 seeds; therefore, at least 1000 pollen grains must land on the stigma and grow down into the ovary. The greater the pollen transfer, the bigger the fruit is likely to be.

b *Note: This question is not about general plant biology – competition for light and nutrients, disease etc. are not relevant.*

Kiwifruit producers would need to understand:

- the need to plant both male and female vines
- how pollination occurs in kiwifruit flowers
- how close/far apart male and female plants should be

- the ratio of male to female plants (is it 1:1, or can one male plant supply enough pollen for many females?)
- how to encourage pollinators such as bees and the impact of wind
- that fertilisation follows pollination to make seeds inside the fruit
- how the number of seeds impacts the shape and size of the fruit.

2 The aim of the producer is to produce quality fruit in high yields and at low production costs. Knowledge about reproduction is crucial to the planning of the orchard and ensuring enough quality fruit is produced. Kiwifruit production begins with choosing a variety to grow, or breeding a new variety using artificial pollination and crossing between varieties. To produce large, well-shaped fruit the producer needs to know about pollination and how the sex of the plants affects the final product. Trial experiments to find out new information have a downside –
financial costs, having to dedicate land to experiments and receive potentially low income from experimental plots. To create a new variety, the producer needs to know about the flower anatomy of the kiwifruit plant and which parents will give the desired offspring. They need to understand that male flowers are found on a different plant to female plants, to understand when the pollen is ripe on the anthers and to understand when the stigma is in a state to accept pollen transfer. Hopefully the two times are synchronised, and pollen production and acceptance occur close to each other. If not, then the producers will need to know how to collect and store pollen until the female flowers are ready. They will need to do trials to know the number of days for which the female stigma is able to receive pollen for maximum seed production. Experiments have already shown that, for current varieties, 1000–4000 seed per fruit produces an acceptable size and quality of fruit. Producers need to have knowledge of how far natural pollinators will travel in order to know how many male plants are needed in an orchard and how far apart they need to be. If a producer only had female plants, there would be no fruit. Having males that are too far away from the females means there will not be enough pollination events to produce large, well-shaped fruit. Too many males or having males too close together is a waste, and excess male plants take up space and other resources that could have taken by females. Without an understanding of the species' reproductive biology there would be less choice of fruit in shops and lower yield, and hence fruit would be more expensive.

Chapter 2: Cell replication

WS 2.1 PAGE 23

1 a Meiosis

b Mitosis occurs anywhere in an organism where more somatic cells are needed for growth or for replacement of cells. Mitosis starts when a zygote divides into two cells and continues throughout the life of the organism.

c Flowering plants: anther and ovary; Animals: testis and ovary

d Haploid cells have half the number of chromosomes compared to diploid cells. If $n = 8$, then the gametes of this organism will have 8 chromosomes. This is the haploid number for this organism. The somatic cells of this organism will have $2n$ chromosomes; that is, 16. This is the diploid number for this organism.

e The amount of DNA in the nucleus must double before the cell divides. If it did not, then the two resulting cells would have half the correct amount of DNA and the next generation of cells would have a quarter of the correct amount. Doubling DNA before mitosis ensures the resulting cells have the correct amount of DNA.

2

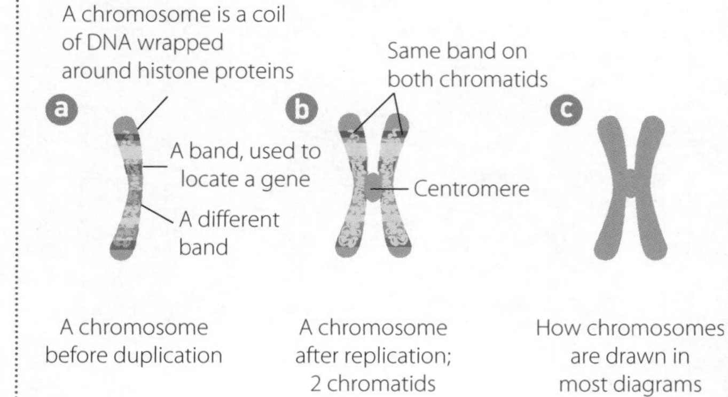

d

Nucleus (usually cannot see the chromsomes)
Cell
Tightly coiled DNA/histone combination
Histone protein (grey)
DNA
Very condensed DNA/histone combination becomes visible in the nucleus = a chromosome
DNA (pink) wrapped around a histone protein
University of Waikato

3 a G_0 is the state where the cell is performing its normal cell functions but not involved in cell division.

G_1 and G_2 are both stages of cell growth. In G_1 the cell is preparing for DNA synthesis. In G_2 the cell is preparing for mitosis, and DNA synthesis has already occurred.

b S phase

c Nerve cells are always in G_0 because they do not divide. Skin cells divide frequently, so they spend much less time in G_0 than do nerve cells.

d The cell mass would increase during S phase because more DNA is being made. At the start of M phase, the cell has double the amount of DNA but after mitosis the DNA has been separated into two cells, and therefore the cell will have less mass.

4 D

5 Cancer cells are the result of faulty cell division. If cell division happens too frequently and cells do not perform their normal function, masses of unspecialised cells can build up, forming a tumour. There are genes that control when a cell should start dividing or stop dividing. Cells with faulty genes are usually detected during the cell cycle and destroyed by a process called apoptosis, in which the cell destroys itself. If a faulty cell is not detected it can go on to produce many more faulty cells by mitosis.

9780170449625

6

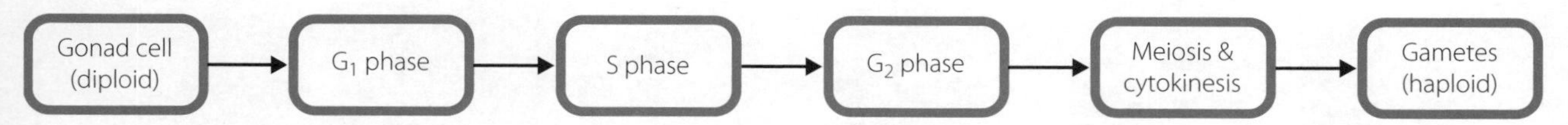

7 Meiosis results in gametes with fewer chromosomes than somatic cells, therefore it reduces the number of chromosomes. It is a process of *division* whereby there is a *reduction* in chromosomes.

8 The purpose of gametes is to join at fertilisation. The male gamete and the female gamete should each have half the normal amount of DNA for the species. When the nuclei of the two gametes join, the amount of DNA in the resulting zygote is doubled, bringing the amount of DNA back to the normal amount for that species.

WS 2.2 PAGE 28

1 a & b

Cytoplasm and organelles duplicate and new cell membranes form around daughter cells

Cytokinesis

Interphase → Prophase → Metaphase → Anaphase → Telophase → Interphase → G_0 phase

Interphase: Growth, DNA synthesis and preparation for mitosis

Prophase: Chromosomes become visible; nuclear membrane breaks down; spindle fibres attach to chromosomes

Metaphase: Duplicated chromosomes line up across the middle of the cell

Anaphase: Chromatids split at the centromere and start to move toward opposite ends of the cell

Telophase: Single chromosomes are now in two bunches; two nuclear membranes form around them

G_0 phase: Some cells enter G_0 phase and don't divide for some time

2 a a: metaphase; b: anaphase; c: late telophase or cytokinesis; d: prophase; e: late anaphase or early telophase; f: interphase

b Photo e shows the chromosomes in two bunches at the far end of the cell, but the nuclear membranes are not visible. Therefore, it is difficult to say whether this is late anaphase or early telophase.

3 a 4

b $n = 2$

c

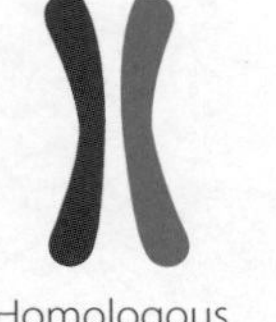

Homologous pair 1 Homologous pair 2

4

Sister chromatids

Centromere

Homologous pair 1 after duplication

Homologous pair 2 after duplication

5 a four; two; low; simplify; easier; nuclear; chromosomes; arranged

b

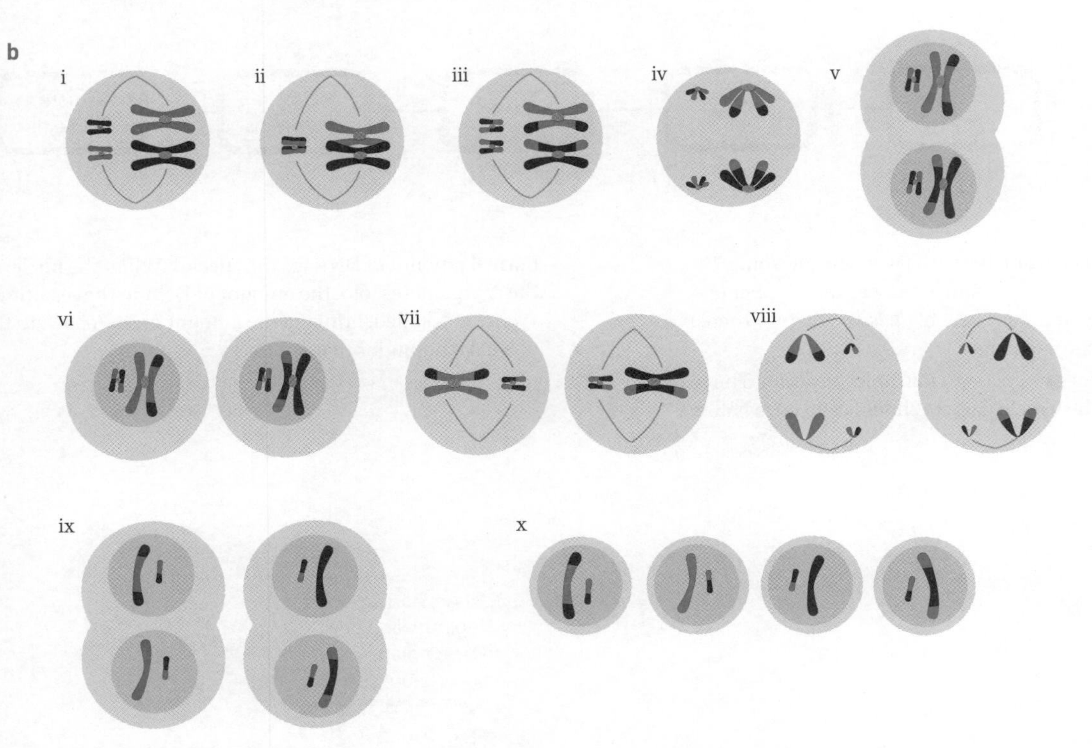

c Answers should refer specifically to the drawings created. The chromosomes in the gametes are not identical. At point ii in the model the homologous chromosomes are on top of each other. There is a chance for parts of one chromatid to swap places with parts of another homologous chromatid. Once the chromosomes move apart at point iii in the model, these pieces are bonded and there is no more opportunity for pieces to swap. The final gametes all have the same number of chromosomes but the combination of alleles on the chromosomes are different.

d

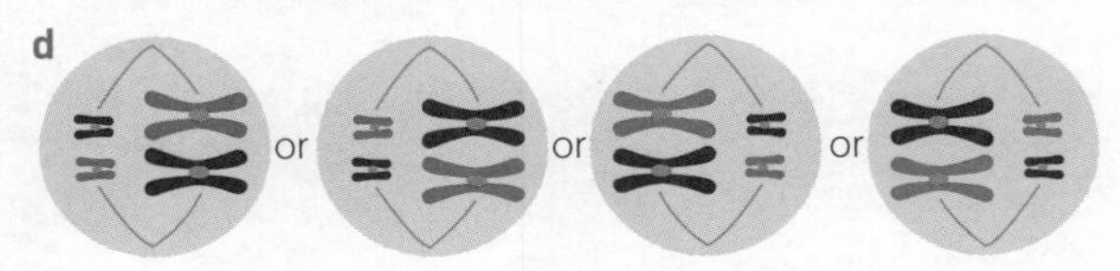

~continue in right column

e *Example answer:* At point iii, the places that I swapped between chromatids could have been different and the gametes would have been different as a result. At point vii, the chromosomes could have been flipped in a different way when I drew them, and the combination of chromosomes in the gametes would have ended up different.

f *Example answer:* The model helped me understand that the decisions made at points iii and vii affect which chromosomes end up in the final gametes. At these two points there are multiple combinations that could have been chosen when I drew the chromosomes on the left or the right or the top and the bottom. This helped me to see the randomness of this part of the process.

WS 2.3 PAGE 32

1 a

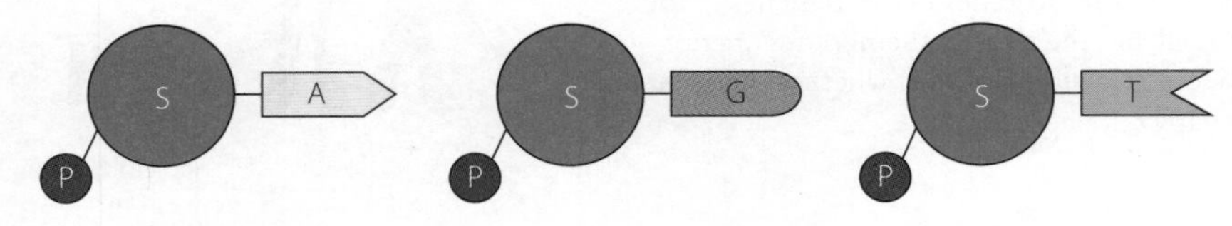

b *Example answer:*

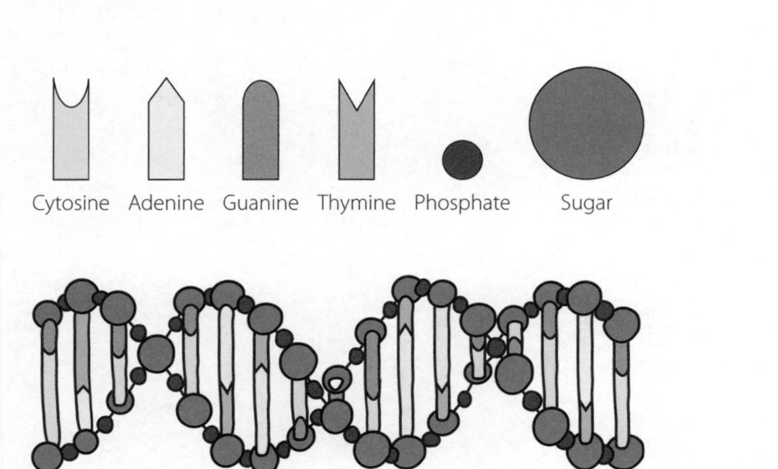

c & d *Students should draw the left part of the diagram below for part c. They should then draw in the remainder of the diagram for part d.*

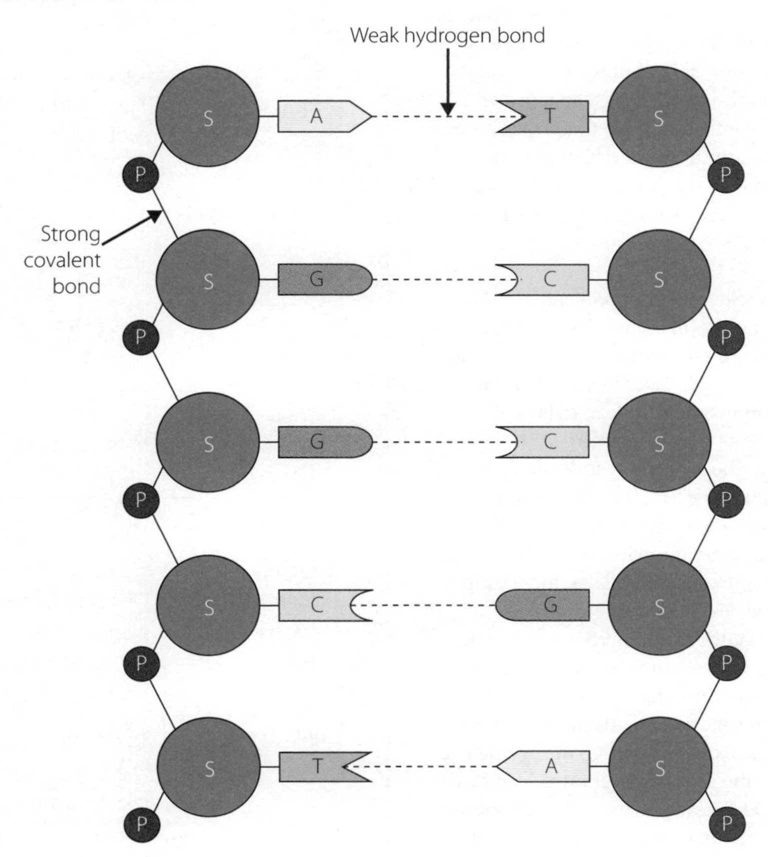

e Drawings like this are models. The intention is to make a difficult concept more easily understood. Most people understand the concept of shapes that match each other, therefore making the idea of complementary bases more easily consolidated. The concept of base pairing is more important to understand than the reasons behind it (i.e. the number of bonds and realistic shape of the molecules).

2 The DNA is arranged so the phosphate and the sugar backbone are on the outside and in contact with fluid, while the nitrogenous bases are in the inner portion of the molecule. The sugar–phosphate backbone of DNA has the most contact with the liquid in the nucleus. It must be hydrophilic. The nitrogenous bases are hydrophobic and have the least contact with water. The helical shape further prevents the nitrogenous bases from coming into contact with the water in the nucleus.

3 a All pieces of DNA in the cell are in the G_1 phase of the cell cycle and the cell is being prepared for DNA replication.

b S phase begins and the enzyme helicase unwinds the DNA helix.

c Helicase also unzips the DNA, separating the two strands. Other enzymes bind to the newly opened DNA and prevent it from zipping back together again.

d There are many nucleotides floating free in the nucleus.

e The enzyme DNA polymerase helps nucleotides join on to the template strands of DNA according to the base pairing rules. The enzyme DNA ligase helps glue the phosphates and sugars together.

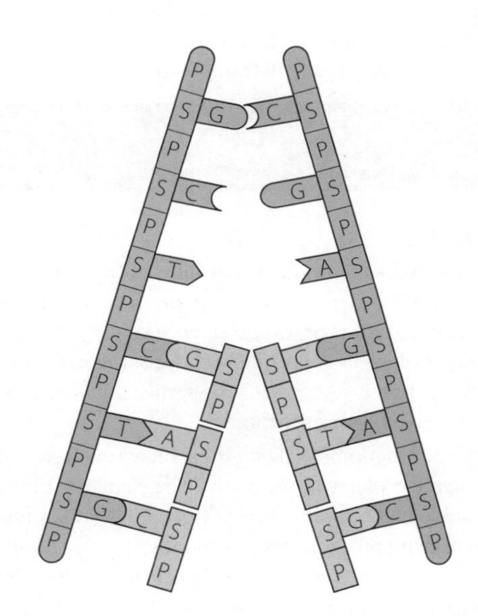

f Two double-stranded DNA lengths are produced, each with one strand from the parent molecule and one new strand. DNA polymerase backtracks to 'proofread' and 'edit' the strand, correcting any base pair errors.

g The two double strands are wound back into helices. The DNA is checked again for errors before the cell enters mitosis.

4 Helicase, DNA polymerase, DNA ligase

5 DNA replication is essential for mitosis, for meiosis and to ensure as few mutations as possible. Early in cell division, the chromosomes replicate and form two chromatids. The genes along the length of the chromosome must also be replicated. This can only occur when DNA is replicated in phase S of the cell cycle. DNA replication contributes to identical, healthy offspring cells in mitosis. Meiosis also relies on DNA replication so that the chromosomes can undertake crossing over and random segregation to create variation in the gametes. During DNA replication the strands of DNA are also checked for errors and repaired, leading to healthier cells.

WS 2.4 PAGE 37

1 The continuity of species relies on exact replication of DNA and cells for normal functioning of organisms. During mitosis, identical cells are necessary for growth. If mitosis is out of control, tumours can develop. Mistakes in DNA replication and reassortment of chromosomes can lead to lack of function in cells, affecting the whole organism. Although these processes operate at the level of the individual, if most individuals had faults like these, the species would be negatively affected. Conversely, there can be benefits to species when mistakes occur in replication of DNA because mutations are a source of new alleles, increasing genetic diversity in the species. Some mutations will be beneficial, and some are neutral, until there is a change in the environment. Meiosis is an example of a cell process where non-exact replication is crucial. The intention is to make cells with half the normal number of chromosomes; by definition, there is never exact replication during meiosis. Independent assortment and crossing over introduce variety into the gametes produced. These processes in meiosis, when coupled with fertilisation in sexual reproduction, ensure there is variety in populations. A species is more likely to survive changes in the environment if populations have genetic diversity. Exact replication and non-exact replication are both important to continuity of the species.

Chapter 3: DNA and polypeptide synthesis

WS 3.1 PAGE 39

1 a Prokaryotes reproduce asexually, so there is no fertilisation. They only have genes from one parent. The eukaryotes that reproduce sexually get two copies of every gene; one each from the male and the female parent. Alleles for the same trait are found on homologous chromosomes.

b All the kingdoms of living things have evolved from an original prokaryotic ancestor. The structure of the DNA double helix, and the role of ribosomes and amino acids in making proteins have been inherited by the eukaryotes. The similarity in biochemistry is evidence for evolution.

2 nucleus; linear; proteins; base; cytoplasm; single; supercoiled; non-chromosomal; plasmids; advantage; conjugation; mitochondria

3

Features	Prokaryotic cells	Eukaryotic cells
DNA structure	Double helix – sugar – phosphate backbone and four nitrogenous bases	Double helix – sugar–phosphate backbone and four nitrogenous bases
Shape of DNA pieces	Circular	Linear
Supercoiling	Yes, around dense proteins	Yes, around histone proteins
Sketch of chromosome		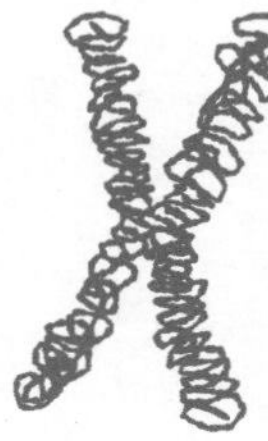
Location of chromosomes	In the cytoplasm	In the nucleus
Number of chromosomes	One	Many
Non-chromosomal DNA	One or more circular plasmids that can be swapped between cells	In mitochondria and chloroplasts; no plasmids
Role of DNA	To provide instructions for protein synthesis	To provide instructions for protein synthesis

4 Nucleotides

WS 3.2 PAGE 41

1 a Carbon, hydrogen, oxygen, nitrogen

b Hydrophilic means the molecule is attracted to water molecules whereas hydrophobic molecules are repelled from water molecules.

c Peptide bonds are what hold amino acids to each other. 'Poly' means many. Therefore, 'polypeptide' refers to many amino acids joined by peptide bonds.

2

Type of protein	Example	What this protein does
Antibody	Immunoglobin G	Protects against infections
Enzyme	Amylase	Digests carbohydrates in the mouth
Structural	Keratin	Provides strength to hair, skin and nails
Functional	Haemoglobin	Helps transport oxygen in the blood
Storage	Ferritin	Stores and controls the release of iron in the body
Contractile	Myosin	Causes muscle cells to contract
Hormonal	Insulin	Regulates blood glucose

9780170449625

3

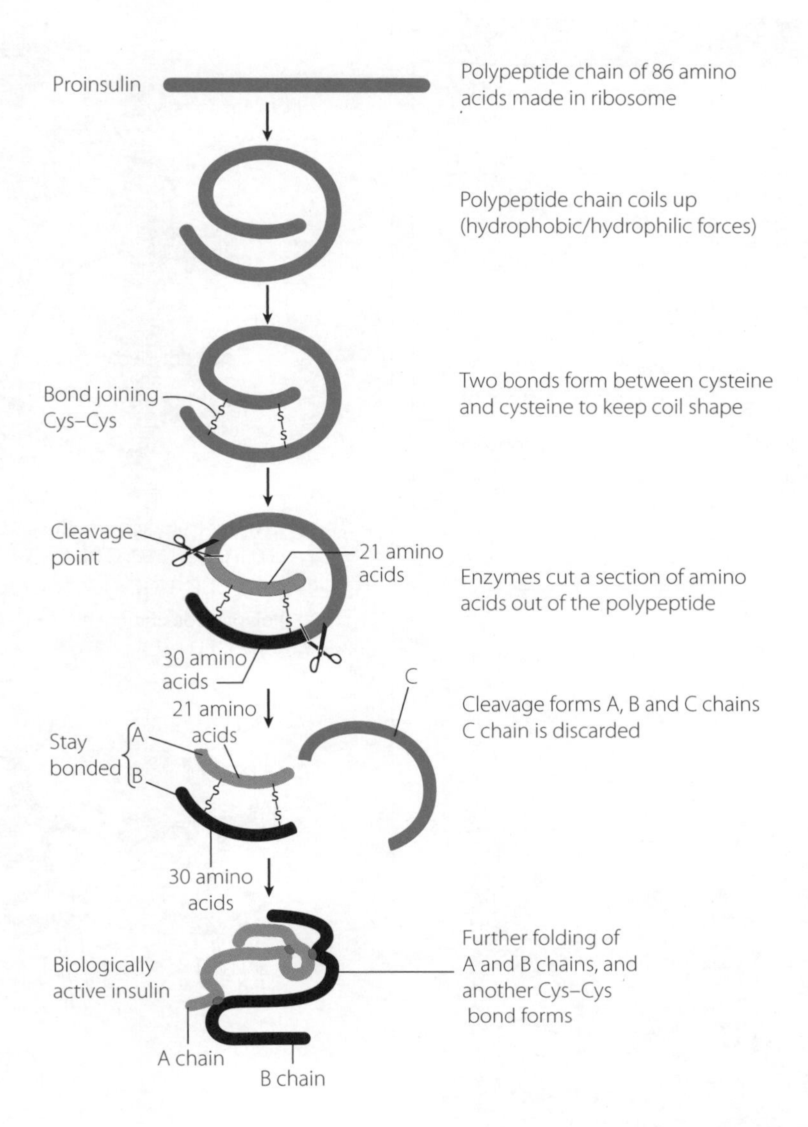

4 Enzymes are proteins that help with chemical reactions. They are chains of amino acids folded into a three-dimensional shape. Each enzyme has an active site where it joins with a substrate like a lock and key:

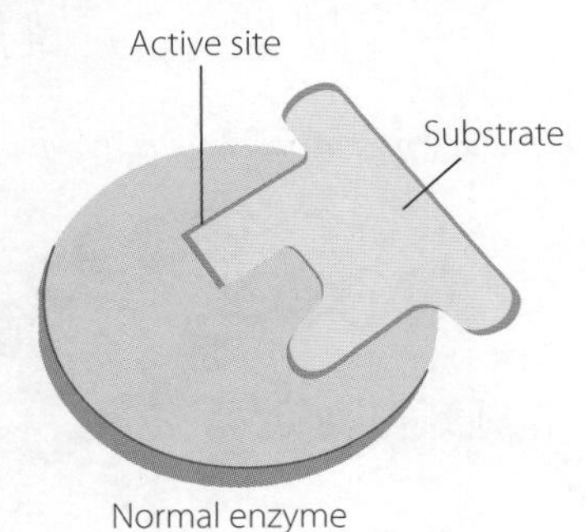

The shape of both molecules is important. A change in temperature or pH can change the shape of the protein even though these things do not change the order of the amino acids in the protein:

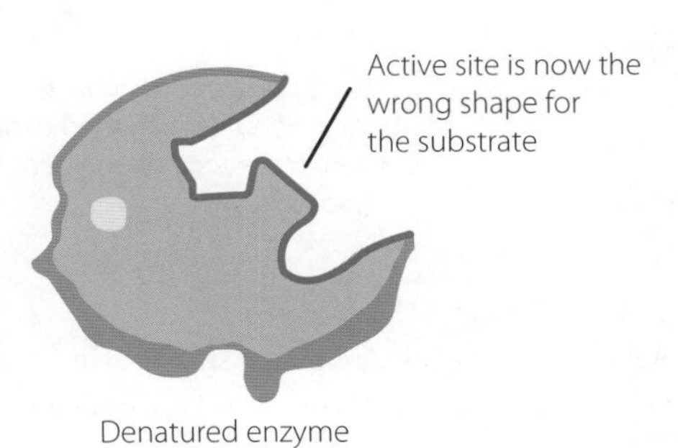

If the active site is the wrong shape, it can no longer fit with the substrate, which means it cannot perform its function.

WS 3.3 PAGE 44

1 a Amino acids
  b Molecules: DNA, RNA and amino acids
    Cellular structures: ribosomes
  c Endoplasmic reticulum and Golgi bodies

2

Features	DNA	mRNA	tRNA
Shape	Double helix	Linear	Twisted into a clover-leaf shape
Strands	Double	Single	Single
Sugar	Deoxyribose	Ribose	Ribose
Bases	A, T, C, G	A, C, G, U (uracil) instead of T	A, C, G, U (uracil) instead of T
Location	Stays in nucleus	Moves between nucleus and cytoplasm	Cytoplasm

3 a ribosomes; big; nuclear; copy; messenger; transcription; cytoplasm

b

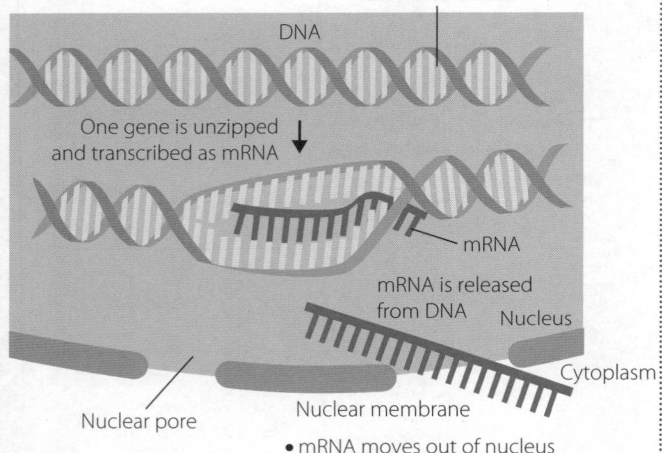

4 a CCU, CCC, CCA, CCG

b Alanine, Asparagine, Arginine, Aspartic acid

c GAG for glutamic acid is only one base different to GUG for valine.

GAA for glutamic acid is only one base different to GUA for valine.

5 Codons are triplets of bases on mRNA, whereas anticodons are triplets of bases complementary to codons, based on base pairing rules. For example, if a codon is UAC, the anticodon would be AUG.

6

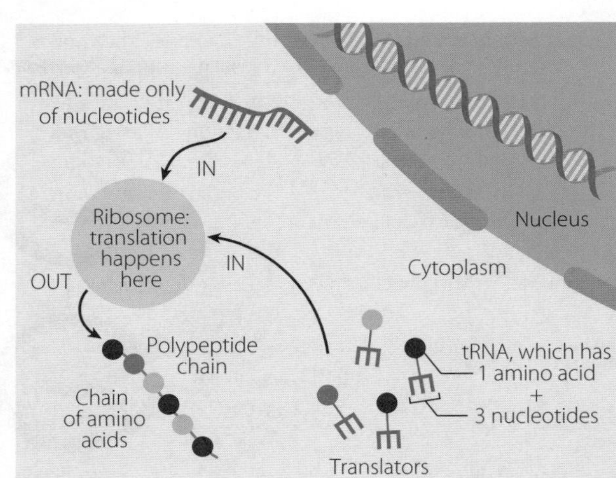

7 a In order from left to right: GTG, GAC, ACG, CCG, AGT, GTG, GAC, CAC, CTT, CGA

b CAC, CUG, UGC, GGC, UCA, CAC, CUG, GUG, GAA, GCU

c Polypeptide chain: HIS- LEU- CYS- GLY- SER- HIS- LEU- VAL- GLU- ALA

d

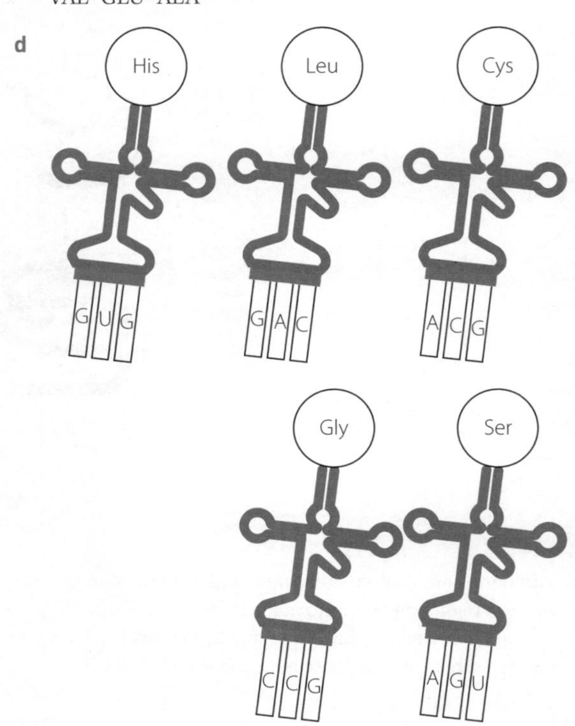

e *The transplanted part of the chain is shown in pink.*

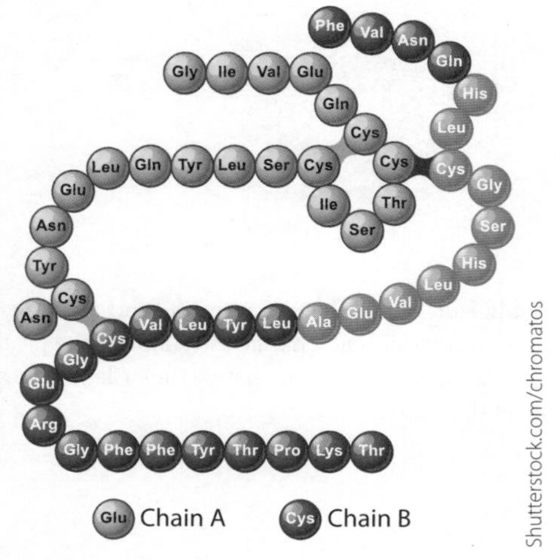

Shutterstock.com/chromatos

f $(10/61) \times 100 = 16\%$

9780170449625

WS 3.4 PAGE 48

1 a Black fur
  b High temperature
  c Black fur only grows on extremities and not on the central part of the body.
  d The *C* gene is turned off when the temperature is above 35°C so the body is white because no black pigment is made. The *C* gene is switched on and the fur contains black pigment if the temperature is 15–25°C. This occurs in the extremities where the rabbit loses the most heat.
  e The *C* gene is active at 15°C; hence the breeder thinks the rabbits will grow black fur all over if they are in a cool room. This experiment is unlikely to be successful. Even if the rabbits are kept in a room at a cold temperature, their body temperature will still be much higher, because they are mammals. Their central body areas will still be too warm for the *C* gene to be turned on. The extremities will grow black fur, but the main torso will remain white.

2 a Male genotype will have male phenotype.
  b High heat of incubation
  c Genotypic males are phenotypic females.
  d High heat turns on two genes that override the male sex chromosomes so eggs hatch as females even though the genotype is male.

3 a Regardless of the temperature at which the embryos are incubated, all the genotypic females are also phenotypic females.
  b At incubation temperature X or below, all the genotypic males are phenotypic males. As the incubation temperature increases, the proportion of males reversing sex and becoming phenotypical females increases. Above temperature Z, all the genotypic males are phenotypically female. There are no males at all.
  c Temperature Y causes half of the embryos that are genotypically male to become phenotypically female.

~continue in right column

  d Temperature Z produces 100% phenotypically female lizards, but equal proportions have the female and male genotype. That is, half of the females are reversed sex genotypic males.

4 PKU results in a child developing a permanent intellectual disability over time because of their inability to stop phenylalanine (phe) building up in the body. The damage is irreversible. If it is known at birth that the baby has the genotype for PKU, and the diet is immediately altered to exclude phe, the typical phenotype does not develop. The special diet needs to continue to allow normal intellectual development. No phe in the diet is a change in the environment that stops the PKU gene being expressed as the PKU phenotype.

Chapter 4: Genetic variation

WS 4.1 PAGE 51

1 Genetic diversity refers to the amount of variation in the genes of a population of a specific species.

2 a

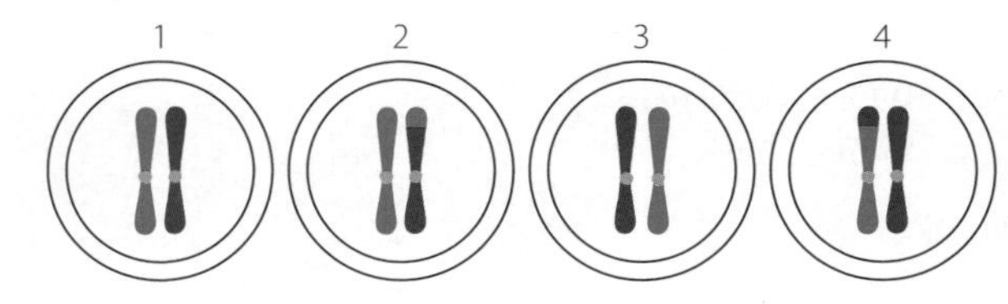

  b 2 or 4
  c Gametes 2 and 4 are likely to be most genetically diverse as crossing over has occurred. This process involves homologous chromosomes exchanging genetic material, and forming a chromatid with a sequence of genes that is different from its parents. This will increase the genetic diversity of the offspring.

WS 4.2 PAGE 52

1

Term	Definition
Gene	A sequence of nucleotides on a specific part of a chromosome that can code for a specific trait
Allele	A variation of a gene on homologous chromosomes
Dominant gene	A gene that will be expressed over a recessive gene
Recessive gene	A gene that will only be expressed in the absence of a dominant gene
Phenotype	The observable expression of a trait resulting from an organism's genotype and its environment
Genotype	The genetic makeup of an organism
Heterozygous	Having two different forms of an allele inherited from an organism's parents
Homozygous	Having two of the same genes inherited from an organism's parents
Codominance	Both phenotypes are expressed individually at the same time; for example, a black and a white fur gene resulting in black and white spots
Incomplete dominance	Both phenotypes are expressed as a mix or a blend; for example, a black and a white fur gene resulting in a grey phenotype

2

Term	Example
Dominant phenotype	Purple flower
Recessive phenotype	White flower
Homozygous dominant genotype	PP
Heterozygous genotype	Pp

3 a

	F	f
f	Ff	ff
f	Ff	ff

With freckles: 50%
Without freckles: 50%

b

	F	F
f	Ff	Ff
f	Ff	Ff

With freckles: 100%
Without freckles: 0%

4

	Y	y
Y	YY	Yy
y	Yy	yy

The allele for short stalks is dominant because, if the allele were recessive and two individuals were crossed, 100% of the offspring would display the trait. The farmer had 25% regular stalks in the first generation; therefore, it is likely that both parents carried the recessive allele and 25% of offspring received the recessive allele from both parents. The Punnett square above with two heterozygous parents resulted in 75% short-stalked broccoli and 25% regular stalked broccoli, adding further credence to the assumption.

5 Sexual reproduction involves the process of meiosis and the combination of haploid gametes from two genetically distinct individuals. The result is a genetically unique offspring. During meiosis, crossing over and independent assortment greatly increase the genetic diversity of the gametes that go on to form the offspring. This could be a disadvantage to a farmer that is trying to replicate a favourable trait as it may not be expressed in all of the offspring.

WS 4.3 PAGE 55

1

Codominance	Similarities	Incomplete dominance
Both phenotypes are expressed individually at the same time; for example, a black and a white fur gene resulting in black and white spots.	Both phenotypes are expressed.	Both the phenotypes are expressed as a mix or a blend; for example, a black and a white fur gene resulting in a grey phenotype.

2 This type of inheritance is codominance. Both of the alleles from the parental generation are expressed in the offspring phenotype equally and distinctly, compared to the blended mix that would be shown in incomplete dominance.

3 a

	T^R	T^R
T^W	T^RT^W	T^RT^W
T^W	T^RT^W	T^RT^W

b

	T^W	T^R
T^R	T^WT^R	T^RT^R
T^R	T^WT^R	T^RT^R

Phenotype: White 50%, Red 50%
Number of offspring: 60, 60

WS 4.4 PAGE 57

1

Symbol	Meaning
	Male – unaffected
	Male – affected
	Male – carrier
	Female – unaffected
	Female – affected
	Female – carrier

2 The statement is true.
If each parent was affected and the allele was recessive, then each parent would be homozygous recessive for the trait. Therefore, all of their offspring would be homozygous recessive.

	t	t
t	tt	tt
t	tt	tt

If the parents were heterozygous for the allele and the allele was dominant, they could each carry the recessive non-affected allele. This would result in a 25% chance that their offspring would not show the trait.

	T	t
T	TT	Tt
t	Tt	tt (not shown)

9780170449625

3 a

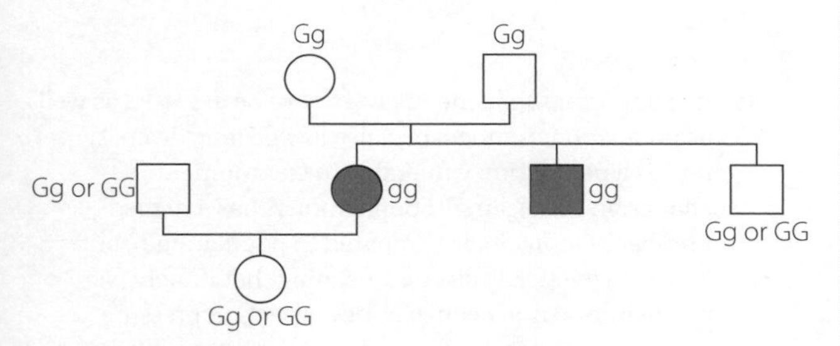

b If two unaffected parents produce offspring that have the trait, the trait is recessive. This is due to both parents being carriers and therefore heterozygous for the trait. As each parent carries a dominant and a recessive allele, only the dominant allele will be expressed; however, they have a 25% chance of producing offspring that are homozygous recessive.

4 a

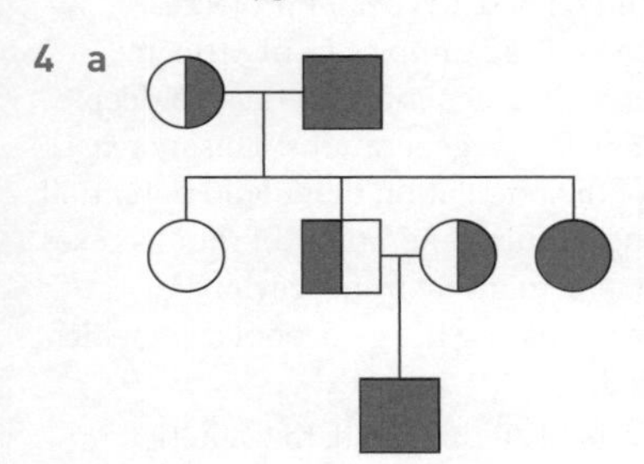

b Recessive

c

	K	k
K	KK	Kk
k	Kk	kk

Probability: 25%

WS 4.5 PAGE 59

1 Male: X^hY, female: X^hX^h

2 X^HX^h

3 Males have only one X chromosome, so it is more likely that they will express sex-linked traits. It is possible that the male's mother and grandmother were heterozygous for haemophilia and so were carriers of the gene. The male, therefore, inherited the recessive gene from his mother on the X chromosome and as a result expressed this trait.

4 a X^hY

b The female offspring receive one X chromosome from each parent. As the father (person A) has only one X chromosome and this chromosome carries the recessive gene (X^h), he will pass this onto his female offspring. The mother (person B) does not have haemophilia and so has passed the dominant gene X^H to her offspring. Both female offspring have the genotype X^hX^H.

c

	X^H	X^h
X^h	X^HX^h	X^hX^h
Y	X^HY	X^hY

Offspring:

Male with haemophilia: 25%

Male without haemophilia 25%

Female with haemophilia: 25%

Female without haemophilia: 25%

5 a

	X^r	X^r
X^R	X^RX^r	X^RX^r
Y	X^rY	X^rY

Female colour blind: 0%

b Each male only has one X chromosome and that X chromosome always comes from the mother. As the mother is homozygous recessive, she will always pass on an X chromosome with the allele for colour blindness to her son, and there will be no dominant allele present.

6 a X^DY and X^dX^d

b The second generation shows all females with the dominant trait and all the males with the recessive trait. This is due to the females inheriting one X chromosome from the mother and one from the father. As the mother is homozygous recessive and the father is dominant, each of the females is heterozygous. As the trait is dominant, it is expressed in each of the females. The males only inherit their X chromosome from the mother and so only have the recessive trait.

WS4.6 PAGE 62

1 I^AI^B

2 a Blood group A

b Blood group B

c Blood group O

3

	I^B	I^I
I^A	I^AI^B	I^AI^I
I^I	I^BI^I	I^II^I

AB blood group: 25%

O blood group: 25%

B blood group: 25%

A blood group: 25%

	I^B	I^B
I^A	I^AI^B	I^AI^B
I^A	I^AI^B	I^AI^B

AB blood group: 100%

	I^B	I^B
I^A	I^AI^B	I^AI^B
I^I	I^BI^I	I^BI^I

AB blood group: 50%

B blood group: 50%

	I^B	I^I
I^A	I^AI^B	I^AI^I
I^A	I^AI^B	I^AI^I

A blood group: 50%

AB blood group: 50%

4 Blood group type O is a universal donor as its red blood cells do not contain any ABO antigens and it can be given to people of all ABO blood group types without antibodies being triggered.

5 Sexual reproduction results in an offspring that is genetically unique from the parents. Therefore, the offspring could have a different blood group type compared to the mother if the male gamete coded for a different blood group type. If the blood from mother and offspring were allowed to mix, antibodies in both the mother and the offspring would attack the foreign blood and this could put at risk the lives of both individuals.

Chapter 5: Inheritance patterns in a population

WS 5.1 PAGE 64

1 A gene pool is the total collection of all of the different genes found within a population of interbreeding organisms.

2 a 2

b 8

c i 8

ii 12

d $q = 8/20$
$= 0.4$
$p = 12/20$
$= 0.6$

e $0.6^2 + 2(0.6)(0.4) + 0.4^2 = 1$
$0.36 + 0.48 + 0.16 = 1$

3 a 80%

b $p = 0.8, q = 0.2$

c $p^2 + 2pq + q^2 = 1$
$0.8^2 + 2(0.8)(0.2) + 0.2^2 = 1$
$0.64 + 0.32 + 0.04 = 1$
32% heterozygous

WS 5.2 PAGE 66

1 *Example answer:*

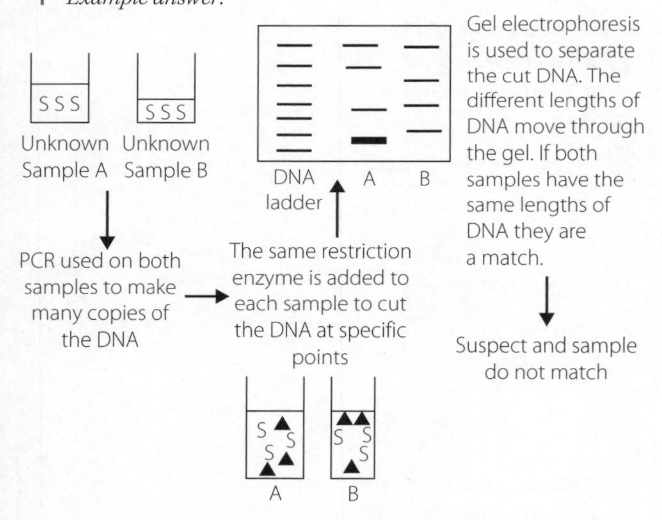

2 isolating; heating; primer; DNA polymerase; nucleotides; complementary; terminate; lengths; dye; electrophoresis; shorter; quickly; laser beam; sequence

WS 5.3 PAGE 68

1 Variation within a population increases the likelihood of traits within a population that will allow individuals to survive environmental changes. For example, a population of plants with varying root lengths may have a better chance of surviving a drought because those plants with long roots would have access to deeper deposits of water than plants with short roots. However, if the members of a population all had short roots, the entire population may die in times of drought.

2 a Population B

b It is important for genetic diversity to be assessed as well as population numbers because low genetic diversity makes a population vulnerable to environmental change. In this example, population A has a higher number of individuals compared to population B, but it is less genetically diverse, meaning that if there were an environmental change or new selection pressure applied, and all individuals were disadvantaged by it, the population may not survive. Traditionally, because population B has fewer individual members, it may have been given a higher conservation status and hence given more protection. However, because population B has greater genetic diversity, it is more likely to be more resilient than population A when faced with a disadvantageous change.

3 The Hemmersbach Rhino Force Cryovault project is gathering DNA, semen and egg samples from African rhinoceroses to preserve their genetic material in a deep freeze for possible IVF in future generations. This increases the genetic diversity of the population that could potentially be lost due to poaching. Future generations of rhinoceroses will have access to genetic material from previous generations, therefore increasing the gene pool from which diversity can be derived.

4 If a species falls below its MVP threshold, the genetic diversity of that species has been significantly reduced and the pool of potential mates shrinks. This can result in inbreeding between closely related individuals, which can magnify detrimental traits in the population. If the genetic diversity is reduced, the species would not have enough variation to successfully adapt to changing selection pressures. Each of these factors could contribute to the continued reduction in size and health of the population until no more individuals of a species can survive.

WS 5.4 PAGE 70

1 a The graph represents the replacement hypothesis because it shows a small group of humans migrating out of Africa around 100 Kya and this small population diversifying throughout the rest of the world.

b The lines represent gene flow or interbreeding between humans of different populations.

c According to the information provided in the graph it appears that the statement is true and Sub-Saharan Africans are more genetically diverse compared to the rest of humanity. Evidence for this in the graph is that the Sub-Saharan Africans have more genetically distinct populations compared to people from the Middle East/Europe, Asia, the Americas and Australia/Melanesia combined. The graph also shows that the small branch labelled Phase III indicates the small population of individuals that migrated out of Africa and colonised the rest of the world. This small population had a limited genetic diversity compared to those left behind in Africa.

2 a Huntington's disease is a genetic neurodegenerative disease that usually affects people around 30–40 years old. It causes the breakdown of nerve cells in the brain and results in death. Early symptoms can include difficulty concentrating, memory lapses, stumbling and loss of coordination.

b Genetic counselling can provide families with the information and support they need in order to make an informed decision about predictive testing for

9780170449625

Huntington's disease. Families that decide to be tested will then be talked through all of the factors to consider based on the likelihood that their offspring will inherit Huntington's disease.

MODULE 5: CHECKING UNDERSTANDING

PAGE 72

1 B

2 C

3 D

~continue in right column

4 The embryo is inside the seed. An embryo develops from the fusion of the male and female gametes. Fertilisation occurs when the sperm nucleus from the pollen joins with the ovum nucleus in an ovule within the ovary. Fertilised ovules become seeds. Therefore, the embryo must be inside the seed.

5 a–vii; b–iv; c–i; d–ii, e–iii, f–v; g–vi

6 Haemoglobin – carries oxygen in the blood
Insulin – regulates blood sugar

7

Process	DNA replication	Protein synthesis
When it occurs	Immediately before cell division	When a cell needs to manufacture a protein product
Type of cell involved in this process	Somatic cells about to undergo mitosis or a gonad cell about to undergo meiosis	Any cell that needs to make protein
Main steps involved	DNA unzips Each original strand gives rise to a complementary strand of DNA as nucleotides join onto exposed base Two double helices produced form each DNA piece	DNA unzips only in the part of a chromosome where the required gene occurs. The gene on one strand of DNA gives rise to a complementary strand of mRNA. The mRNA moves into the cytoplasm. tRNA helps attach the correct amino acids together to make a protein.
Result	All chromosomes within a cell are replicated.	Protein products are produced in the cell under the instructions of a gene.

8

	Differences	Similarities
Mitosis	Results in identical daughter cells with same genetic makeup as the parent cell Makes cells with diploid number of chromosomes Makes somatic cells Involves one round of division	Both are types of cell division. Both involve chromosomes being replicated and divided between cells. Both start with a diploid cell.
Meiosis	Results in four genetically unique gametes Makes cells with haploid number of chromosomes Only occurs in gonads Involves two rounds of division	

9 DNA is a double-stranded helix, whereas RNA is single stranded. DNA contains the nitrogenous bases thymine, adenine, cytosine and guanine, whereas RNA contains the nitrogenous base uracil in place of thymine (the other three bases are the same for RNA as for DNA). The sugar that forms part of the sugar–phosphate backbone in DNA is deoxyribose sugar, whereas in RNA it is ribose sugar. It is the sugar that gives them their names: DNA is deoxyribonucleic acid and RNA is ribonucleic acid.

10 a Male 3 must be heterozygous for the gene because he has produced offspring without the gene. As his wife is not affected and the gene is dominant, she must be homozygous recessive. If he was homozygous dominant, all of their offspring would be heterozygous and so be affected. Male 8 and female 10 are both unaffected, supporting this statement.

b The gene is not sex-linked because female 10 inherited her X chromosome from male number 3, who has the trait. If the gene were sex-linked, she would show the trait.

11

Blood type	Genotypes
A	I^AI^i, I^AI^A
B	I^BI^B, I^BI^i
AB	I^AI^B
O	I^iI^i

12 If the allele is recessive it would only be expressed if two recessive alleles were combined, so the majority of combinations would result in the allele not being expressed.

13 Endangered species have a very low number of individuals, and hence have limited genetic diversity. Through the use of DNA profiling, the best and most diverse pairings could be made to ensure maximum genetic diversity in the resulting offspring. By increasing genetic diversity, the overall health of the species can be increased and then maintained.

MODULE SIX: GENETIC CHANGE

REVIEWING PRIOR KNOWLEDGE PAGE 76

1 amino acids; nucleotides; codon; translated; polypeptide; protein

2 a

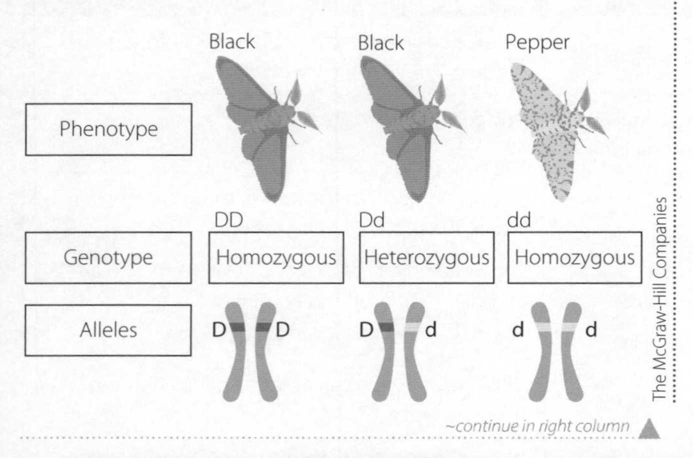

~continue in right column

b Dominant

c Variation: The mutation resulted in the population being both pepper coloured and black.

Selection pressure: As soot from factories darkened the habitat of the peppered moth, they were more visible to predators.

Adaptive advantage: Black moths were more camouflaged and therefore more likely to avoid predation.

Survival and reproduction: The black moths were more likely to survive and breed successfully.

Change in population: Eventually the black allele became more common in the gene pool.

3 C

4 C

5 Genes are parts of DNA. A gene provides the instructions for making a polypeptide. One or more polypeptides folded into a 3D shape makes a protein.

6

Cell division	Role	Outcome
Mitosis	Cell division used for growth, repair and asexual reproduction	Produces two genetically identical daughter cells
Meiosis	Cell division used to create gametes (sex cells)	Produces four genetically different daughter cells

7

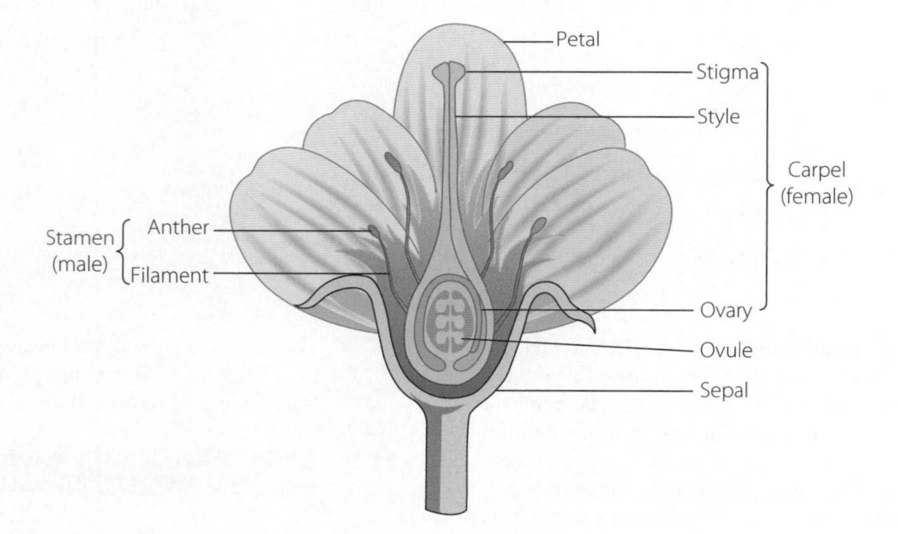

8 Biodiversity is the number, relative abundance and genetic diversity of organisms in an area. It encompasses variation at the genetic, species and ecosystem levels.

9 Bioethics is important because it allows for careful consideration of the impacts of a procedure or other technologies on an individual and/or on society. Bioethics allows for appropriate decisions to be made based on information from balanced arguments.

9780170449625

Chapter 6: Mutation

WS 6.1 PAGE 78

1

Mutagen	Classification	Mutagen	Classification
Sunlight	Electromagnetic radiation	Viruses	Naturally occurring
Cigarettes	Chemical	Transposons	Naturally occurring
Processed meat	Chemical	X-rays	Electromagnetic radiation

2 a Each enzyme has a specifically shaped active site that is designed to bond with a complementary substrate. Thymine dimers change the shape of the DNA molecule and so prevent the enzyme RNA polymerase from bonding and correctly reading the template DNA strand.

~continue in right column

b

Process	Description	Diagram
Photoreactivation	UV light causes two thymine bases to covalently bond together, forming a thymine dimer. The enzyme photolyase detects the thymine dimer and bonds to the site. Photolyase uses visible light to break the covalent bond and repair the dimer. DNA is returned to healthy condition and normal complementary pairs are restored.	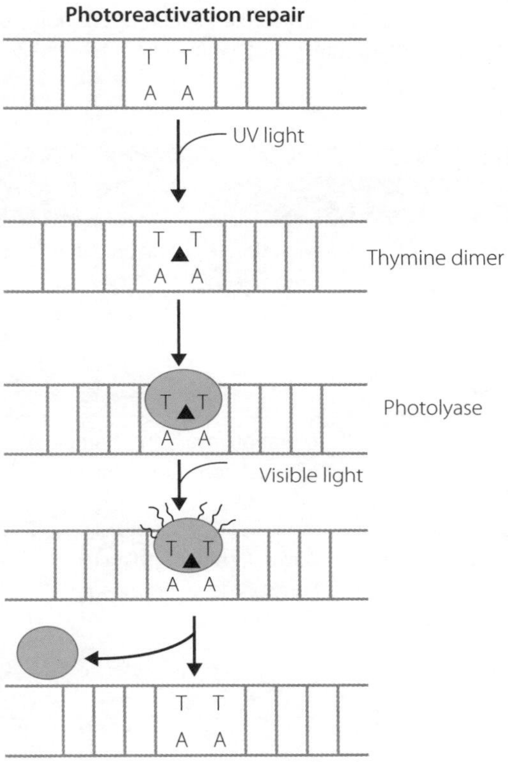 **Photoreactivation repair** T T / A A UV light T T / A A — Thymine dimer T T / A A — Photolyase Visible light T T / A A T T / A A
Nucleotide excision repair	Dimer recognised by DNA polymerase and DNA section is cut. Dimer is removed by nuclease. Gap in DNA is filled by DNA polymerase. Gap is sealed by DNA ligase.	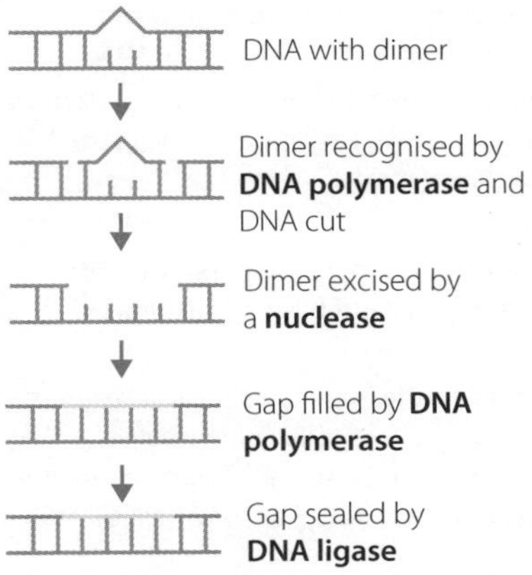 DNA with dimer Dimer recognised by **DNA polymerase** and DNA cut Dimer excised by a **nuclease** Gap filled by **DNA polymerase** Gap sealed by **DNA ligase**

c Photoreactivation requires the enzyme photolyase, and, although this enzyme is found in bacteria, archaea and eukaryotes, it is not found in placental mammals. Therefore, the process of photoreactivation cannot take place in humans.

WS 6.2 PAGE 80

1 a A substitution point mutation will lead to a change in the codon sequence of the transcribed RNA. This change could result in a different amino acid being coded for during translation and in turn could alter the polypeptide chain being formed. This alteration of the polypeptide chain could result in a change of shape of the final enzyme's active site, preventing the enzyme from bonding with its intended substrate.

b i A frameshift mutation in gene 2 would result in the flower being pink because gene 1 will not be affected by a frameshift mutation. Only genes coded for after and including gene 2 will be affected.

ii A chromosomal inversion would flip gene 1 and gene 2 and place the coding bases in reverse order. This would likely result in neither gene 1 nor gene 2 correctly coding for their corresponding enzyme. This would result in the flower remaining white.

~continue in right column

2 a Missense: A change in one of the nucleotides results in a change to the amino acid used in the final polypeptide chain.

b Silent: A change in the nucleotide results in the same amino acid being coded for in the polypeptide chain.

c Frameshift insertion: A nucleotide is inserted, which changes the codons for each subsequent amino acid in the mRNA.

WS 6.3 PAGE 82

1 a Duplication mutation

b Inversion mutation

c Deletion mutation

d Insertion mutation

e Translocation mutation

2 During meiosis II, specifically anaphase II, sister chromatids fail to separate. This results in both chromatids going into one daughter gamete and none to another gamete. This means that one gamete would have an extra chromosome (trisomy), whereas one gamete would be missing a chromosome (monosomy).

3

Type of mutation	Description of mutation	Description and example of syndrome in humans	Chromosome affected
Trisomy	Mutation results in an extra chromosome	Down syndrome Physical features are altered, resulting in a smaller head with a flattened face; excessive flexibility Intellectual disabilities of varying degrees	21st pair
Monosomy	Mutation results in a gamete with a chromosome missing	Turner syndrome Swollen hands and feet, high palate, short stature, flat feet, droopy eyelids	23rd pair: sex chromosomes
Sex-linked trisomy	A male is born with an extra copy of the X chromosome: they are XXY	Klinefelter syndrome Small testicles or penis, taller than average height, weak bones, low muscular development, increased belly fat, enlarged breasts	23rd pair: sex chromosomes

WS 6.4 PAGE 84

1 The mutation is a point mutation because only one nucleotide is being substituted.

2 This mutation cannot be inherited because it is stated that the mutation is somatic. It will therefore not be passed on through gametes to the offspring.

3 Gene 1 + Gene 3 + Gene 4

4 Scientists can discount gene 2 because person 1 has a mutated version of that gene and has a 0% likelihood of developing cancer.

5 A substitution of a nucleotide in gene 2 may not result in an increased likelihood of developing cancer because the codon may result in the same amino acid being coded for in the resulting polypeptide chain. This is known as a silent mutation.

6 From the data in the table, mutation in gene 3 will result in the largest increase in the likelihood of developing cancer and therefore this gene should be the focus of the scientific research. In person 1, gene 2 is mutated, but it is shown that this does not increase the likelihood of developing cancer. In person 2, genes 1, 2 and 4 are mutated, and it seems that, when mutated, genes 2 and 4 increase a person's likelihood of cancer by 30%. In person 3, the likelihood of cancer increases by 60% when a mutation occurs in gene 3, therefore indicating that gene 3 has the most influence on cancer development.

WS 6.5 PAGE 86

1 Genetic drift can occur due to random events. In the diagram, a significant portion of the population is killed randomly with a swatter. This random sample contained a large portion of insects with one trait. This then led to genetic drift that was not selected for by natural selection, because the death of those individuals was completely random. As the traits were not selected for, they may not be

favourable. This could be the case with a natural disaster in an ecosystem or widespread land clearing.

2 isolated; colonised; founder; alleles; genetic drift

3 *Points for:* Both models use colours to clearly indicate the allele frequency in a population.

The models clearly show how alleles in new populations can be different from those in the original population.

Points against: The models are oversimplified and do not indicate dominant and recessive alleles.

The models do not show alternative allele combinations that could have occurred due to bottlenecks or genetic drift.

Judgement: The models are effective at comparing the two genetic processes because they clearly show each process in a simplified way.

4 *Answers will vary. Example answer:*

1 Count the number of individual alleles (lollies) in the original population and the number of each colour. Each colour represents a different allele.
2 Record the original allele frequency for the entire population.
3 Mix the individuals in the flask.
4 Let five alleles through the neck of the flask. The remaining alleles have been struck down by a disaster that has caused the population to dwindle.
5 Record the colours of the five alleles.
6 Calculate the allele frequencies in this new population.

~continue in right column

7 Repeat the process of letting five individuals through a bottleneck two more times.
8 Recording the allele frequencies in each new population.

Chapter 7: Biotechnology

WS 7.1 PAGE 88

1 a Yes
b No
c Yes
d No
e No
f Yes
g Yes
h Yes
i Yes
j Yes
k Yes
l No
m No

2 a Making a product using living things is part of the definition of biotechnology. The possum cloak itself is a product made from an animal. In the making of it, other intermediary products are involved. These include making thread from animal tendons or plant fibre, making pins from wood, bone and echidna quills, and using shell and bone as tools to etch artwork into the skin. The paint made from tree sap and ochre is also a biotechnology. Stone tools and ochre alone are not made from living things.

b

Positive social implications	Negative social implications
Cultural connection; positive emotions; stories and knowledge shared during workshops; product valued for ceremonies; better relationships across generations	People who are not of Aboriginal or Torres Strait Islander descent may not appreciate the effect of connection to culture and therefore not support the continuing practice
Positive ethical implications	**Negative ethical implications**
Using a feral animal rather than just killing them with no use	Possums not from local area; are animals humanely treated? Are the other aspects of the original process used (e.g. are kangaroos killed for their tendons?)?

3 a Primary producer
b Crossing
c Species
d Domestication
e Cultivation
f Breed
g Variety
h Strain

4 Merino: fine wool; Dorper: meat; Corriedale: meat and wool; East Friesland: sheep milk

5 The original varieties each had advantages and disadvantages. By breeding them together to create Yandilla, a variety with the best features of both parents was created – it was good for milling and baking, and was rust resistant and drought tolerant. One more cross, with Purple Straw, added the characteristic of strong straw to the next generation. The resulting variety was named Federation, and it had all the features of Yandilla with the addition of strong straw. Selective breeding was crucial for developing varieties of wheat suited to growing in Australia.

6 Hybrids are often sterile, so the only way to continue that variety is to continuously breed the same parents together and then collect and plant the seeds. The farmer will have to control the pollination to make sure only the desired parents breed together. This is unlikely to be a possibility for the primary producer. If sterility is not a problem and the hybrids can breed together, the resulting offspring will not all display the desired features of the parent. Therefore, the seed collected from this generation will produce very few offspring with the hybrid vigour and the desired features expected. The Punnett square that follows looks only at one feature, and shows only a 50% chance of the offspring (Gg) of heterozygous parents displaying the same characteristic as the heterozygous parents. When you consider that there may be many features being chosen when crossing is planned, there is little chance that the offspring will have the hybrid condition for every desirable feature.

	G	g
G	GG	Gg
g	Gg	gg

WS 7.2 PAGE 92

1 a Animal pancreases are the source of insulin. The insulin product is made using the processes of extraction and

purification. Since this is a product made from living things, it fits the definition of biotechnology.

b Type 1 diabetes was known as a disease before the hormone insulin was discovered, but most people died within a few years of disease onset because the cause of the disease was not known, and hence there was no available treatment. As such, the discovery of the hormone insulin and its relevance in the presentation of Type 1 diabetes (where the pancreas of a person with the disease does not produce enough insulin) was vitally important to society. Understanding the cause of the disease was the first step in developing a treatment for it. Knowledge of extracting and purifying insulin from animal pancreases was also vitally important, and was the next step in developing a treatment.

This knowledge was crucial to society because it meant that individuals with the disease would not necessarily die from it. Those individuals could continue to be, or grow up to be, productive members of society. The industries involved in sourcing animal insulin, making synthetic insulin and distributing insulin also support jobs for people in society.

c A person with Type 1 diabetes has only two options: use insulin treatment, or die. In the case of this disease, a religious waiver would likely be considered, given that the alternative to treatment is death. If a person has strong moral (rather than religious) convictions about using products derived from animals, they would need to make the ultimate choice between life-saving treatment or death. This could be a harrowing decision for someone in this situation.

2 a Proinsulin is a single polypeptide composed of many amino acids that is made in the ribosomes. Insulin is the final protein formed after cleavage and discarding of a section of amino acids, leaving an A and B chain folded into a 3D shape with bonds between cysteine molecules.

b The vector is the piece of DNA placed where the foreign gene is inserted. This is a plasmid from a bacterium. The plasmid cannot do anything unless it is in a cell. This is the host cell. The vector is the modified plasmid and the host is the cell containing the vector. The host might be a bacterium or a yeast cell.

c Step 2 – inserting *INS* gene into a DNA vector and into a host cell

d

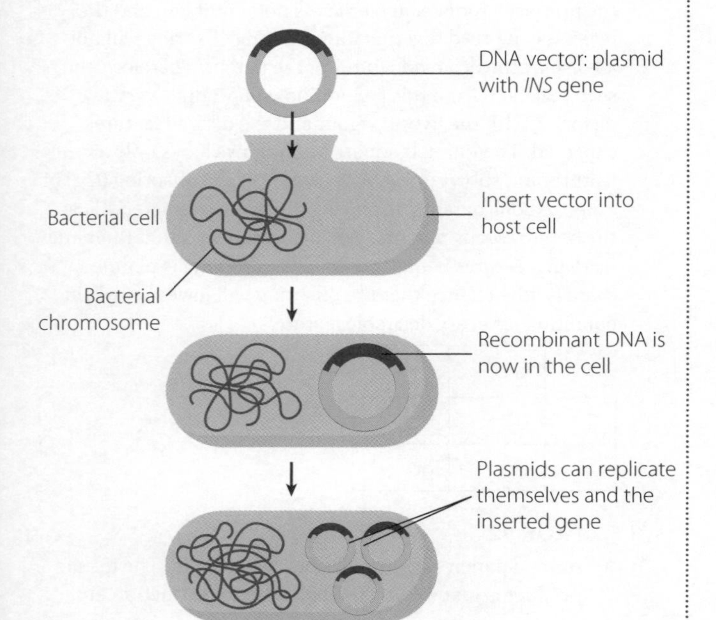

e Asexual reproduction of the bacteria cells by binary fission to make many more cells with the recombinant plasmid.

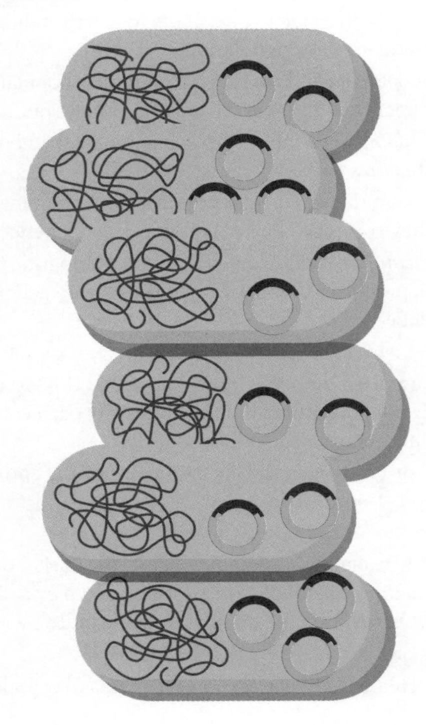

f Yeast are eukaryotic cells and have ribosomes, endoplasmic reticuli and Golgi bodies, which are all involved in the full process of cleaving and folding the final insulin protein. Bacteria have only ribosomes, so they can make the polypeptide but might not be able to complete the processing of the protein's 3D shape.

3 a To make recombinant DNA, you must be able to cut pieces of DNA in the lab. This is what restriction enzymes do. Different enzymes will cut DNA at different places, so scientists can choose which restriction enzyme to use to cut a specific gene out of a DNA strand. Therefore, there could be no recombinant technology without knowledge of restriction enzymes.

b A restriction enzyme would need to cut the human DNA at the point where the insulin gene is located. A restriction enzyme would also need to cut a gap in the bacterial plasmid. Ligase would be used to glue the piece of DNA into the plasmid, so the complementary bases remain attached.

9780170449625

4 a

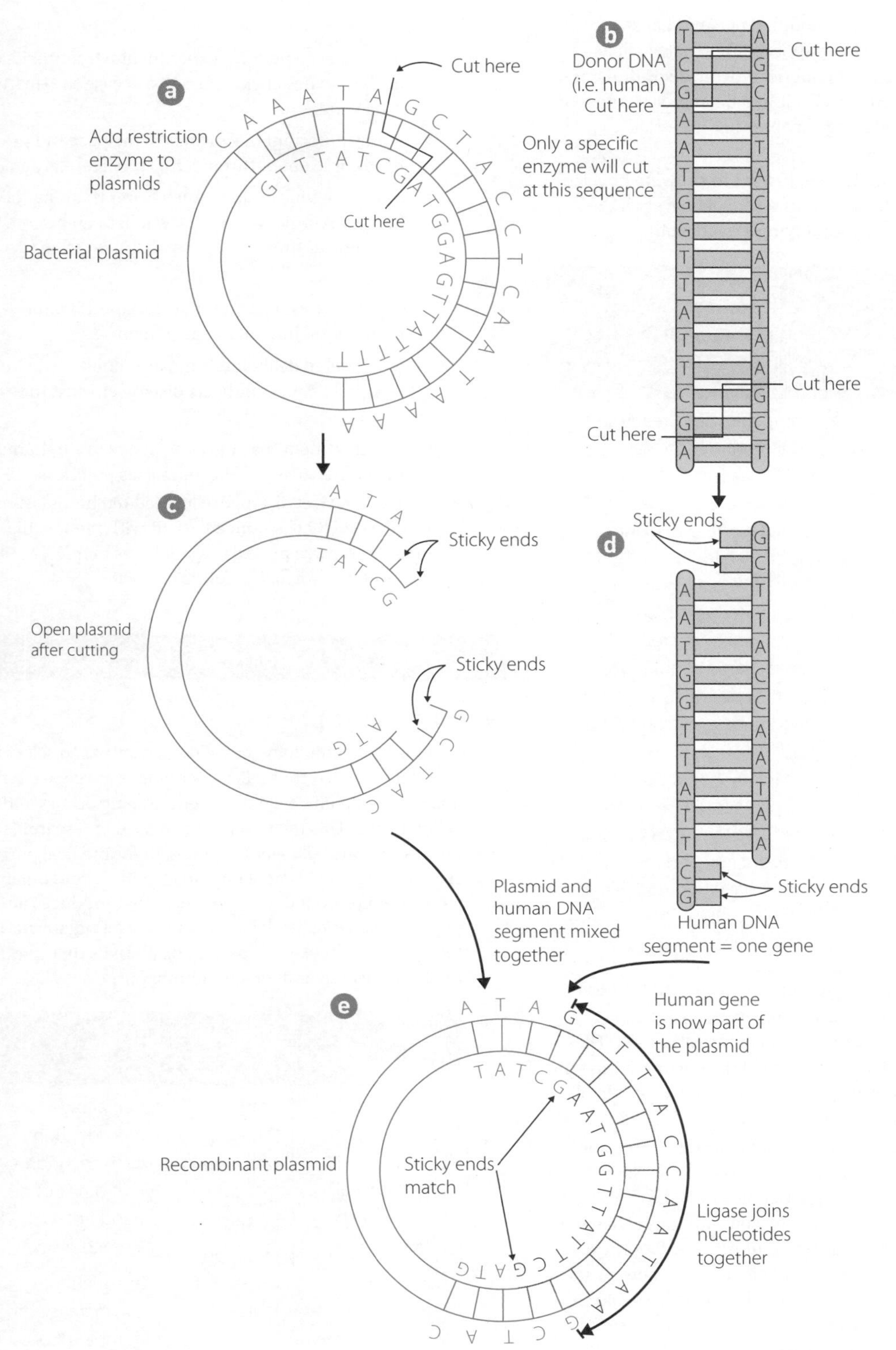

b The same restriction enzyme cut the DNA and the plasmid, leaving the same sticky ends. On one end of the plasmid cut there were two exposed bases – C and G – and the end of the gene had G and C exposed. The bases at the other end of the plasmid cut and the other end of the gene are exactly the opposite. This means sticky ends match up as per the base pairing rules. The sticky ends will all be attracted to each other so the gene fits easily into the gap in the plasmid.

c The enzyme requires a TAT sequence on *both* sides of the plasmid strand in order to make a cut that creates sticky ends. The plasmid does not have TAT locations that will allow a complete cut through the plasmid strand to open up two sticky ends. There is also only one TAT sequence on the human DNA strand. No complete cut will be made. No gene will be spliced out of the strand.

d Models are intended to make difficult ideas easier to understand.

Benefits of this model: It makes us look at the way restriction enzymes work by considering the base sequences on the DNA strands and the creation of sticky ends. It helps us understand the idea of why the gene easily inserts into the plasmid. The model helps with definition of a recombinant plasmid and the role of ligase.

Limitations of the model: This model focuses on modifying a bacterial plasmid by adding a gene from human DNA into the plasmid; it does not show any of the steps involved with obtaining the DNA and plasmids or getting the modified plasmid back into a bacterial cell, or what happens after that, such as plasmid replication. The model only involves drawings. It would have been better if it had been more hands-on, such as using modelling clay or other materials to make it more fun.

Judgement: The model is sufficient in showing how to modify a plasmid, but it does not really cover all the aspects of genetically modifying a bacterium.

5

Benefits of using recombinant plants to produce human insulin	Disadvantages of using recombinant plants to produce human insulin
• Seeds or leaves could produce high levels of proinsulin. • It is easy and inexpensive to asexually reproduce the GM plants. • Seeds containing insulin could be stored for a long time until required. • Patients could possibly ingest the insulin from the plant rather than inject it.	• Companies are already set up to continue to make insulin via fermentation; it could be costly to set up manufacturing of plant-based insulin. • A lot of land and water is needed to grow plants compared to microbes. • These plants might make proinsulin, which would still need to be processed into insulin. • The initial production of the GM plant could be costly. • Seeds are the result of sexual reproduction and may not always have the modified gene because of meiosis and fertilisation.

6 a Stem cell research and gene editing with CRISPR-Cas9

b Prior to publishing a research paper, the paper is scrutinised by a group of experts in the same field. These experts are considered the peers of the person writing the paper. A peer-reviewed paper is one of the most reliable sources of information on a subject.

c CRISPR has the ability to snip out a harmful mutation and replace it with a 'normal' DNA sequence. Gene editing could potentially lead to the eradication of genetic diseases. If only wealthy communities can access this technology, then harmful mutations will only exist in poorer communities. This could lead to a new social norm. Potential parents could feel pressure to avoid the conception of embryos or foetuses that carry harmful genetic mutations. The less wealthy may feel obliged to stretch themselves financially to comply with these expectations. There could be more discrimination against or stigma surrounding disabled people and those living with the conditions that could have been 'corrected'. In addition, if CRISPR was used for genetic enhancements, further inequalities will occur between rich and poor.

d For:

- It is better to perform experimental treatments on mice than directly on humans, in case the treatment is harmful.
- Many important discoveries in the past and present have come about through this process; e.g. penicillin.
- The mouse life cycle is much faster than that of a human, so embryo-to-adult studies can be completed in a shorter time.

Against:

- It is cruel to subject animals to experimental treatments that might harm them.
- It is cruel to deliberately make a mouse diabetic or give them cancer or heart disease etc. just to test treatment.
- The treatment may never be applicable to humans, so the harm caused to the animals is pointless.

e Maxwell, K.G. et al. (2020) 'Gene-edited human stem cell-derived β cells from a patient with monogenic diabetes reverse pre-existing diabetes in mice.' *Science Translational Medicine* Vol. 22, no. 540.

Chapter 8: Genetic technologies

WS 8.1 PAGE 101

1 Cloning is a technology that allows scientists to select for particular desired traits. Two techniques are gene cloning and whole-organism cloning. Gene cloning occurs at the cellular level. This technique can be used in research, to understand the behaviour of genes, or for medical purposes, such as the production of human insulin. Whole-organism cloning is a reproductive technology that produces an organism genetically identical to the target organism. This technology is focused on producing animals that may assist medical research and for agricultural purposes.

2

Selective breeding	Similarities	IVF
Technique is easily utilised by farmers geographically close to each other Relatively inexpensive when compared to other reproductive technologies	Produces fertile offspring Allows for the selection of desired traits	Can allow reproduction between individuals that may not be geographically close to each other In agriculture, can increase the number of offspring with desired traits produced in a year; use of surrogate mothers means that reproduction rate is not limited by gestation period of the targeted mother

9780170449625

WS 8.2 PAGE 102

1

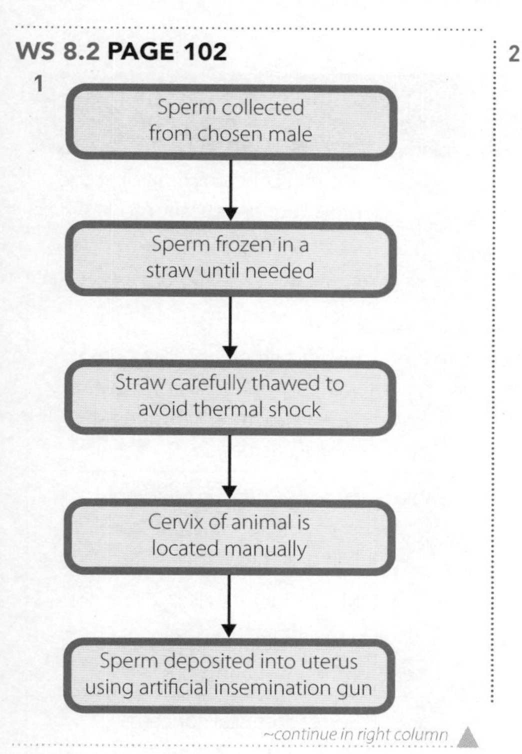

~continue in right column

2

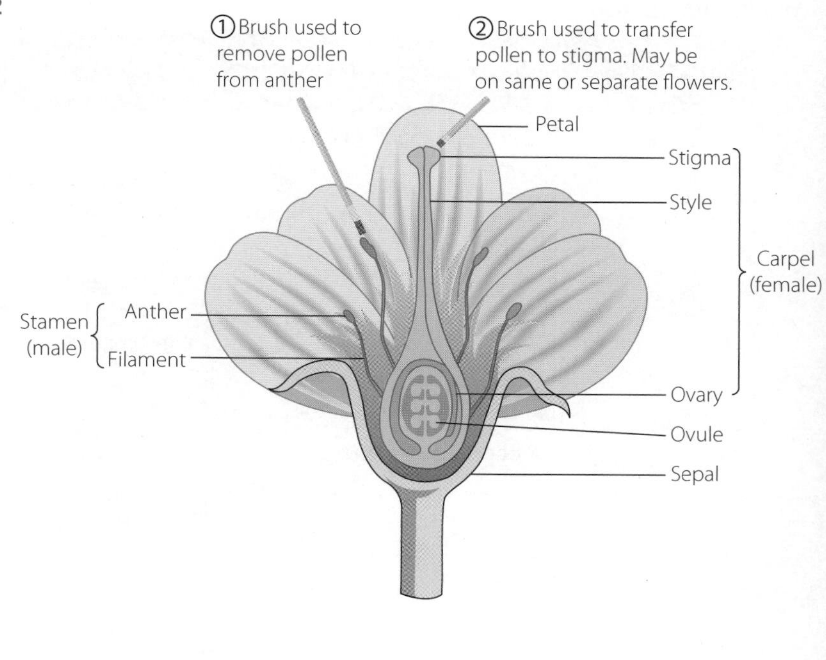

3

Artificial insemination	Artificial pollination
Reproductive technology used with animals; e.g. racing horses Sperm can be frozen, so a male can produce offspring after he has died Can be used to assist with conservation efforts; has been used with gorillas and sharks	Reproductive technology used with plants; e.g. production of hybrid corn with increased growth rate and yield Can reduce the fruit and seed yield compared with insect pollination

WS 8.3 PAGE 104

1

Gene cloning	Similarities	Whole-organism cloning
Occurs at the cellular level Produces identical copies of a gene Used for biotechnology research	Produces copies of genetic material	Produces a genetically identical whole organism Considered to be reproductive technology A form of asexual reproduction

2 *Answers will vary. Highly regarded answers will have subheadings to address all parts of the question: cloning description, advantages, limitations, judgement.*

WS 8.4 PAGE 105

1 Restriction enzymes are proteins that recognise short DNA sequences and cut the DNA at these specific locations. In the diagram, this location is between G and A in the sequence GAATTC. Once cut, these enzymes create sticky ends, which are single strands of DNA that can pair with complementary bases on other DNA strands.

2

Embryonic stem cells — Lacks detail. These cells are genetically engineered. Production of knock out mice requires inactivation of a target gene.

↓

Put into blastocyst — Stem cells are injected into blastocyst in uterus of host female.

↓

Mouse gives birth

— Step missing here. Question asks for how these mice are used in medical industry. Knock out mice can be used to study disease symptoms and treatments, e.g. inactivation of tumour suppression gene and understanding cancer treatments.

WS 8.5 PAGE 106

1

Industry	Genetic technology example	Benefit/s of the named technology	Limitation/s of the named technology
Agricultural	Genetically modified Atlantic salmon	Fast growing Puts less pressure on wild stock, which are significantly reduced in number Fewer greenhouse gas emissions as fish can be grown close to the consumers; reduces transport	Societal concerns regarding the growth of genetically modified organisms Some fears over increased risk of allergic reaction Risk of salmon escaping into the wild; e.g. during a storm event when structures housing fish may be damaged
Medical	Monoclonal antibodies	Highly reproducible and scalable Can stop cancer cells dividing, help with drug delivery or activate the immune system	Ethical issues surrounding production inside the stomach lining of mice May induce an allergic reaction Other side effects: fever, vomiting, headaches
Industrial	Potato plants that produce the starch needed to make paper and textile products	Offers environmentally friendly solutions to producing high-demand products Saves water and energy Potato starch is high quality and produced in high yield per hectare	Societal/ethical concerns regarding the production of genetically modified organisms

2 *Answers will vary.*

3 a Fungi that can more effectively degrade the polymer will decrease the optical density of the polymer.

b *Possible answers:* the amount of polymer tested, the amount of fungi added, the same environmental conditions such as atmospheric temperature, ensuring that the conical flask was sterile before adding the fungi and polymer

c *Answers will vary. Example answer:* If the conical flask was not sterile, it would not be possible to determine if any degradation of the polymer was due to the target fungus, or any other microbes that may have been present at the time the investigation was conducted.

d The species *Pestalotiopsis microspora* strain E3317B and *P. microspora* E2712A would both potentially be most useful for bioremediation. As shown in the graph, *P. microspora* E2712A decreased the optical density of the polymer the fastest after 8 days, down to 0.1 (600 nm), and after 15 days *P. microspora* E3317B had the lowest optical density at around 0.02 (600 nm).

e *Answers will vary. Possible answers:* Investigating how much of each fungus is needed to be effective/most efficient/degrade the polymer the fastest, to determine which environmental conditions create the most favourable conditions for polymer degradation or to investigation whether other types of polymers can be degraded using these fungi.

WS 8.6 PAGE 108

1 Bt corn has been engineered to be resistant to insect attacks, and as a result the need for spraying of pesticides is reduced. This is shown in the graph by a significant reduction in the use of insecticides between 1996 and 2010, from 0.2 kg/ha to less than 0.01 kg/ha. Spraying these chemicals can result in the death of many species of invertebrates, which in turn can impact soil health and limit the ability of some plants to grow, and remove food sources from food webs. It is therefore likely that planting Bt corn will increase the local biodiversity because only target pest species will be impacted by the crop.

2 a Sunny Creek Farm: 23; Sunny Creek National Park: 18; Windy Valley Farm: 29; Windy Valley National Park: 18

b

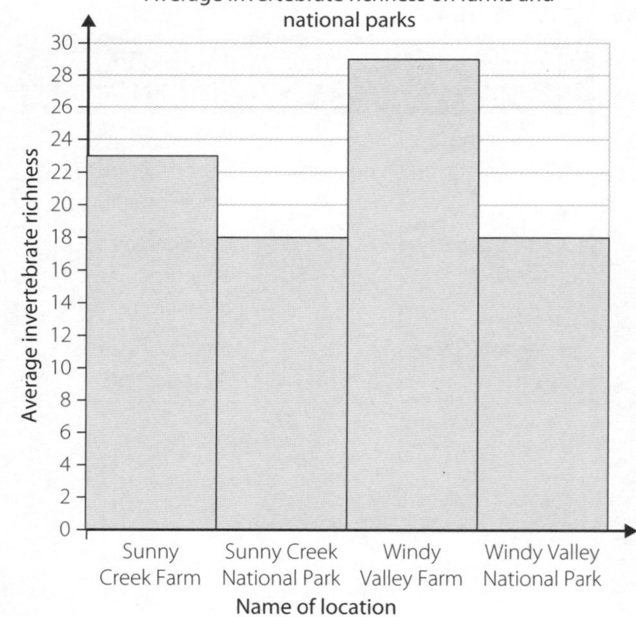

9780170449625

c The trend in the graph shows that invertebrate Order richness was highest on the farms, when compared with the local national parks. For example, the average invertebrate Order richness on Sunny Creek Farm was 23, compared with a richness of 18 in Sunny Creek National Park. One possible reason for this may be due to the disturbance on farms, such as land clearing, which creates available niches for the invertebrates to occupy. Between the two farms, Windy Valley Farm had the highest richness at 29 invertebrate Orders. This may be due to the fact that Windy Valley Farm was growing non-genetically modified wheat.

WS 8.7 PAGE 110

1 a Lack of access to the technology for people who cannot afford it, some may not agree with the harvesting of cells from the dead animal, issues may arise from the use of surrogate mothers

b According to the information provided, the three companies listed are located in the USA, South Korea and China. This may be due to the fact that these countries have a higher proportion of people that are able to afford the $100 000 cost of the technology. The expense of the technology will limit how much the industry is able to grow, as size of the available customer base is limited.

c Despite the fact that cloning costs around $100 000 it is likely that the size of the cloning industry will increase in five years' time. According to the graph, cloning is becoming more accepted by the general public, with 40% of those surveyed in 2018 supporting the cloning of animals, compared with 29% in 2002.

2 *Answers will vary. Data must first be determined to be accurate and reliable before validity can be assessed. Data cannot be valid if it is not accurate and reliable. It is also important that the data are relevant, and that any referenced primary data were collected using appropriate scientific methods.*

MODULE 6: CHECKING UNDERSTANDING PAGE 112

1 A substitution of one nucleotide due to a mutation could result in a silent mutation. This is when the altered codon still codes for the same amino acid and so no change is made to the resulting polypeptide chain. Therefore, there is no phenotypic change.

2 The founder effect occurs when a small number of individuals of a species are isolated from the original population. The alleles present in this small number may not be representative of the alleles present in the overall population, and some allele combinations may be missing altogether. The diagram below shows how two members of the new 'founder' population do not possess the allele that the majority of the original population possess.

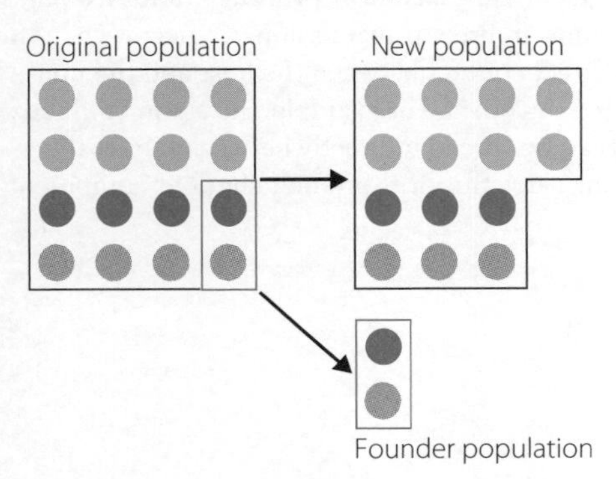

3 C

4 To cut DNA at specific sites in specific sequences

5 Yeast is a microbe involved in making bread, beer and wine. Yeast turns the sugars in flour or fruit into alcohol + carbon dioxide + energy. Fermentation satisfies the definition of biotechnology because it is using a living thing (e.g. yeast) to make a product.

6

Artificial insemination	Similarities	Artificial pollination
Sperm is collected from chosen animal and introduced into selected females	Gives control over the breeding process; allows for selection of desired traits	Stamen is removed from selected flower and pollen dusted onto selected stigma

7 Whole-organism or reproductive cloning involves creating a genetically identical whole organism from a somatic cell. This process is used to select for particular desired traits. As only a few parent animals can be cloned, genetic diversity is decreased as all clones will have the same genetic material.

8 a Xenotransplantation involves using organs from other animals for human transplant. Transgenic animals, such as pigs, are engineered to have proteins known as surface markers on their cells that are complementary to human receptors and this method inhibits organ rejection.

b The graph shows that the number of people on organ transplant waiting lists increased from 23 198 in 1991 to 121 272 in 2013. This number of people far outweighs the number of available organ donors, 14 257, in 2013. The use of xenotransplantation would help to remove pressure from organ transplant waiting lists and likely increase life expectancy for patients on the waiting list.

MODULE SEVEN: INFECTIOUS DISEASE

REVIEWING PRIOR KNOWLEDGE PAGE 114

1 Both terms are used to describe cellular structure. Eukaryotic refers to cells that have membrane-bound organelles, such as a nucleus. Prokaryotic cells do not have membrane-bound organelles.

2 Unicellular organisms, such as amoeba, consist of just one cell. Multicellular organisms, such as a banksia tree, exist as many differentiated cells working collaboratively.

3 Disease refers to any condition that prevents an organism from performing its normal functions.

4

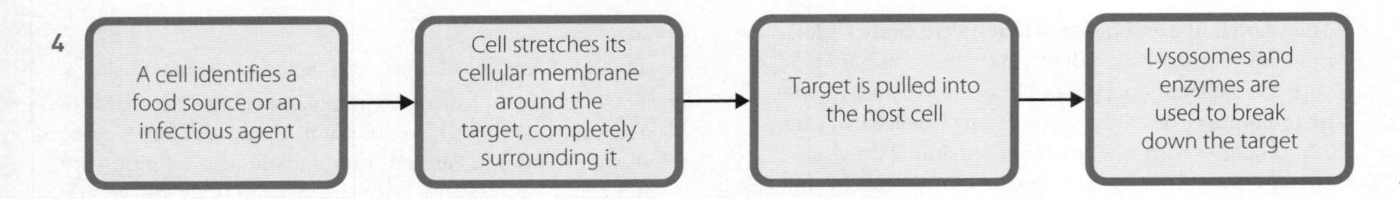

Chapter 9: Causes of infectious disease

WS 9.1 PAGE 115

1

Pathogen	Structure	Approx. size	Example disease
Prion	Abnormal protein	~253 amino acids in length	Mad cow disease
Virus	Genetic material (DNA or RNA) covered by a protective protein coat known as a capsid	~20–400 nm	Acquired immune deficiency syndrome (AIDS) caused by the human immunodeficiency virus (HIV)
Bacterium	Unicellular prokaryotic organisms that have a cell wall	~0.5–5 micrometres	Whooping cough caused by *Bordetella pertussis*
Fungus	Eukaryotic organisms, cells have a cell wall Can be unicellular or multicellular	~2–50 micrometres	Tinea caused by tinea pedis
Protozoan	Unicellular eukaryotic organisms	~1–150 micrometres	Cryptosporidiosis caused by *Cryptosporidium parvum*
Macroparasite	Multicellular invertebrate animals that can be either endo- or ectoparasitic	~1 mm – 30 cm, although very wide range of sizes	Mange caused by mites

2 Cellular pathogens are living, have true cells and include bacteria, fungi, macroparasites and protozoa. Both bacteria and fungi have cell walls. Non-cellular pathogens are non-living and consist of parts of a cell, such as prions (composed of proteins) and viruses (genetic material covered by a capsid).

3 *Example answer:*

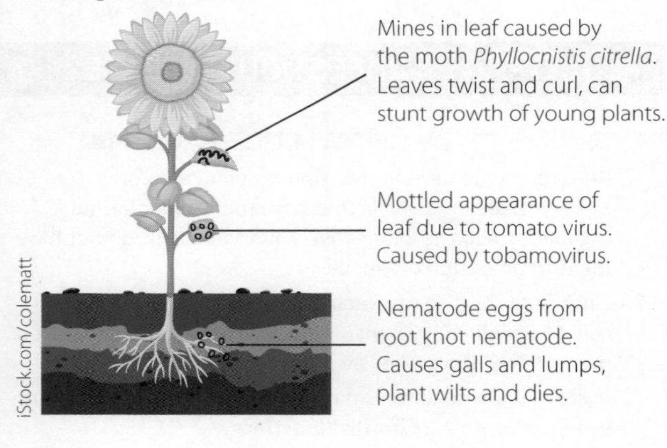

4 *Answers will vary. Highly regarded answers will include the name of the pathogen, the names of the hosts, a description of the pathogen life stages and how the pathogen is able to move between hosts.*

WS 9.2 PAGE 117

1 a Indirect
b Direct
c Indirect
d Direct
e Vector
f Direct
g Indirect

2 a *Examples will vary. Example answer:* Direct transmission of infectious disease occurs when there is actual contact between the host and a non-infected organism. For example, the equine influenza virus orthomyxovirus A can be spread between horses through contact, specifically through fluids from nasal secretions. Indirect transmission can occur when there is no direct contact between the host and the non-infected organism. For example, the equine influenza virus can be spread indirectly between horses if humans carry the virus on their shoes or equipment.

9780170449625

3 a

Disease name	Dengue fever
Pathogen name and classification	Dengue virus from the family *Flaviviridae*. There are four serotypes of the virus that cause dengue (DENV-1, DENV-2, DENV-3 and DENV-4).
Disease transmission	Vector-borne via mosquitoes, usually female *Aedes aegypti* mosquitoes. Some evidence of maternal transmission.
Symptoms	Severe, flu-like symptoms for 2–7 days. Seldom results in death. Symptoms include high fever, severe headache, muscular pain, nausea and vomiting, and a rash. Some people develop severe dengue fever, which can result in potentially fatal complications such as organ impairment or severe bleeding.
Treatment	There is no specific treatment for dengue fever. Paracetamol can be taken for symptoms. Ibuprofen should be avoided. Medical treatment, such as fluid replacement, should be sought for severe dengue fever.
Location of 2016 epidemic	*Answers will vary. Locations include:* The Philippines, Brazil, Malaysia, Solomon Islands, Burkina Faso
Number of cases and number of deaths during the 2016 epidemic	*Answers will vary. Example answer:* In the Philippines, 211 108 suspected cases of dengue were reported and 1019 deaths.
Management and prevention strategies used to control the epidemic	*Answers will vary. Example answer:* • Vaccination • Avoid getting bitten by mosquitoes, especially if you have the disease (e.g. window screens, insect repellents, long-sleeve clothing) • Education of the community • Spraying of mosquito-prone areas

b *Answers will vary. Example answer:*
World Health Organization (2020) *Dengue and severe dengue.* [online] Available at: <https://www.who.int/news-room/fact-sheets/detail/dengue-and-severe-dengue> [Accessed 9 April 2020].

c *Answers will vary. Data must first be determined to be accurate and reliable before validity can be assessed. Data cannot be valid if it is not accurate and reliable. It is also important that the data are relevant, and that any referenced primary data were collected using appropriate scientific methods.*

4 a Over the 15-year period from 2001 to 2016, the number of cases per year in Korea increased from 6 to 313.

b Between the years 2012 and 2016, the highest number of cases occurred during summer, with an average of 30 cases in July and an average of 37.6 cases in August.

c The data presented suggest that the optimum temperature for dengue transmission is around 30°C. The biting rate, the probability of infection and transmissible rate are all highest at around 30°C.

d It is likely that in the future dengue epidemics will become more common in Korea under climate change. The data provided suggest that the possibility of dengue infection is highest from around 25°C and the transmission rate is highest between 25°C–30°C.

WS 9.3 PAGE 122

1 a To investigate microbial growth in three different sources of water

b Dependent variable: Number of colonies and/or colony morphology
Independent variable: Source of water

c *Answers will vary. Possible answers:* Amount of water, incubation temperature, incubation time, amount of agar, size of Petri dish, inoculation equipment and technique

d *Answers will vary. Highly regarded answers will provide a direct link between the variable selected and what would happen to the results if the variable was not controlled.*

e No treatment should be applied – the plate should have sterile agar only.

f An experimental control removes the effect of the independent variable. If any microbial growth appears on the control plate, then it would show that the agar was contaminated and therefore the results on the other agar plates must be discounted.

g *Answers will vary. Example answer:* The agar plates should be incubated at 25°C to prevent the growth of any human pathogens.

h i Selected mark: 3

ii The method was provided in steps.
It outlines the process in general terms.
It does not include repetition or treatment of experimental control.
It does not describe the equipment needed to inoculate the agar plate or how to inoculate the plate.
It does not specify incubation time.
It does not describe how to count the colonies or where to record the data.

2 *Answers will vary.*

WS 9.4 PAGE 125

1 a
- Marshall had an endoscopy to confirm he was negative for *Helicobacter pylori*.
- He drank a suspension containing *H. pylori*.
- After five days he developed symptoms of gastritis: bloating, decreased appetite, bad breath and vomiting water.
- An endoscopy confirmed severe active gastritis.

- He took antibiotics after 14 days; bacteria were no longer visible in culture.

b Koch's postulates state that a microorganism must be present in the host with the disease and must be able to be isolated and cultured. When a sample of the pure culture is inoculated into a healthy host, that host must develop the disease and, when the microbe is taken from the second host, it must be identified as the same pathogen present in the original host. In this example, Dr Marshall is acting as the first host and the gerbil is acting as the second host. Both developed the same disease and the same bacterium was present, therefore fulfilling the postulates.

2 a Dependent variable: Growth of microbes

Independent variable: Presence or absence of the swan neck

b *Answers will vary. Example answers:* Size of flask, amount and type of broth, time flasks were boiled for

c *Answers will vary. Example answer:* Increasing the amount of broth may have made more nutrients available for microbial growth, and therefore may have encouraged additional growth. It was therefore essential that the amount of broth was kept constant.

~continue in right column

d Microbes will grow in the flask without the swan neck, while no microbial growth will be present in the swan-necked flask.

e Pasteur could have repeated his investigation three times.

f Pasteur's investigation supported the theory that the microbes present in air needed to be present for decay to occur and that microbes could not spontaneously generate.

WS 9.5 PAGE 127

1 a Yellow spot, stripe rust and *Septoria nodorum*

b *Answers will vary according to disease selected*

c i Root and crown fungi

ii Northern

iii 72.8%

iv 38.3%

WS 9.6 PAGE 130

1 Adhesion is the first step in establishing infection. For example, viruses can use surface proteins to 'stick' to host cells. Invasion involves entry of the cell. For example, some bacteria use enzymes to break down host cell contents.

2

Adhesion structure	Type of pathogen	Adhesion or invasion?	Labelled diagram and/or description: how does it work?
Surface proteins	Virus	Adhesion	Adhere to host cell surface receptors 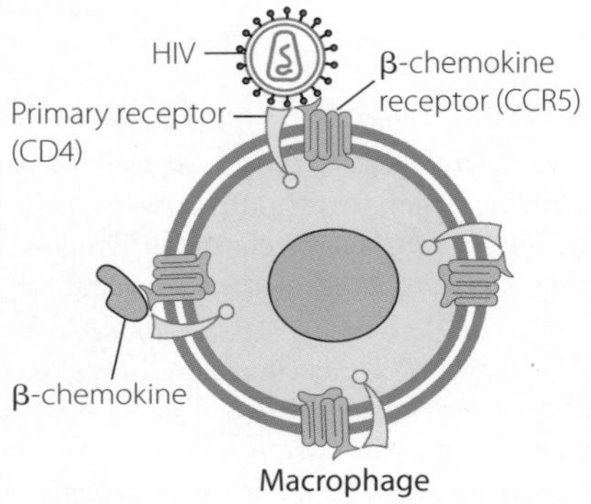 HIV; Primary receptor (CD4); β-chemokine receptor (CCR5); β-chemokine; Macrophage Source: *Molecular Biology of the Cell*, 4th edition. Alberts B, Johnson A, Lewis J, et al. New York: Garland Science; 2002.
Pili	Bacteria	Adhesion	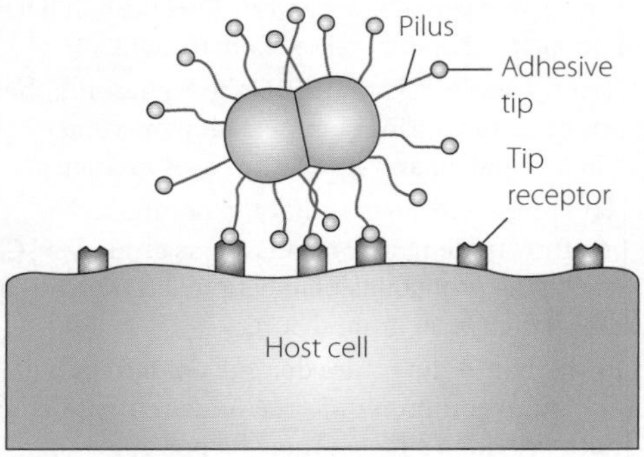 Pilus; Adhesive tip; Tip receptor; Host cell *Microbiology* by Gary Kaiser, Libretexts, licensed under Creative Commons 4.0, https://creativecommons.org/licenses/by/4.0/ Pili are thin protein tubes that extend from the bacterial cell. Pili allow the bacterial cell to adhere to the host cell.
Fimbriae	Bacteria	Adhesion	Fimbriae are thin protein tubes that extend from the bacterial cell. They tend to be shorter and more numerous than pili. These structures allow the bacterial cell to adhere to the host cell.
Microtubule protrusion	Protozoan	Invasion	Microtubule protrusion into the host cell allows the protozoan to enter the host cell.
Secretion of hydrolytic enzymes	Fungi	Invasion	Damages host cells and provides nutrients to the fungus

3

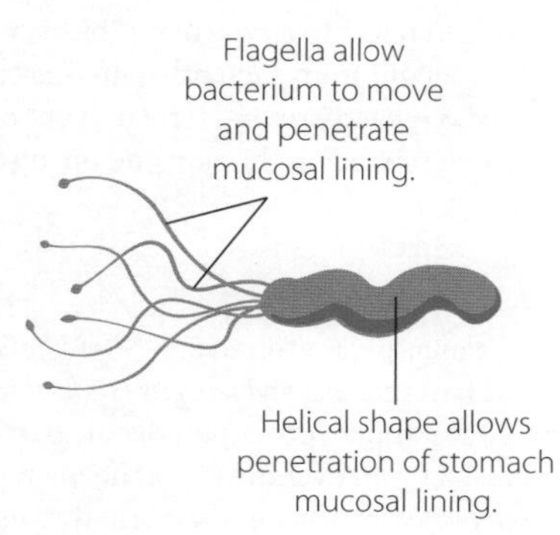

4 *Answers will vary. Example answer:* The influenza virus can be transmitted indirectly when a host sneezes and sends droplets into the air. An influenza virus infection can cause irritation of mucous membranes, which results in sneezing, and therefore increases the chances of the virus being transmitted to a new host.

Chapter 10: Responses to pathogens

WS 10.1 PAGE 132

1 a–iii; b–vi; c–iv; d–ii; e–i; f–v

2 Xylem vessels are hollow tubes that allow water to move in one direction through a plant from root to leaf. Unlike phloem vessels, which are joined by sieve tubes, these vessels are more like hollow straws, which would make it easy for a pathogen to spread rapidly throughout the plant. Tyloses block the xylem vessels and therefore limit the spread of the disease. The tylose also blocks the flow of water to any areas beyond the barrier and therefore would kill any part of the plant beyond the tylose.

3

Chemical plant defence	Named example	Mode of action
Hydrolytic enzymes	Lysozymes	Digest cell wall of bacteria
Alkaloids	Nicotine, caffeine	Inhibits or activates enzymes and affects metabolic reactions
Terpenoids	Menthols	Found in mint and have antibiotic and antifungal properties
Phenols	Tannins	Bind to insect salivary digestive enzymes and eventually kill insects
Defensive peptides	Defensins	Inhibit protein channels in pathogen cell membrane, resulting in cell death

WS 10.2 PAGE 134

1

Term	Definition	Type(s) of pathogens
Biotroph	An infection that does not kill the host but hinders its usual function	Bacteria and viruses
Necrotroph	An infection that results in the death of the host	Fungi

2 a The main symptoms of the disease are yellowing and wilting of leaves and splitting of stems.

b Xylem carries water from the roots throughout the plant. If the xylem is damaged, water cannot pass through the plant and so cannot reach areas where it is needed for photosynthesis and to fill vacuoles.

Phloem moves sugars and nutrients around the plant. If these vessels are damaged, aerobic respiration, growth and reproduction of cells will be limited.

3 If all of the plants have very similar genes then they will have limited genetic diversity. This means that if one plant is affected by the fungus, then all the plants will be. The disease will easily spread through entire regions and cause an economic disaster for farmers.

4 a: Healthy cell

b: Fungal pathogen attaches to the cell

c: Fungal pathogen sends hyphae into the cell through cell wall

d: Antimicrobial molecules inside the cell are triggered, signalling the cell wall to seal off the infected cell

e: Cell wall increases in thickness, adding lignin and more cellulose to the structure, sealing the pathogen inside

f: Infected cell dies and releases the signalling molecule methylsalicylic acid, which triggers healthy cells to produce antipathogenic chemicals

WS 10.3 PAGE 136

1 Left box: Environmental bacteria falls onto the stratum corneum and begins to grow

Centre box: Outer layer of the stratum corneum begins to shed

Right box: Outer layer of the stratum corneum is fully shed and replaced by the layer of cells underneath

2 a

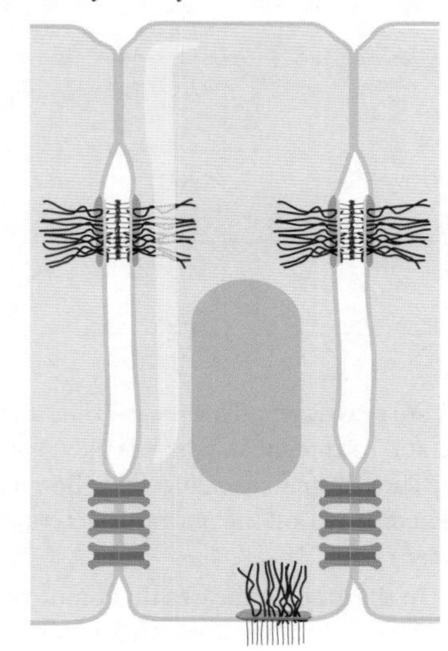

Cell junctions are tightly packed together in the mucous membranes, which restricts the entry of pathogens into the body.

b & c

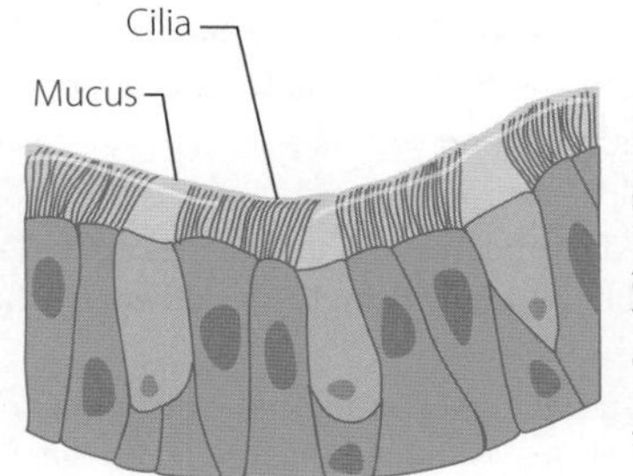

Alamy Stock Photo/Science History Images

Mucus is a liquid substance produced by cells that line the mucous membranes. It keeps the cells moist and collects dust, pollen and pathogens. It works in conjunction with the cilia structures to move trapped particles out of the lungs.

Cilia are muscular hair-like structures that line the respiratory system. The cells to which they are attached beat to move a layer of mucus up and out of the lungs, where it can be swallowed or coughed up.

3 a

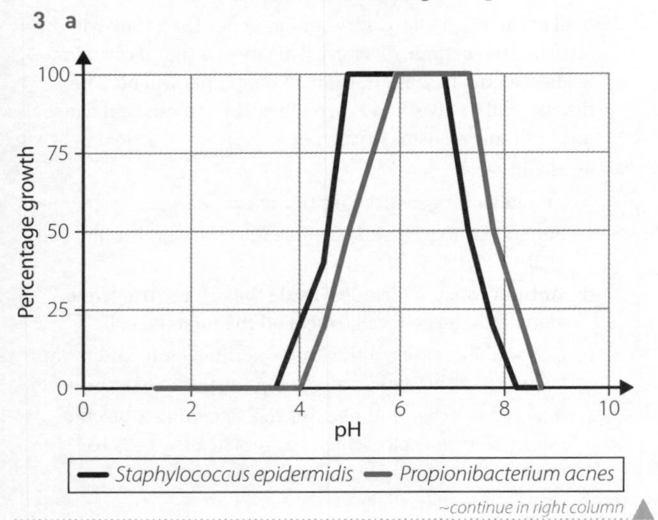

~continue in right column

b Commensalism refers to a type of symbiosis where both parties benefit from each other. In this case the *Staphylococcus epidermidis* bacteria prevent other bacteria from growing on the skin and are also given a place to grow.

c pH 5–7

d pH 6–7.5

e The skin's regular pH is around 5.4–5.9. At this pH the commensal bacteria *Staphylococcus epidermidis* can thrive and outcompete other pathogenic bacteria. Acne is caused by an overgrowth of the bacteria *Propionibacterium acnes*. This bacteria thrives at a higher pH of 6–7.5. If regular soap is used on the skin it increases the pH of the skin, making it less hospitable for *Staphylococcus epidermidis* and more hospitable for *Propionibacterium acnes*. Therefore, it would be more beneficial to use a slightly acidic, specialised face wash.

WS 10.4 PAGE 139

1

	Location	How does it prevent infection?
Saliva	Mouth	Saliva is slightly alkaline and contains lysozymes that can help to break down bacteria.
Tears	Eyes	Tears wash eyes and contain lysozymes.
Urine	Urethra	Urine cleans and flushes the lower urinary tract. It is sterile and its acidity inhibits growth of bacteria.
Sebum and sweat	Skin	Sebum and sweat create a waterproof layer of fatty acids across the skin. They maintain the slightly acidic pH of the skin at around 5.5 pH.
Gastric juices	Stomach	The stomach wall secretes gastric juice, which contains hydrochloric acid and protein-digesting enzymes that destroy pathogens in the stomach.

2 a Vultures are scavengers. This means that they regularly eat dead and rotting meat, which is filled with harmful bacteria. Their stomach acid needs to be very acidic (i.e. low pH) in order to kill the harmful bacteria before it makes them sick.

b The data in the graph indicates that omnivores and carnivores have a stomach pH range of about 1.2–3.5, whereas herbivores have a stomach pH of 4 and above. As human stomach pH is around 1.5, this suggest that humans are biologically 'meant' to eat meat.

WS 10.5 PAGE 141

1 Body is breached: The skin or other physical barrier is penetrated and pathogens enter the body.

Vasodilation: Blood vessels widen, increasing blood flow to the area.

Increased permeability: The walls of capillaries become more permeable, and fluid from the blood enters the interstitial space.

Phagocytes: Phagocytes leave the blood and enter the interstitial spaces to fight invading pathogens.

Diagrams will vary.

2

Symptom	Cause
Redness	This is due to vasodilation and increased circulation in the area. The redness is a result of an increase in red blood cells in the area.
Heat	As blood flow increases, more heat is brought to the area in the blood.
Swelling	Swelling is due to the increased permeability of the blood vessels, resulting in increased fluid in interstitial spaces.
Pain	Pain is due to swelling and the release of chemicals such as prostaglandins and serotonin.

3 a Phagocytes are only effective against pathogens that have entered the body. This means that the pathogens have already made their way past the skin or other physical barrier. Physical and chemical barriers could be classed

9780170449625

as the first line of defence, whereas phagocytes are the second line of defence against pathogens.

b

Macrophages	Similarities	Dendritic cells
Found in the circulatory system and can be released into the interstitial fluid when inflammation occurs	Break down pathogens through the process of phagocytosis Can release cytokines to recruit other immune cells into the area	Antigen presenting to activate the adaptive immune response, involved in the activation of T-cells Found in tissues that come into contact with the outside environment, such as mucous membranes in the nose, stomach, lungs and intestines

c A: Invading bacterium is identified as 'non-self'

B: Phagocyte begins to change shape, completely surrounding the invading bacterium with its membrane

C: Lysosomes inside the phagocyte release digestive enzymes into the vesicle containing the bacterium

D: The bacterium is destroyed, harmless particles are released from the phagocyte and the vesicle moves to become part of the plasma membrane

4 a When macrophages sense pathogens they release cytokines. These cytokines stimulate the hypothalamus in the brain to produce prostaglandins, which trigger a fever.

~continue in right column

b Enzymes are proteins that catalyse all reactions inside cells. These enzymes have an optimum temperature range where they function at their best. If temperatures are increased too much past the optimum temperature range, the enzymes may denature, which results in a change of shape to their active site. If many enzymes are denatured and can no longer catalyse reactions, cell functions could be hindered, resulting in illness and even death.

Chapter 11: Immunity

WS 11.1 PAGE 144

1 innate; chemical; physical; phagocytosis; inflammation; fever; penetrate; adaptive

2

Innate	Similarities	Adaptive
Targets a specific area of the body to protect Responds in the same generalist way to all pathogens Does not have a memory of past infections Present at birth	Uses chemicals and cells to kill pathogens	Can fight throughout the whole body at once Has a specific response to an individual pathogen Remembers specific pathogens and how to fight them Develops over a person's lifetime

3 The immune system uses antigens to trigger an immune response. Without the presence of an antigen, the body's immune response would not be triggered and therefore a pathogen would be able to attack and kill a host organism.

4

Type of cell	Where do these cells develop and mature?	Types of cells found in the body	How does this cell destroy pathogens?	Humoral or cell-mediated?
B cell	Develop: Bone marrow Mature: Bone marrow	Plasma cells Memory cells	Uses antibodies to bind to antigens to attack the pathogens from the outside	Humoral
T cell	Develop: Bone marrow Mature: Thymus	Cytotoxic T cells Memory T cells Helper T cells Suppressor T cells	Binds to pathogens and releases chemicals and enzymes to destroy the pathogen from the inside	Cell-mediated

WS 11.2 PAGE 146

1 Antibodies each have a specific shape, just like enzymes. The binding site of an antibody suits the exact shape of the antigen that it corresponds with; therefore antibodies can only bind with one type of antigen.

2 If B cells with autoantibodies are allowed to enter the blood stream of the organism, they will begin to attack 'self' cells. This could lead to major damage of organs and tissues, severely compromising the health of the individual.

3

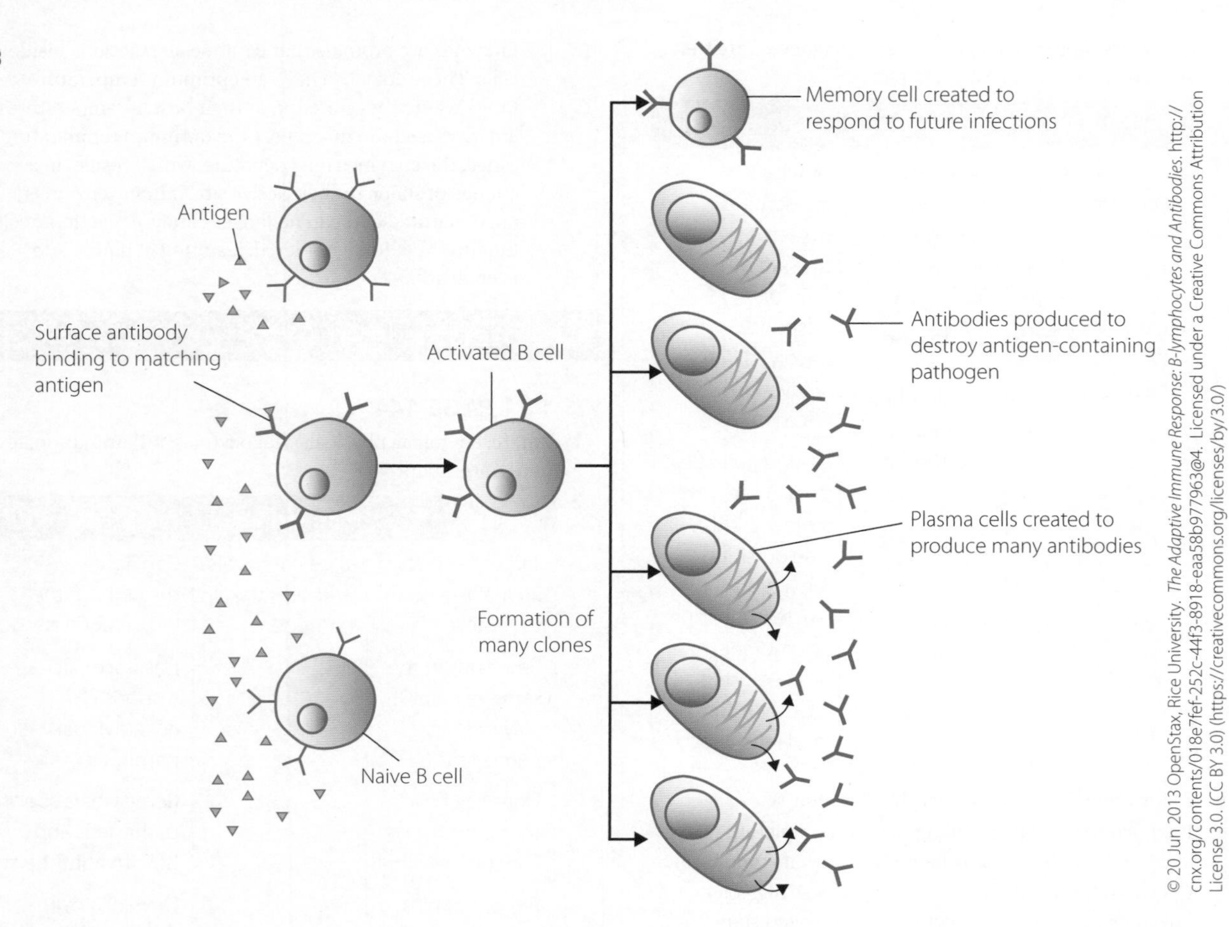

4 Each B cell has a specific antibody on its surface that corresponds to an antigen found on a pathogen. These antibodies can inactivate or destroy invading pathogens. If a naive B cell is not exposed to an antigen it will die within a few weeks. Memory cells are formed if the naive B cell is triggered, and they remain in the body long after the pathogen has been destroyed to allow a much more rapid response on subsequent infection compared to the initial infection.

5 a

Phase 1
Phase 2
Phase 3
Antibody levels in blood (arbitrary units)
100
80
60
40
20
0
0 10 20 30 40
Days since exposure to pathogen

b Antibodies are part of the third line of defence in the body, meaning that they are not triggered until a pathogen has overwhelmed the first and second lines of defence. First, a pathogen has to penetrate the physical and chemical barriers in the first line of defence. It then has to survive the biochemical process of fever and overwhelm the phagocytes brought to the site of infection by inflammation in the second line of defence. A pathogen then has to come into contact with a B cell with the corresponding antibody matching its antigen. This B cell replicates and begins to produce antibodies into the blood. This final process is represented around day 6 on the graph.

6 a The first exposure shows antibodies increasing from day 10, peaking at day 20 and then decreasing. The second exposure shows antibody levels increasing from around day 4, peaking around day 10 and being sustained for longer. This is due to acquired immunity and memory B and T cells. The second exposure triggered a much faster, stronger and prolonged response as the body had already been exposed to the pathogen.

9780170449625

b

First exposure	Similarities	Second exposure
Antibodies start to be produced around day 10 Antibodies peak around day 20 Antibody quantity begins to reduce after peak	Both exposures show an increase in production, a peak in production and a decline	Antibodies start to be produced around day 4 Antibodies peak around day 10 Antibody quantity continues at a high level for around 15 days after peak

WS 11.3 PAGE 150

1

Process description	Diagram of process
Neutralisation of bacterial toxins: antibodies bind to bacterial toxins, blocking the action of the toxin	
Agglutination: antibodies bind antigens on the surface of the cell and form antigen–antibody complexes, activating phagocytes and the complement cascade, leading to antigen/cell destruction	
Opsonisation: bound antibodies tag pathogens for destruction by phagocytes	

Source: Pranzatelli MR, *The immunopharmacology of the opsoclonus-myoclonus syndrome.* Clinical Neuropharmacology 19(1): 1-47, 1996

2 a Newborn babies have not been exposed to pathogens, so their adaptive immunity has not developed. Theoretically, they could be exposed to many pathogens each day that could overwhelm their immune system. By taking antibodies from their mother via colostrum, the baby will be protected from many diseases for a short time. This boost in immunity allows the baby to slowly develop their adaptive immunity over time.

b The baby is not being exposed to antigens, and therefore its B cells are not being activated, so they will not produce memory B cells. Therefore, the baby will not build up an immunity to disease. The antibodies given by the mother will protect the baby while they are present (this is known as passive immunity), but will not offer long-term protection.

WS 11.4 PAGE 152

1 a Step 1 depicts a macrophage undergoing phagocytosis of an infected cell. Macrophages are part of the innate immune response. Antigens from this infected cell are then presented on the surface of the macrophage. Step 2 then depicts the same macrophage presenting the invading pathogen's antigen to the memory T cells and helper T cells, triggering the adaptive cell-mediated immune response.

b Step 4 shows killer T cells identifying an infected cell by binding with the antigen–MHCI complex. The killer T cell then releases chemicals into the cell, which triggers cell death.

c Step 2

2 Helper T cells bind with the antigen–MHCII complex on the surface of antigen-presenting macrophages. This then stimulates them to release cytokines. Cytokines trigger the production of cytotoxic T cells and memory T cells, as well as triggering B cells to replicate and differentiate into plasma and memory cells.

3

Cytokines	Similarities	Foreign antigens
Cytokines are produced by body cells to signal infection and trigger both T cells and B cells	Trigger an immune response	Foreign antigens are present on pathogens and trigger B cells and T cells

Chapter 12: Prevention, treatment and control

WS 12.1 PAGE 154

1

Factor	Description of how factor influences the spread of disease	Scale of impact (local, regional, global)
Waste disposal	Poor disposal of household waste or sewage can encourage the spread of disease; e.g. dysentery or cholera spread through contaminated water.	Local
Antibiotic misuse	Widespread misuse of antibiotics has resulted in the development of antibiotic resistant bacteria around the world.	Global
Regional geography	Local topography can encourage or discourage population movement or settlement. People in isolated areas are less likely to contract diseases.	Regional
Overcrowding	People living in overcrowded situations are more likely to transmit diseases.	Local
Food trade	The movement of contaminated food can spread infectious disease, such as hepatitis A or salmonella.	Regional
Seasonal climate variations	Rainfall and temperature levels can influence vector home ranges. For example, mosquitoes require warm, moist environments; people living in these areas are more susceptible to diseases such as malaria and dengue fever.	Regional
Population migration	Population movement can significantly increase the spread of disease and make disease control measures difficult to implement.	Global
Local cultural practices	Some practices do not follow western hygiene standards. For example, exhuming dead relatives can spread diseases such as the plague.	Local

2 a Map 1 shows the distribution of antibiotic use around the world in 2010 and map 2 shows the resistance of the bacteria *Klebsiella pneumoniae* to the antibiotic carbapenems.

b Global regions characterised by high antibiotic use tend to also be the areas where the antibiotic resistance is found. For example, some parts of southern Europe, central Europe and south-eastern Europe, plus Turkey, have a high use of antibiotics at around 15–20 000 defined daily doses per population of 1000 and this region also has a high carbapenems resistance ranging from 25–75%.

9780170449625

WS 12.2 PAGE 156

1

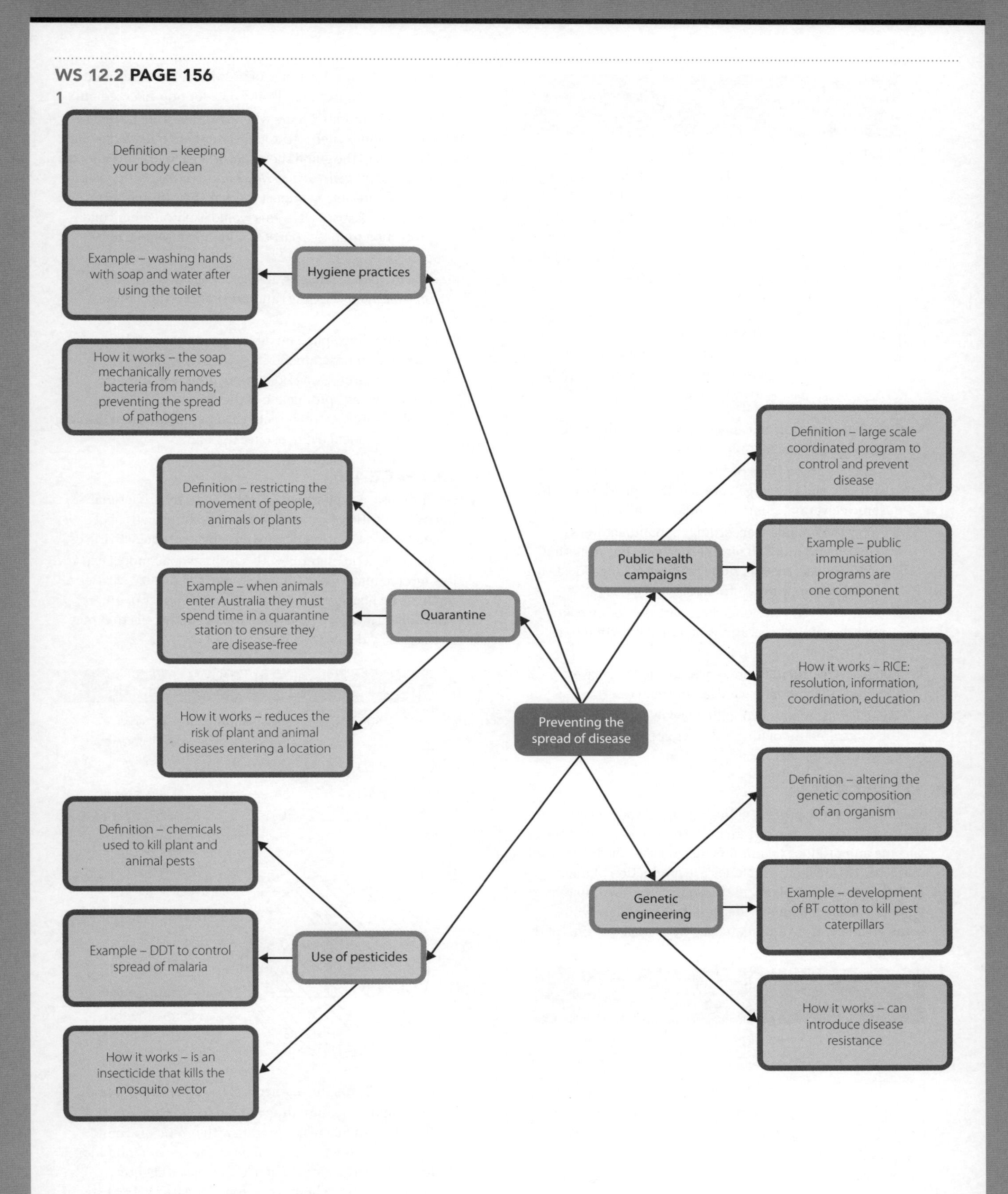

2 a

Risk assessment		
Identify risk	**Assess risk**	**Control risk**
Flint 500	It is harmful if inhaled and can cause skin irritation.	Wear gloves, safety goggles and a mask when applying. Wash skin with soapy water if it comes into contact.
Sun	Working in the hot sun can cause sunburn and dehydration.	Wear a hat and long-sleeved clothing. Apply sunscreen.

1 Prepare the fungicide in a spray bottle with water, according to the manufacturer's instructions.
2 Select 18 grape vines of approximately the same size and age, and growing in areas with approximately the same elevation, slope and sunlight availability.
3 Apply the fungicide according to manufacturer's recommendations to nine of the vines, ensuring that each vine receives the same amount of fungicide. Leave the other nine vines untreated.
4 Observe all 18 vines on a daily basis for four weeks, recording any observations of fungal growth into a results table.

b Reliable: Treatments were repeated: nine vines were given the fungicide and nine were left untreated.
Valid: Environmental conditions were controlled and the same amount of fungicide was applied to the experimental vines.

WS 12.3 PAGE 158

1 Vaccination is the process of introducing a vaccine into the body. The immune system in the body will react to the vaccine and produce an immune response to the introduced antigens. This response includes the production of memory cells that confer immunity in the body. There are different types of vaccines: some are delivered via injection, while some are made from attenuated, or weakened, cultures of the pathogen.

2

Passive immunity	Similarities	Active immunity
Short lived: weeks to months A person is given antibodies, rather than acquiring them; e.g. newborn babies from their mother	Provide protection against antigens Involve obtaining antibodies	Long lived: can be life-long Antibodies produced from exposure to a disease or via immunisation

3 a *Answers will vary. Example answer:* They could have incubated the bacteria in the different milks at the same temperatures.

b In each type of milk tested, the percentage of Kato III cells that tested positive was always lower when breast milk from an infected woman was used. For example, in column 3, the percentage of positive cells was around 45% for the infected milk and 68% for non-infected milk.

c Passive immunity occurs when antibodies are passed to an individual, such as through a mother's milk to her child. The data in the graph show that when a mother is infected with a pathogen, such as *Helicobacter pylori*, antibodies in her milk are able to impact the way that this pathogen can adhere to gastric cells – this would likely offer her child protection from this particular pathogen were it to be passed on.

4 a The incidence and prevalence of cheesy gland based on treatment (dipping) and prevention (vaccination) methods

b Recommended plan: no dipping and the complete vaccination program. Both of these have the lowest infection rates. Using no dipping produced the lowest average prevalence of the disease at 14%, and implementing the complete vaccination program resulted in a 3% incidence rate of cheesy gland.

WS 12.4 PAGE 160

1 An antibiotic is a pharmaceutical used to treat bacterial infections. Antibiotics work by slowing the growth of or killing bacteria. Antibiotics are not effective against viruses.

2 A virus is genetic material with a protein coat and, because it is not technically alive, it cannot be 'killed'. Also, viruses replicate by hijacking the replication machinery of other cells. Killing a virus would therefore kill body cells that may be needed by the organism.

3 a

Variable	Graph A	Graph B
Independent	Time (days)	Diameter of inhibition zone (mm)
Dependent	Temperature (°C)	Diameter of inhibition zone (mm)

b

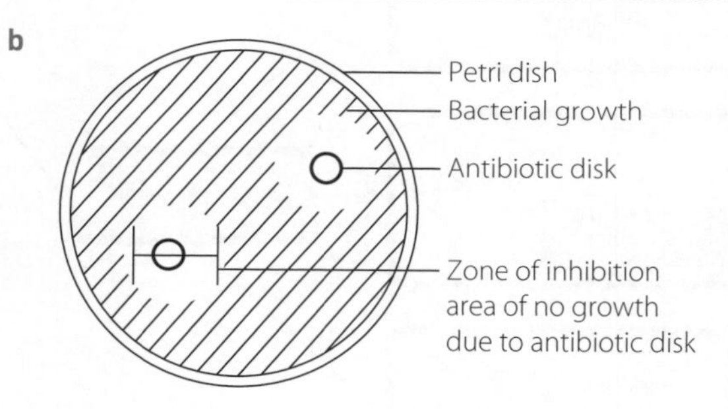

c Graph A shows the diameter of the zone of inhibition over time for garlic extract stored in the dark and at 4°C. Both treatments of garlic start with a zone of around 22 mm in diameter. Over 10 days, the zone of inhibition for garlic extract stored at 4°C decreased slightly to around 21 mm, whereas the zone for garlic extract stored in the dark decreased more quickly down to around 11 mm after 10 days.
Graph B shows the zone of inhibition for garlic extract heated at different temperatures. At about 22°C the zone of inhibition is around 19 mm, and this decreases as temperature increases, with the zone at around 15 mm at 80°C and no zone of inhibition for garlic treated at 100°C.

d The results of the study suggest that the best way to prevent the growth of *E. coli* is to store garlic extract at 4°C and then heat treat the garlic extract at 22°C. These

9780170449625

treatments resulted in the largest zones of inhibition, at around 22 mm and 19 mm, respectively.

WS 12.5 PAGE 162

1 An epidemic is an outbreak of an infectious disease in a community at a particular time. A pandemic is an outbreak of an infectious disease that affects an entire country or the whole world.

2 a Zika virus, member of *Flavivirus* family

b Vector: *Aedes* mosquito, usually *Aedes aegypti*. It can also be transmitted through organ transplantation, blood transfusion or sexual contact, or from mother to foetus during pregnancy.

c 3–15 days

d Most people don't develop symptoms. Symptoms are mild and include headache, fever, rash, conjunctivitis, muscular pain and feeling unwell. Symptoms tend to last 2–7 days.

e There is no available treatment.

3 a Graph 1 shows the number of confirmed and expected cases in Brazil from the end of 2015 until the beginning of 2017. Case numbers peak at the beginning of 2016, with suspected cases much higher than confirmed cases.

Graph 2 shows the number of Zika cases in the 1000s during 2016, across the five Brazilian regions. Cases were highest in epidemiological week 10.

b i 21 000 cases

ii 11 000 cases

iii 500 cases

iv South

4 a
- Vector control: 20 million homes were fumigated as the mosquito *A. aegypti* breeds successfully in urban areas, especially in artificial containers such as bottle caps. The mosquito also hides in dark places inside homes.
- The military and community health agencies collaborated to educate and collect data in remote areas. Education included advising residents about how to reduce available mosquito breeding grounds, anti-Zika carnival parades and establishment of cleaning-up Saturdays.
- Institutions such as the Fundacao Oswaldo Cruz conducted disease research.
- Authorities requested the US to support the development of a vaccine.
- Release of GM mosquitoes to breed with the wild mosquitoes and produce sterile offspring. One neighbourhood reported an 82% reduction in mosquito population using this method.
- People were not allowed to be blood donors if diagnosed with Zika.

b *Answers will vary. Data must first be determined to be accurate and reliable before validity can be assessed. Data cannot be valid if it is not accurate and reliable. It is also important that the data are relevant, and that any referenced primary data were collected using appropriate scientific methods.*

c Criteria: For environmental management strategies to be effective, they must contribute to reducing the number of Zika virus cases in Brazil.

Control measures implemented:
- 20 million homes were fumigated to kill the vector, the mosquito *A. aegypti*
- Education of the community – visiting remote areas, carnivals and clean-up days
- Publically funded disease research, including the development of a vaccine
- Release of GM mosquitoes resulted in an 82% reduction in mosquito population
- Prohibition of being a blood donor if diagnosed with Zika – disease known to be transmitted via blood transfusion

Evidence of success:

The graphs show that after a peak of confirmed Zika cases in epidemiological week 10 in 2016, Zika virus cases in Brazil significantly reduced by week 20, with low numbers of cases present for the rest of 2016.

Judgement: The environmental management strategies were successful in controlling the 2015 Zika virus epidemic in Brazil.

WS 12.6 PAGE 166

1 Disease incidence refers to the number of new cases in a specific time, whereas the prevalence is the number of all cases of disease. Health management authorities can use these data to make decisions about how to distribute resources to provide healthcare to the sick.

2 a The prevalence of dengue increased from just below 20 000 in the year 2000 to around 138 000 in 2001. Prevalence then decreased until 2004 after which it started to increase again until 2008, when prevalence was around 90 000. Numbers fluctuated between 2008 and 2011.

b Incidence and prevalence of dengue are closely related – as prevalence increases, incidence increases, and as prevalence decreases, incidence also decreases.

c

Year	Incidence per 100 000	Calculated incidence (%)
2000	25	0.025
2003	100	0.1
2006	75	0.075
2009	90	0.09

d i 5–9 year olds, and 10–14 year olds

ii 45–54 year olds, and the 65+ age group

iii 5–9 year olds

iv 130 per 100 000

v 2000

3 a While the incidence in both provinces fluctuated between 2010 and 2013, the incidence of dengue fever in Lakhonpheng district is consistently higher when compared to Manchakhiri district. For example, in 2010 the incidence in Lakhonpheng district was around 2649, compared with around 170 in Manchakhiri district.

b Yes, the socioeconomic data does support trends seen in the dengue incidence data in the two areas. Graph 3 shows that dengue fever in Lakhonpheng district is consistently higher when compared with Manchakhiri district. Households in both rural and suburban areas in the former province were characterised by higher room occupancy, larger proportion of households classified as poor and most houses made of wood.

WS 12.7 PAGE 169

1

Health management category	Implemented during 1918 pandemic?	Details of strategies implemented	Comment on effectiveness of strategy/strategies
Hygiene	Y	Good personal hygiene, use of disinfectants Some doctors and nurses wore masks when treating infected patients Some places promoted fresh air, sunlight	Applied unevenly – reduced effectiveness Evidence from a Boston hospital suggests strict hygiene standards prevent transmission to staff and patients
Quarantine	Y	Isolation and quarantine used in some locations	Applied unevenly – reduced effectiveness
Vaccination	N	Not available	N/A
Public health campaigns	Y	Limitations of public gatherings Dance halls closed in some cities Some streetcar conductors told to keep car windows open in rain Some areas moved court cases outside In the USA, congress approved funding to recruit additional doctors and nurses Conversion of schools, homes and other buildings to makeshift hospitals Some places ordered residents to wear masks Schools, churches, libraries and theatres closed in some places Advice to stay inside and not shake hands Regulations passed to ban spitting	Applied unevenly – reduced effectiveness
Pesticides	N	Not relevant	N/A
Genetic engineering	N	Not available/not relevant	N/A
Pharmaceuticals	Y	Antivirals not available Antibiotics to treat secondary bacterial infections not available Some doctors prescribed aspirin (a newly patented drug) at lethally high doses	Overdose of aspirin was fatal to some

WS 12.8 PAGE 170

1 Leaves can be chewed or boiled.

2

Compound	Medicinal properties
Tannin	Acts as an astringent, which contracts tissues; useful for bathing wounds and treating coughs
Oil	Antifungal and antibacterial properties; can be used to treat sore throats (e.g. tea tree oil)
Latex	Contains enzymes that can clean wounds and remove warts; can also hold wounds together; some can be poisonous
Mucilage	Hydrates and protects the skin
Alkaloid	Can be both poisonous and therapeutic; examples include caffeine, morphine, quinine and codeine; can be used for pain relief

9780170449625

3 *Answers will vary. Example answer:* Tea tree was used in a number of ways, including crushing the leaves and inhaling the vapour to treat headaches, brewing the leaves to help with sore throats or applying to wounds a poultice made from the leaves.

4 *Answers will vary. Example answer:* Tea tree oil is a commercially available product that is distilled from the leaves of *Melaleuca alternifolia*. It is used for cleaning, to treat head lice, to treat viral infections such as *Herpes labialis* and for its antibacterial properties through wound care or to treat gingivitis.

5 Points for:
- The application is allowed under Australian law.
- Large-scale commercialisation of *Pittosporum angustifolium* leaf extracts would allow many more people to have access to the health benefits of the plant.
- Possibility of employment of local Aboriginal and Torres Strait Islander peoples.

Points against:
- The application does not align with the values of Australian Aboriginal customary law.
- Under this law it is considered that the language would belong to all people, and so should not be owned by just two people.
- The lack of consultation with Aboriginal Elders was very disrespectful.
- The patent and trademark will prohibit Indigenous peoples and other businesses from using the plant's products. In turn, there will be a loss of employment and decreased availability of medicinal benefits to current users.

Judgement:

The trademark should not have been filed. The applicants did not consult with the traditional land-owners and the application did not align with the values of Australian Aboriginal customary law.

MODULE 7: CHECKING UNDERSTANDING PAGE 172

1 C

2 C

3 C

4 A

5 An infectious disease is caused when a pathogen invades a host. *Examples will vary.* For example, the disease whooping cough is caused by the bacteria *Bordetella pertussis* and results in the host experiencing a runny nose and noisy bouts of coughing, as well as gagging and vomiting.

6 Many pathogens are known to live in water. If a source of drinking water was to become contaminated with pathogens, then it may result in the transmission of an infectious disease. For example, the disease cholera is caused by the bacterium *Vibrio cholerae*. Symptoms can include diarrhea, vomiting and muscle cramps. Testing water samples for the presence of such pathogens can prevent the spread of disease.

7 The zone of inhibition refers to an area on an agar plate that is without microbial growth. It shows that a substance on the plate contains antimicrobial properties. The larger the zone, the stronger the antimicrobial properties of the substance.

8 Local factors relate to the neighbourhood, village or city scale and can include the cultural beliefs of the local people, or the quality of local infrastructure. For example, a locality with a poor waste management system is more likely to experience disease outbreaks. A build-up of household waste increases the risk of bacterial contamination, which may lead to gastrointestinal illness, or skin or lung infections.

9

Antivirals	Similarities	Antibiotics
Work on viruses Inhibit viral development inside cells, but do not kill the virus	Control infectious diseases caused by pathogens	Work on bacteria Some can kill the bacteria, while others inhibit growth

10 Point for: Tylose formation blocks the xylem vessels and minimises the spread of pathogens throughout the plant.

Point against: All parts of the plant connected to the xylem above the tylose formation will die due to a lack of water.

Judgement: Although part of the plant will die due to the tylose formation, the rest of the plant will be protected from the pathogen and so it is an effective strategy in maintaining the overall health of the plant.

11 Stomach acid is a chemical barrier that helps to prevent infection. If the potency of the stomach acid is reduced with the use of antacid tablets then the chemical barrier is compromised and could lead to infection.

12 a An antibody–antigen complex is formed, which activates naive B cells and begins replicating plasma cells and memory cells.

b Cell X is a memory B cell.

c Antibodies are produced, which bind with foreign antigens fighting the pathogen.

13 Phagocytes recognise the 'non-self' antigen on the bacterium surface. This causes the phagocyte to engulf the bacterium. Once engulfed, the bacterium is exposed to enzymes that break down the invading bacterium. Some of the bacterium's antigens are then presented to T cells, which activate the adaptive immune response.

MODULE EIGHT: NON-INFECTIOUS DISEASE AND DISORDERS

REVIEWING PRIOR KNOWLEDGE PAGE 175

1 a Temperature and pH

b During denaturation, the active site of the enzyme changes shape. Prior to denaturation the active site was the perfect shape for a specific substrate, but afterwards, because it has changed shape, it can no longer bind with that substrate, making the enzyme inactive.

2 Increasing an organism's temperature increases its metabolic processes up to a point. Past this point metabolic processes dramatically decrease. This point is the temperature at which enzymes denature. As enzymes catalyse all metabolic reactions, they can no longer perform their function and therefore all metabolic processes cease.

3 A mutation is an error in DNA replication resulting in incorrect base pairing and a change in DNA base sequence.

4 Genetic engineering involves the manipulation of genetic material to change the genome of an organism.

5

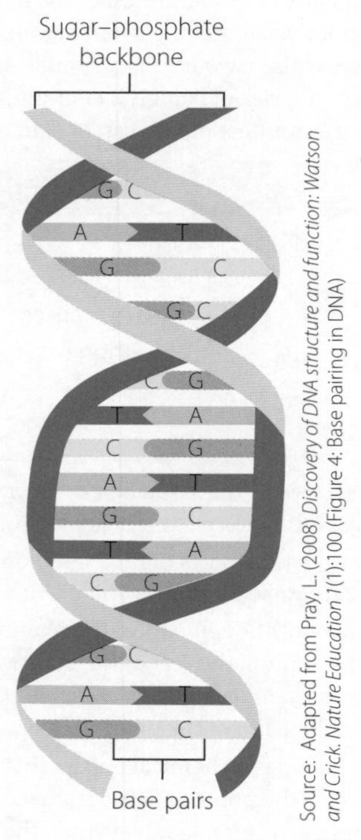

~continue in right column

6 B
7 C
8 C
9 A
10 C
11 A

Chapter 13: Homeostasis

WS 13.1 PAGE 177

1 e Retina is protected from bright light; c CNS detects bright light and sends message to effectors; a Looking at a bright light; d Pupils contract and eyelids blink; b Photoreceptors triggered

2 a The graph shows a regular fluctuation of levels increasing and decreasing around a central point. The central point is the ideal conditions inside the body; this could be temperature, water level or many other conditions. For this response the focus will be on temperature. As the internal temperature reaches an upper tolerance level, receptors trigger a response (such as sweating) to counteract the increase. This then leads to a fall in temperature. When the fall reaches a lower tolerance level, receptors trigger a response (such as shivering) to increase the internal temperature.

b *Example answer:*

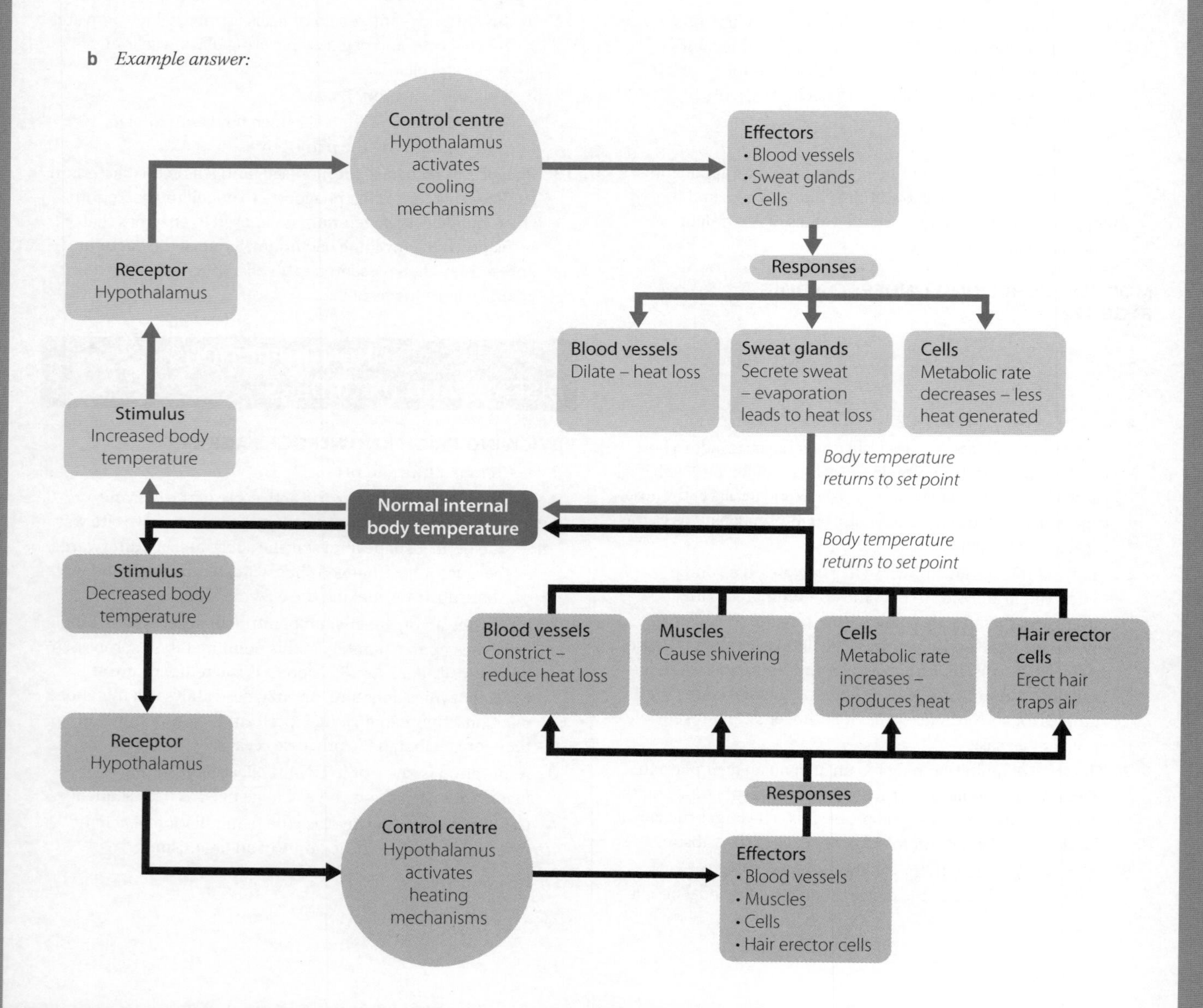

9780170449625

WS 13.2 PAGE 179

1 *Animal examples will vary.*

Response	Animal example	Response to hot or cold	Description of mechanism for cooling or heating
Sweating	Primates	Hot	Sweat glands release water onto the surface of the organism. The water evaporates, reducing the surface temperature of the organism.
Shivering	Horses	Cold	The contraction of muscles increases metabolic processes and so increases the heat generated inside the organism.
Contraction of pili muscles	Cat	Cold	As pili muscles contract, hair or fur on the surface of the organism stands erect. This traps a layer of air between the skin and the environment, insulating the organism from the cold.
Vasodilation	Human	Hot	Blood is brought to the surface, allowing heat to radiate away from the animal.
Licking forearms	Kangaroo	Hot	Forearms contain many blood vessels near the surface. As saliva evaporates, the blood is cooled.
Curling up	Dog	Cold	Reduces surface-area-to-volume ratio, preserving heat inside the organism and limiting exposure to the cold environment.

2 Licking forearms is a behaviour employed by kangaroos to lower body temperature and so could be classified as a behavioural adaptation. However, for the evaporation of saliva to effectively cool the organism, there has to be a large

number of dilated capillaries near the surface of the skin in order to cool the blood. Dilated capillaries is a physiological adaptation. Therefore, licking forearms can be classed as both behavioural and physiological.

3

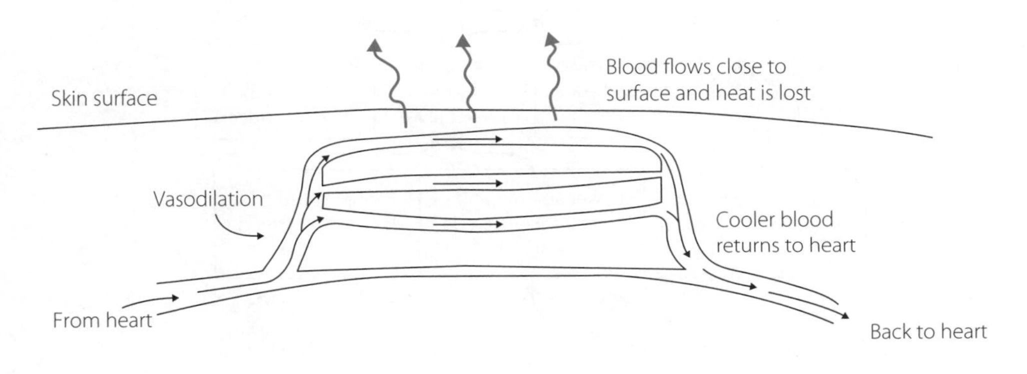

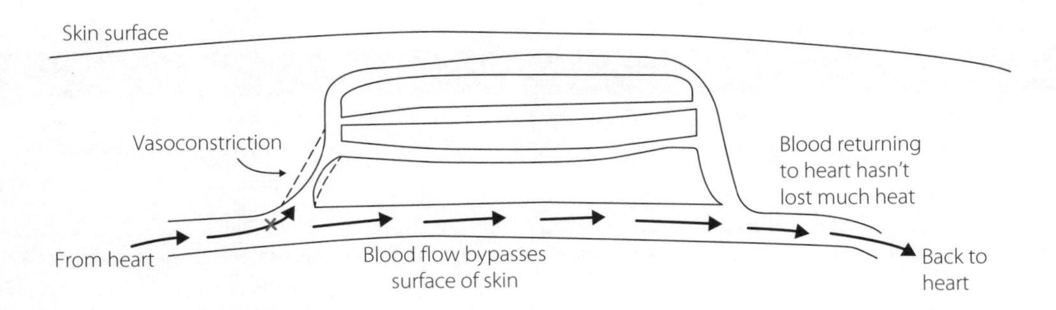

WS 13.3 PAGE 181

1 Too little glucose in the blood means that cells will not receive enough glucose to produce energy via respiration. Too much glucose in the blood can cause hyperglycaemia and damage blood vessels, which can lead to heart and kidney disease.

2 Aliyah would have had low blood sugar levels after sleeping all night. She then would have had a spike in her blood sugar as her body began to absorb glucose from her breakfast. While running, her cells would have required more glucose for respiration and so she would have shown a decrease in blood sugar.

3 a Insulin binds to the outside of cells and increases the absorption of glucose from the blood. This reduces blood sugar levels.

b Glucagon stimulates the liver to release glucose from stores. This increases blood sugar levels.

4 a & c

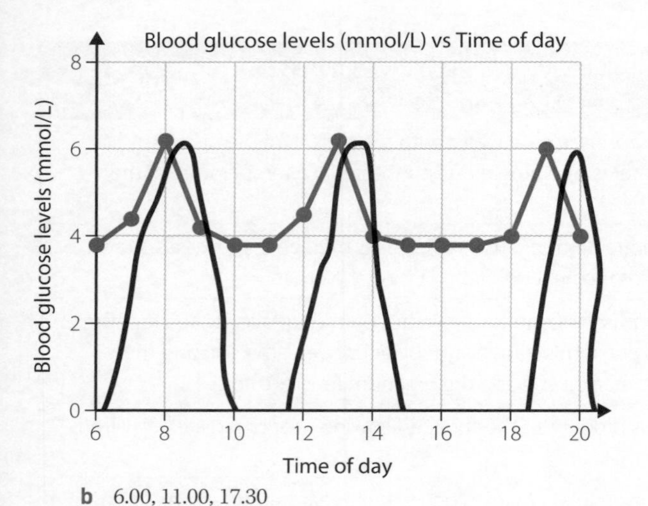

b 6.00, 11.00, 17.30

~continue in right column

WS 13.4 PAGE 183

1 Physiological: Urine is concentrated by countercurrent exchange in the loop of Henle. This returns more water to the blood of the rat and reduces the waste in urine.

Behavioural: The animal remains in a cool burrow during the day. This prevents excess evaporation from occurring.

2 The loop of Henle is the site of water reabsorption in the kidney. The longer the loop of Henle, the more water can be reabsorbed by the animal in its kidneys. Beavers live in aquatic environments and do not have to conserve water. This is why their loop of Henle is short. Rabbits live in many different environments, from grasslands to tundra, and so they have a longer loop of Henle in order to absorb a lot of water if needed. Kangaroo rats are desert animals, so water conservation is necessary for their survival. This has led to the development of a large loop of Henle in this mammal.

3

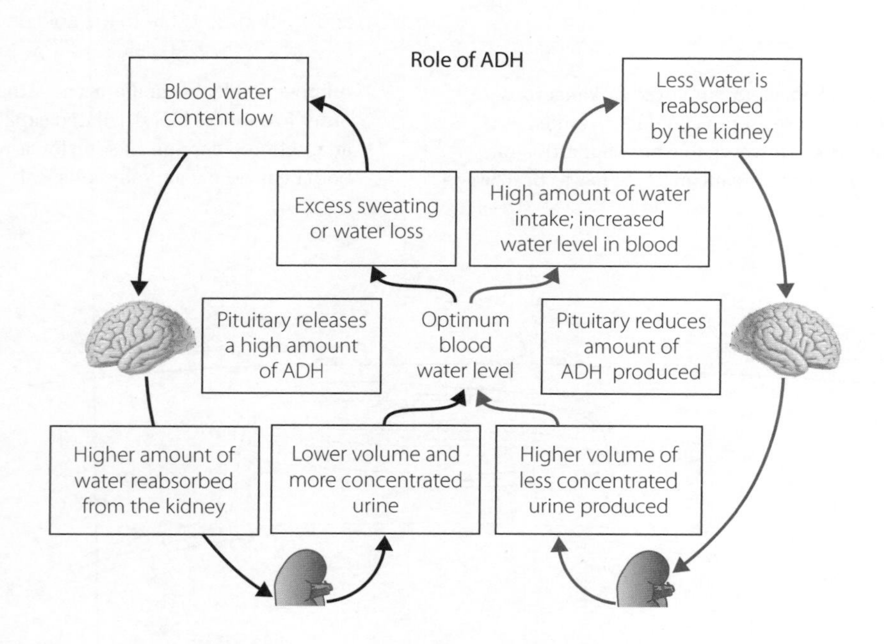

WS 13.5 PAGE 185

1 a *Example answer using* Acacia*:*

Adaptation	Description	How it limits water loss
Rolled leaves	Leaves are rolled inwards, keeping stomata sheltered.	This creates a humid microclimate, trapping moisture inside the rolled leaf and reducing the concentration gradient, therefore reducing evaporation.
Thick, waxy cuticle	Leaves have a thick, waterproof outer layer.	This protects the leaf from heat and also prevents transpiration from occurring.
Low growing	Plants grow low to the ground, which limits their exposure to the wind and increases shade.	Less exposure to the wind reduces transpiration, and shade keeps the plant cool, also reducing transpiration.
Opening stomata at night	Stomata open only at night.	Night is cooler than daytime in a desert, so less transpiration takes place.

9780170449625

b *Example answer using* Spartina alterniflora:

Adaptation	Description	How it limits water loss
Tissue partitioning	Salts can be concentrated in particular leaves, which then drop off the plant.	The plant moves excess salts into spare leaves that act as a waste container. In this way, the plant can maintain osmotic levels.
Salt excretion	Certain glands and bladders can fill with salt.	These glands and bladders remove salt from the inside of the plant and excrete it onto the outer surface, maintaining osmotic levels inside the plant.
Root exclusion	The structure of the root can prevent 95% of the salt from the soil from entering the root system.	This prevents water loss as the plant is able to control the levels of ions in the roots.

~continue in right column

WS 13.6 PAGE 187

1 a A change in action potential allows an electrochemical impulse to pass from the dendrite along the axon of a neuron. The myelin sheath acts as an insulator that increases the efficiency and speed of the impulse.
A damaged myelin sheath would result in slower electrochemical impulses, or the impulses could leak to other neurons close by, triggering an unwanted response.

b MS is an autoimmune disease, so the body is essentially attacking itself. An antigen–antibody complex is formed with 'self' antigens, and triggers B cells and T cells to attack non-foreign cells in the body.

2 Stimulation X is more immediate and short-lived, which indicates that it is a nervous stimulation. Stimulation Y takes longer to begin and lasts for longer, which suggests that it is hormonal because hormones remain in the bloodstream for long periods of time.

Chapter 14: Causes and effects

WS 14.1 PAGE 189

1 Genetic diseases result from mutations in chromosomes or genes. Mutations can alter the coding for a protein, and the changed or missing protein may cause a disease. Additionally, errors that occur during cell division can affect the number of chromosomes in a cell to create chromosomal abnormalities.

2

Category of disease	Name of disease	Cause of disease	Symptoms
Genetic	Down syndrome	Three copies of chromosome 21	Developmental delay, intellectual disability, distinct facial features
Environmental exposure *Answers will vary.*	Asbestosis	Inhalation of asbestos fibres	Inflammation, scarring, stiffening in lung tissue, breathing difficulty, cough, chest pain
Nutrition	Bulimia nervosa	Complex psychological factors	Extreme weight loss due to excessive strenuous exercise, self-induced vomiting, taking laxatives and appetite suppressants
Cancer	Melanoma	Skin cells divide uncontrollably due to DNA changes in genes that control cell division; often due to Sun exposure	Symptoms often vary Include change in size, colour, shape and texture of a mole, or a new lump appearing on the skin

3

Undernutrition	Similarities	Overnutrition
Can be caused by lack of minerals, vitamins or proteins Can be linked to psychological disorders (e.g. bulimia)	Occurs due to lack of balance and incorrect nutrients A form of malnutrition	Caused by consuming more kilojoules than energy expended Common in the developed world Can cause significant health impacts (e.g. type II diabetes)

4 a People aged 85 and older
b People aged 15 years and younger
c Australian Capital Territory
d Northern Territory
e *Answers will vary.* DALY measures can be used to compare the impact of disease on populations from different regions or countries, or the influence of other risk factors such as socioeconomic status. This can help management authorities to effectively distribute health care resources.

WS 14.2 PAGE 191

1

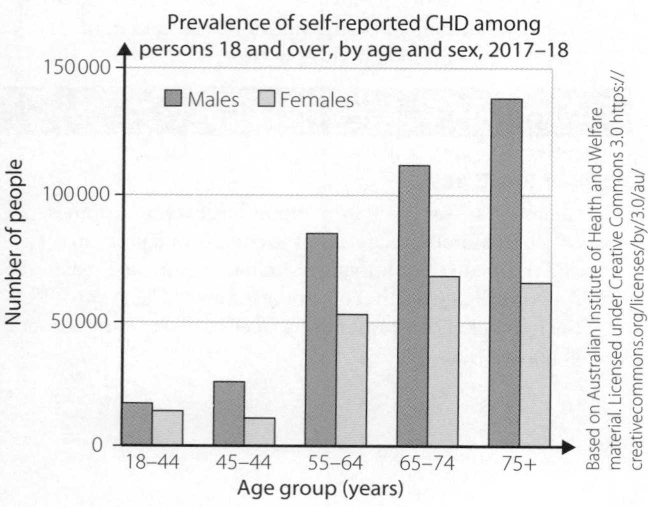

2 *Answers will vary.*

3

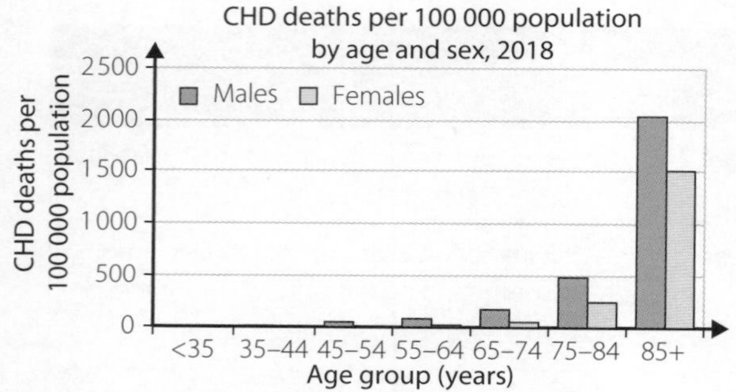

Based on Australian Institute of Health and Welfare material. Licensed under Creative Commons 3.0 https://creativecommons.org/licenses/by/3.0/au/

Chapter 15: Epidemiology

WS 15.1 PAGE 193

1 a Queensland
b 56.2 years
c 5.9 per 100 000
d 25.9 per 100 000
e 1610.3

2

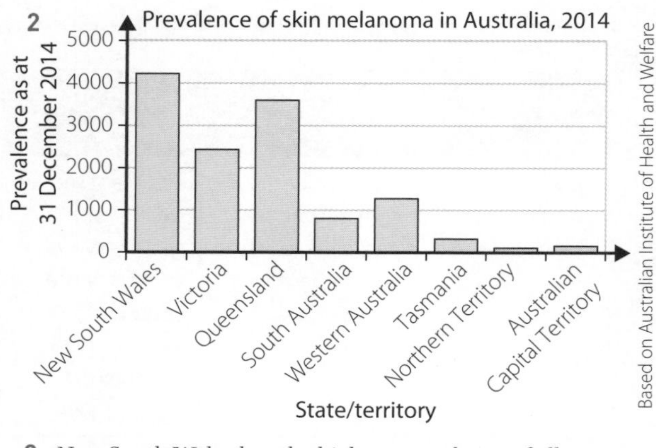

3 New South Wales has the highest prevalence of all states and territories, at around 4200, followed by Queensland with a prevalence of around 3600 cases. Australia's territories have the lowest prevalence of melanoma, with the Northern Territory at 77 and the ACT at 156 total cases of melanoma.

WS 15.2 PAGE 195

1 a *Answers will vary. Example answers:*
Bariatric surgery: surgery to reduce the size of the stomach, which limits its food-carrying capacity and limits food intake to small portions that make the person feel full
Intragastric balloon: the non-surgical insertion of a specialised balloon into the stomach, which limits its food-carrying capacity and limits food intake to small portions that make the person feel full
Gastric electrical stimulation: the insertion of a device that applies electrical stimulation to the stomach, modulating gastric motility (stretching and contractions of the stomach muscles) and altering hunger/appetite-related signalling to the brain.
b *Answers will vary. Answers should include points for the use of the source, points against the use of the source and a judgement about the reliability of the source. Criteria for reliability include currency of information, author's credentials, lack of bias, source of information (e.g. .edu, .org) and reference to first-hand data.*

2 a *Answers will vary. Example answer:* Does the use of antibiotics affect the diversity of gut microbiota?
b *Answers will vary. Example answer:*
Independent variable: Ingestion of antibiotic (could be extended to include the use of various types of antibiotics)
Dependent variable: The number of species of bacteria found in stool samples
c *Answers will vary. Example answer:* It is only possible to culture bacteria aerobically, and most bacteria found in the colon grow anaerobically.
d *Answers will vary. Example answer:* For an investigation to be valid, it needs to be a fair test with all variables controlled where possible. The inability to effectively grow most of the bacteria found in the colon means the results will not be a true reflection of the impact of antibiotic use on the diversity of bacteria.
e *Answers will vary. Answers should include a description of the treatment method (either the use of prebiotics, probiotics or faecal transplants) and a clear link for how this treatment will help to treat or manage obesity.*

WS 15.3 PAGE 196

1 a Fay would first need to conduct a descriptive study. She would need to collect data from people with and without cancer, including age, sex, occupation and exercise regime, among other things.

9780170449625

b

Source of error	Type of error	Impact on validity
Volunteer bias: employees invited to participate	Systematic error: selection bias	Those that volunteer may have vested interest in the study
Healthy worker bias: only current employees contacted	Systematic error: selection bias	Past employees may have also been impacted by ill health, or even have already passed away due to ill health
Loss to follow-up bias: some respondents changed jobs during the study	Systematic error: information bias	Reduced data set as not all subjects present at the end of the study
Sampling bias: only one airline included	Systematic error: selection bias	A wider sample that included staff from other airlines should have been included
Small sample size	Random error	The sample size is too small to correct for statistical errors that can result due to differences in the respondents

c Fay's method is not valid because the design incorporates too many errors. The design allows for many systematic errors such as volunteer bias and healthy worker bias. These errors result in an incorrect estimate of the impact of exposure to cosmic ionising radiation on rates of cancer.

2 *Answers will vary. Example answer:*

- If researchers can determine the length of exposure required to cause breast cancer, employers can put in safeguards, such as limiting the number of hours flown each year, to reduce the risks of developing breast cancer.
- If people are aware of the health risks when undertaking employment as aircrew, they may be more likely to undergo regular health checks. This proactive health care approach can allow for early detection of cancer and increases the chances of successful treatment outcomes.

~continue in right column

Chapter 16: Prevention

WS 16.1 PAGE 198

1

Criterion	How it was implemented in the NBCSP
Evidence base for action	Program was designed using data from the Australian mortality database, bowel cancer diagnosis data from cancer registries and health statistics from people invited to participate in screening.
Package of evidence-based interventions	Eligible Australians (e.g. citizens, those with Medicare or veterans affairs cards) are added to the NBCSP register. People between the ages of 50 and 74 are sent an information kit and are offered screening every two years.
Effective performance management including monitoring, evaluation and program improvement	Monitoring and program evaluation reports are regularly conducted and published; for example, the *Analysis of bowel cancer outcomes for the National Bowel Cancer Screening Program* is available online.
Public and private sector partnerships	The program is managed by the Australian Department of Health, in partnership with state and territory governments. Patients are encouraged to talk with their GP or pharmacist about how to purchase a screening test. Bowel cancer research is conducted by the Medical Research Future Fund and the Colorectal Surgical Society of Australia and New Zealand.
Communication of accurate information	Information is available online from Bowel Cancer Australia, Cancer NSW (NSW Government) and Cancer Screening (Australian Government). Includes background scientific information, information and brochures for health professionals and monitoring reports.
Political commitment	In 2006 the Minister for Health announced funding for the start of the campaign. In the 2014 federal budget an additional $95.9 million was committed over four years to accelerate the program.

2 The graph shows that in the four years following a positive bowel cancer diagnosis, nearly 100% of patients in the screen-detected category who received a positive diagnosis for the disease survived. The survival rate for individuals with bowel cancer who were not screened by the NBCSP decreased over the four years; for example, the non-responder group had a survival rate of around 80% after four years. These data suggest that early detection of bowel cancer in at-risk groups allows for more effective treatment measures and increases the chances of survival after receiving a positive diagnosis. As the aim of the NBCSP was to actively recruit people for early detection of bowel cancer, these data support the judgement that the program was a success.

WS 16.2 PAGE 201

1 SCID-X1 is caused by a mutation in the gene called IL2RG, which is important for the development and function of lymphocytes.

2 Babies born with this disease get many infections in the first few months of life due to their weakened immune system. The individuals have a lack of T cells, natural killer cells and functional B cells.

3 This name comes from a famous case of a boy who was born with the disease in 1971 and had to spend most of his time in a sterile plastic bubble while waiting for a bone marrow transplant.

4 The patient's bone marrow is collected. An altered version of the HIV virus is used to deliver a working copy of the IL2RG gene into the bone marrow cells. The altered cells are then reintroduced back into the patient.

5 Patients need to be regularly monitored for side effects and to ensure that the changes to the immune system are long lasting. It takes time for the adaptive immune response to build after the gene therapy.

6 Other treatments include a bone marrow transplant from a suitable sibling donor – but this only works for less than 20% of patients. Using bone marrow from non-sibling donors comes with the risk of introducing other diseases and ineffective reconstitution of the immune system. A study investigating the effectiveness of gene therapy treatment found immune system development in all eight patients treated.

7 The genetic engineering of the IL2RG gene into the bone marrow cells is an effective strategy to treat SCID-X1 disease. Research into the application of this technology found an immune system response in all patients. While patients need to be regularly monitored and it takes time for the adaptive immune response to develop, these issues are more inconveniences rather than limitations and are not enough to prevent the use of the technology. SCID-X1 disease is a life-threatening condition, and with the application of genetic engineering patients are able to develop immune system function that will significantly extend their life. This treatment also comes with fewer risks and is more successful than the traditional bone marrow transplants used to treat this disease.

8 a *Answers will vary.*

b *Answers will vary. Answers should address the fact that information from secondary sources can be said to be accurate if there is consistency across multiple sources. To be accurate, the resource should also be reliable. Secondary data are reliable if not biased, are current and refer to first-hand statistics. Other considerations also include whether the author is appropriately qualified and whether the information comes from a respected source; for example, a website containing '.gov' '.org' or '.edu'.*

Chapter 17: Technologies and disorders

WS 17.1 PAGE 203

1 Congenital hearing loss is present at birth; progressive hearing loss is acquired and becomes worse over time.

2 a Males: Mild = 70%; Moderate = 18%; Severe = 3%

Females: Mild = 38%; Moderate = 9%; Severe = 1%

b There are approximately twice as many 80–89-year-old males in each category of hearing loss compared to females.

c In 2017 the following trends applied:

The percentage of people with hearing loss increases with age for both males and females. For males, the numbers are relatively low from birth to 49 years, and then there is a significant increase in percentage of males with hearing loss. The same trend is shown in females, but the increase in percentage in cases occurs after 59 years.

3 The number of Australian males and females with hearing loss is expected to more than double between the years 2017 and 2060. The number of females in 2017 was less than 1.5 million but in 2060 is expected to be almost 3 million. The number of males is expected to increase from approximately 2.2 million to almost 5 million from 2017 to 2060.

9780170449625

4 a *Example answer:*

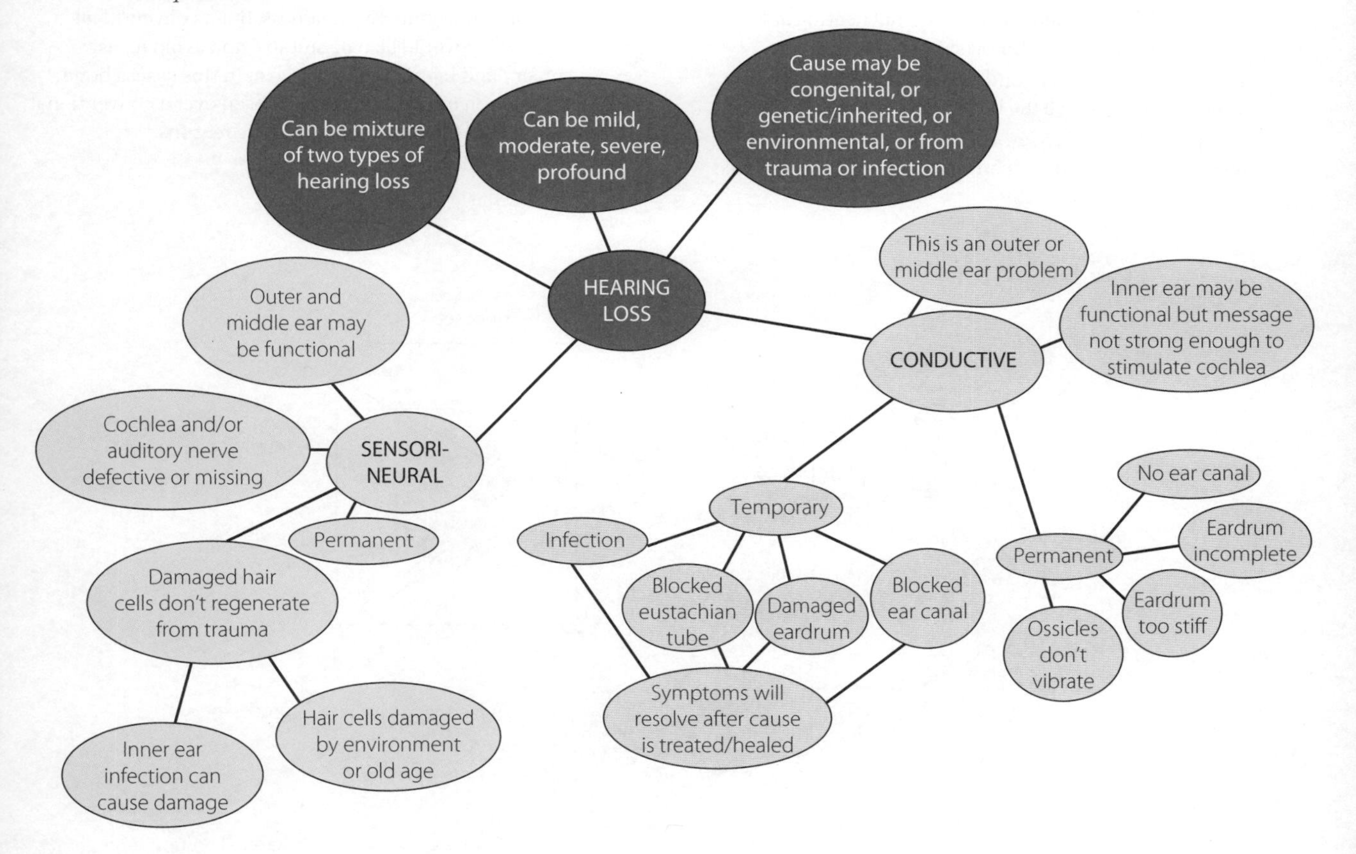

b **i** conductive; temporary

ii sensorineural; permanent; progressive

iii conductive, permanent

iv permanent; progressive

v sensorineural; permanent

vi conductive; permanent; progressive

vii conductive; permanent; progressive

5 a Ensuring the ear canal is sealed would direct all the sound toward the eardrum from the speaker and prevent sound escaping from the ear canal.

b
- The user might not want people to see their device.
- Miniature hearing aids would be more expensive than behind-the-ear aids.
- The ear mould or mini device may be uncomfortable in the ear canal.

~continue in right column

- Small devices may be easily lost.
- A person will not be able to hear if the battery on their aid goes flat.
- Extraneous noise may interfere with understanding conversations.

c Deformed ear canals may make the placement of 'in the canal' aids impossible. If the opening of the ear canal was normal, an aid with an ear mould would be used. If there was no opening to the ear canal a hearing aid could not be used. Excessive wax build-up would reduce the effectiveness of all hearing aids because conduction of sound would be reduced. The components of 'in the canal' aids may become clogged with wax, further reducing their effectiveness. A hearing aid sitting in the opening of the canal or behind the ear may be the best option in this case.

6 a

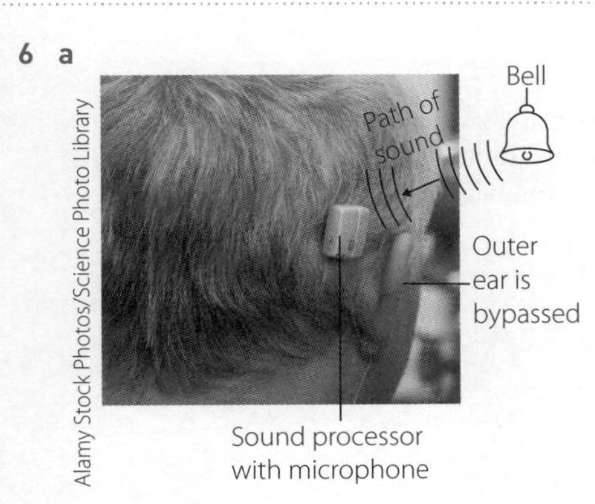

Alamy Stock Photos/Science Photo Library

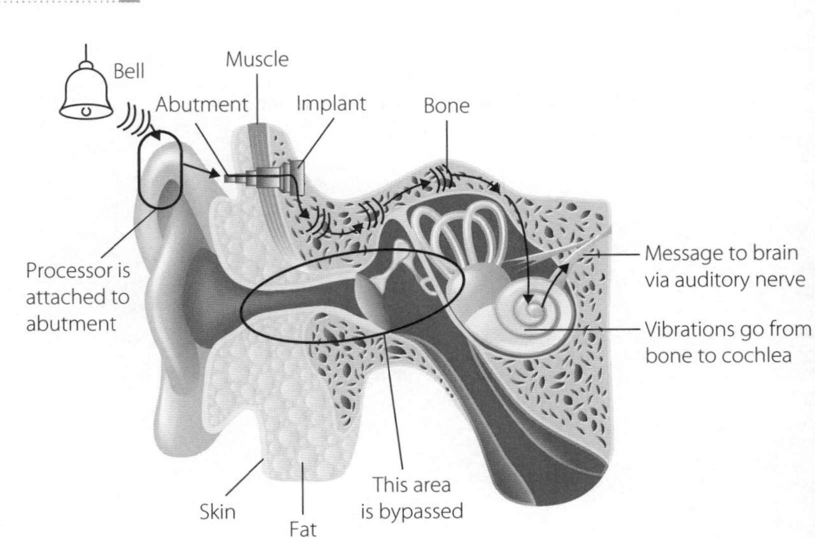

Shutterstock.com/La Gorda

b **i** S; Sound can travel from the implant through bone, bypassing the ear canal. A conventional hearing aid would be unsuitable in this case.

ii U; This is a temporary condition and invasive surgery is unnecessary. Remove the bead and hearing will return.

iii U; A person must have an auditory nerve to send messages to the brain. A conventional hearing aid would not work in this case, either.

~continue in right column

iv S; Sound bypasses the outer and middle ear, which are the cause of conductive deafness, but a conventional hearing aid would also be suitable and avoid invasive surgery and higher monetary cost. In this case, a bone conduction implant might be chosen over a conventional hearing aid for comfort or aesthetic reasons.

7 a & c

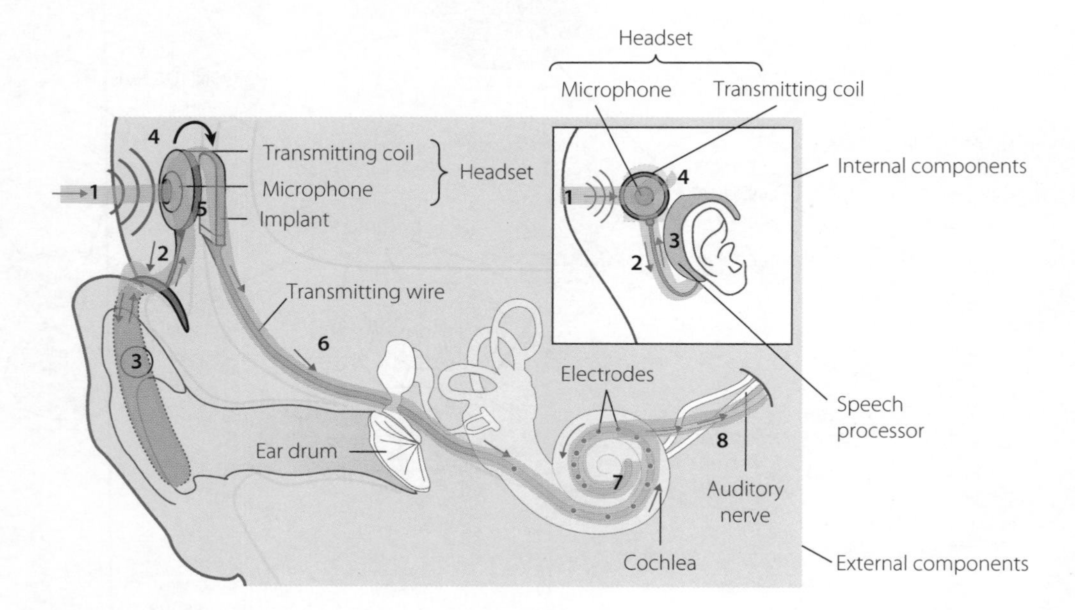

b The order is: 8, 4, 3, 5, 2, 1, 6, 7

8 a

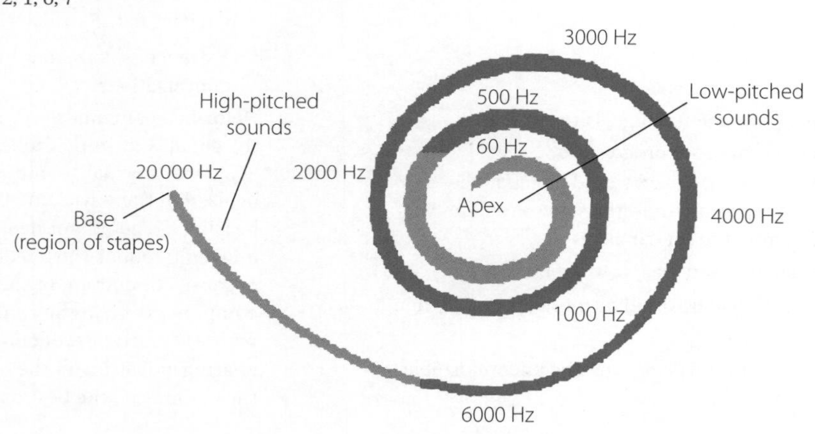

b High frequency sounds from 8000 to 20 000 Hz and low frequency sounds from 250 to 50 Hz

c The highest and lowest musical instruments, music will not sound the same, bird song, high pitched children's voices, low rumbling sounds such as trucks

WS 17.2 PAGE 210

1

Myopia: Can't focus on distant objects

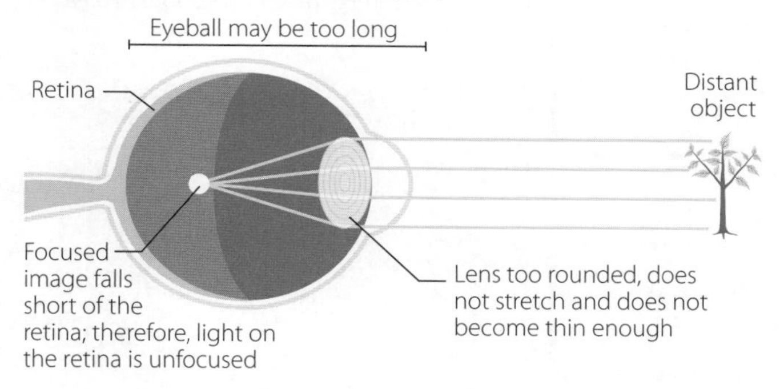

Hyperopia: Can't focus on close objects

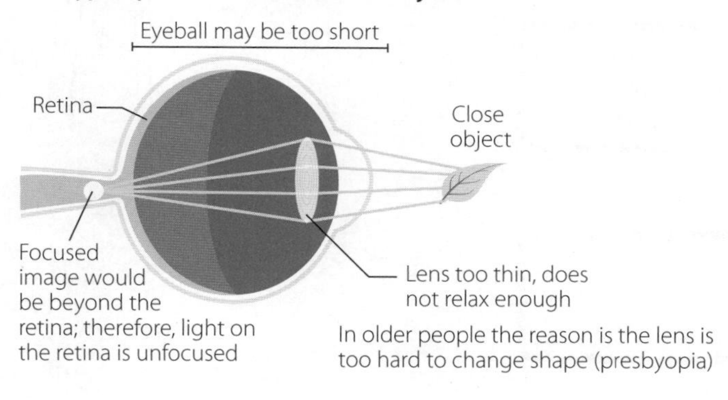

Note: Details of suspensory ligaments and ciliary muscles could also be included in student answers.

2 focused; in front of; blurry; diverged; toward the back; focused; behind; blurry; converged; focused; forward

3 hyperopia; convex; converged; convergence; convergence; focused; myopia; concave; diverged; divergence; divergence; focused

4 *Answers will vary. Sample answer:*

	Benefits	Limitations
Spectacles	• Successfully correct hyperopia or myopia • Usually comfortable • No risk of damage or infection with use • Easily cared for	• Cannot see clearly if not wearing • Can be expensive • Lenses may need to be updated approximately yearly • Easily broken or misplaced • Difficult to use during some activities; e.g. sport • It is obvious the wearer has a vision problem
Contact lenses	• Successfully correct hyperopia or myopia • Can be worn without other people knowing the wearer has a vision problem • Good to use during some activities; e.g. sport • Some types can be worn longer than a day	• Cannot see clearly if not wearing • Can be expensive • Lenses may need to be updated approximately yearly • More easily lost or damaged than spectacles • Ongoing cleaning and disinfecting of lenses • Must have good hygiene to use • Risk of damage or infection with use • Most users can only wear for 8–12 hours

5 A person with myopia cannot see distant objects clearly. The 'before' picture shows that the person's eyeball is too long; the distance between the cornea and the retina is too far. The focused image falls in front of the retina. Therefore, the image on the retina is blurry. After LASIK surgery the cornea is flatter. This shortens the distance from the cornea to the retina so that the focused image correctly falls on the retina and the image is clear to the person.

6

Essential steps to a LASIK procedure:

A mechanical surgical tool called a microkeratome, or a femtosecond laser, is used to create a thin, circular flap in the cornea.

↓

The hinged flap is folded back to access the underlying cornea (called the stroma).

↓

An excimer laser is used to reshape the corneal stroma by removing microscopic amounts of tissue from the cornea.

↓

The corneal flap is laid back in place, where it adheres to the corneal stroma without stitches.

↓

The cornea now more accurately focuses light on the retina for improved vision.

~continue in right column

WS 17.3 PAGE 213

1 urinary; nitrogenous; water; constant; nephrons; correct; urine; ureters; urea; nitrogenous; excess

2

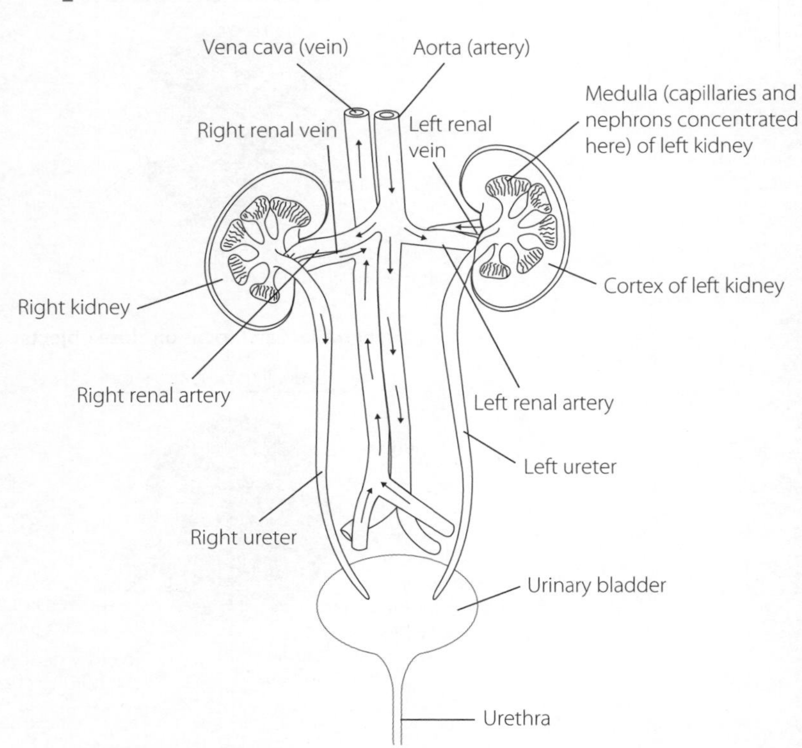

3 Acute kidney disease comes on quickly and may or may not leave permanent damage. Chronic kidney disease develops over time, gradually getting worse, and is not reversible.

4

Common cause of CKD	What occurs in the kidney
Polycystic kidneys	• Fluid-filled cysts develop. • The growth of cysts causes the kidneys to become grossly enlarged, with the cysts replacing functioning kidney tissue.
Hypertension	• Uncontrolled high blood pressure can cause arteries around the kidneys to narrow, weaken or harden. • Damaged arteries are not able to deliver enough blood to the kidney tissue.
Diabetes	• High blood glucose can damage the capillaries in the kidneys. • When capillaries are damaged, they can't be involved in filtering the blood properly.
Glomerular diseases	• Damage occurs to the glomeruli, part of the nephron, letting protein and sometimes red blood cells leak into the urine. • They can occur on their own or because of hypertension or diabetes.

5 *Answers will vary. Sample answer:*
Excess nitrogenous compounds in the blood: The kidney is supposed to remove the nitrogenous compounds so that they move into the urine. If the kidney is not functioning, these stay in the blood and cause the pH to go up.

Too much water is retained: One function of the kidney is to make sure the right amount of water goes out in the urine. If kidney is not functioning correctly, the water stays in the body and accumulates.

9780170449625

6 a The correct order is: 5, 4, 2, 1, 3, 5

b

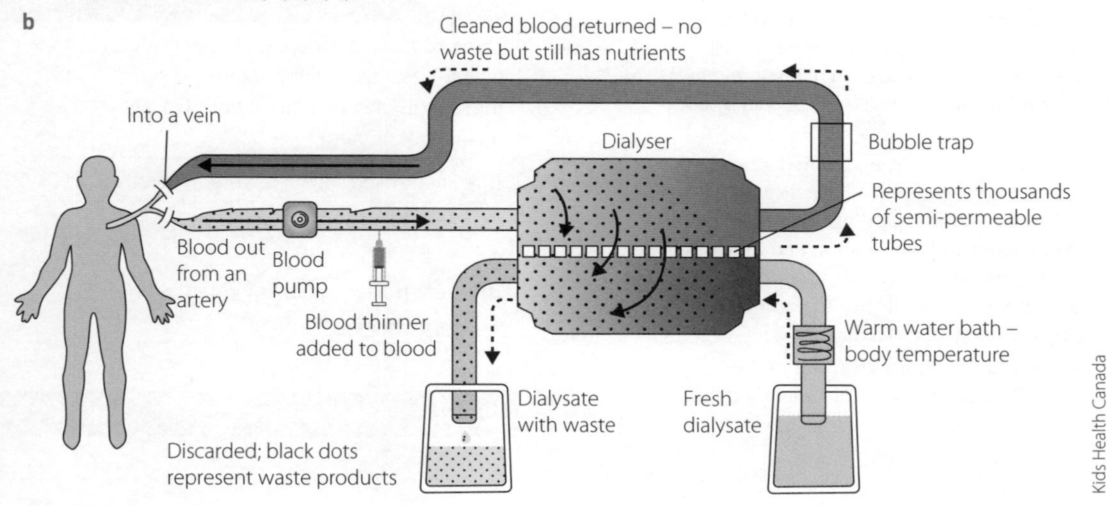

c **i** So the heart doesn't have to pump the blood through the machine

ii To prevent blood clots from forming in the machine and going back into the body

iii To prevent bubbles going back into the body

~continue in right column

iv So the blood doesn't cool down in the dialyser; if it did, cool blood would go back into the patient

v To create greater surface area for diffusion of waste

vi So new blood is constantly exposed to the dialysate

7

	Haemodialysis at a clinic	Haemodialysis at home
Positives	• Blood is filtered of wastes • Cost is covered by government • All equipment and supplies are at clinic • Nurses are on hand if something goes wrong • Less likely to have infections • Experts set up and monitor the procedure • Not at risk of power outage	• Blood is filtered of wastes • Cost is covered by government • Do not have to go to clinic 3 times per week for 6 hours • Can do dialysis when it suits – even while asleep • Can live far from a clinic • More likely to be able to work full-time if dialysis is done when asleep
Negatives	• May fear needles and blood • May not feel well between treatments • Cannot go on a holiday without booking into a clinic at the destination • Not as good as a kidney transplant • Must go to clinic 3 times per week for 6 hours each • Need to live reasonably close to a clinic • Not likely to be able to work • Patient needs to restrict fluids and certain foods in the diet all the time	• May fear needles and blood • May not feel well between treatments • Cannot go on a holiday without booking into a clinic at the destination • Not as good as a kidney transplant • Patient or a family member must be trained to set up and monitor procedure • More likely to have infections • At risk of power outage • Equipment and supplies permanently take up space in your home • No nursing staff present • Patient needs to restrict fluids and certain foods in the diet all the time

WS 17.4 PAGE 217

1 *Answers will vary. Highly regarded answers will thoroughly address the criteria listed in question 2.*

MODULE 8: CHECKING UNDERSTANDING PAGE 219

1 Cellular processes such as respiration require enzymes to lower the activation energy needed for chemical reactions to occur. These enzymes function best under specific conditions, including temperature and pH. In order for an organism to function normally, it is important for all cellular chemical reactions to progress at an optimal rate. This is aided by a constant internal environment.

2 D

3 B

4 Epidemiological studies are used to determine the causes and impacts of disease on populations. Understanding patterns of disease incidence and prevalence can assist health care authorities with the allocation of resources and infrastructure to locations most at risk or impacted by a disease.

5 Descriptive epidemiological studies are the first study conducted when investigating the cause of a disease. These studies allow for data collection regarding which section of the population is impacted and the location and time of the disease outbreak. Intervention studies are used to test the effectiveness of a treatment or public health campaign with the aim of reducing the incidence of a disease.

6 Systematic errors such as sampling bias result in the incorrect estimation of the cause of the disease.

7 D

8 A

9 B

10 B

11 Hearing aids make sounds louder. They are completely outside the body and do not artificially stimulate the cochlea. The statement implies that hearing aids cannot help with any other type of deafness, such as sensorineural deafness. This is not accurate. A person with profound sensorineural deafness would not benefit from simply making the sound louder. However, a person with a milder level of sensorineural hearing loss would still benefit from amplifying the sound. A louder sound travelling down the ear canal, through the eardrum and ossicles and to the cochlea would be able to make the hair cells move and send a message to the brain via the auditory nerve.

Practice examination

SECTION I PAGE 221

1 B
2 B
3 B
4 A
5 D
6 C
7 D
8 A
9 D
10 B
11 D
12 B
13 C
14 B
15 B
16 D
17 C
18 B
19 A
20 B

SECTION II PAGE 227

21 a Number of cells

Criteria	Marks
Correct answer	1

b *Examples include:* only one species of bacteria is used, set quantities of nutrient, same pH, same temperature, always use sterile media, always inoculate the medium at time zero.

Criteria	Marks
Three controlled variables written in a way that conveys the quality is kept the same throughout the incubation time, e.g. 'set quantity of nutrient'	3
Three controlled variables written as above	2
One controlled variable written as above *OR* Three controlled variables written as one-word answers, e.g. 'nutrient'	1

c 2 hours

Criteria	Marks
Correct time in hours	1

22 a i

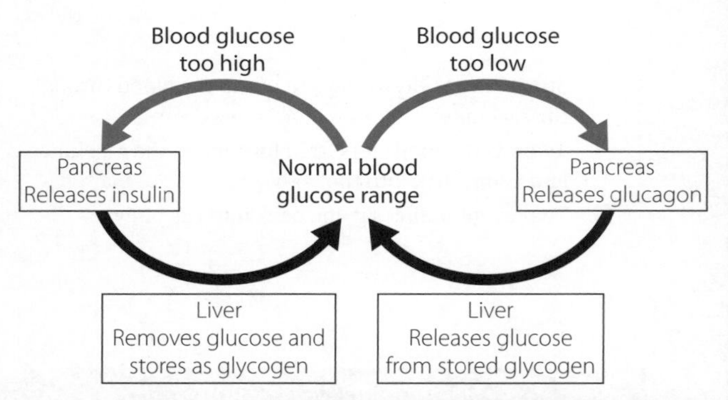

Criteria	Marks
All details complete and correct, including reference to release and storage	4
Hormone release correct for each organ but no reference to stored glycogen	3
Two or more errors	1–2

ii Negative feedback *or* homeostasis

Criteria	Mark
Correct answer	1

b Functioning kidneys should 1) remove nitrogenous wastes from the blood and 2) balance the amount of water and salts in the body.

Someone with loss of kidney function will have high levels of urea and other wastes built up in their blood, which cause toxicity. They will also have excess water in the blood and tissues. Excess water puts pressure on organs such as the heart and lungs. Dialysis involves blood being removed from an artery into a dialysis machine, removal of wastes, excess salts and water, and the filtered blood being returned to the patient. Dialysis needs to be performed 3–4 times a week for about 5 hours. The patient benefits because the toxic wastes are removed and their heart and lungs function better without the excess fluid in the body. Without dialysis, people with no kidney function die, and people with part function are very unwell with poor quality of life.

9780170449625

Criteria	Marks
Two functions of kidney correctly identified Loss of kidney function linked to at least two symptoms of poor health At least two features of dialysis outlined Dialysis linked to at least two health outcomes for patient.	5
Two functions of kidney correctly identified Loss of kidney function linked to poor health Purpose of dialysis outlined Dialysis linked to health outcomes for patient	3–4
Some relevant information about kidney function Loss of kidney function linked to poor health *OR* Some relevant information about dialysis	1–2

c *Acceptable examples must be genetically modified organisms that make a product; they may be currently available or in the research phase.*

Bacteria (or yeast) are genetically modified with a human gene to produce the hormone/protein insulin. The human gene is inserted into a bacterium, which reproduces in culture and insulin is extracted and purified.

Society needs insulin produced because:

- insulin-dependent diabetic patients will die without daily insulin injections
- diabetics who have access to insulin can be productive members of society
- where possible, scientific research tries to help humans with medical needs.

Benefits of making insulin with GM bacteria:

- Industrial fermentation can produce much more insulin than can be extracted from cow and pig pancreases.
- Vegans or people of certain religions who need insulin do not have to be concerned about using animal-sourced insulin.
- The process makes human insulin (not bovine or porcine insulin), so it is more authentic than non-GMO insulin. The only other way to get human insulin would be to extract it from deceased humans.

Criteria	Marks
Logically presented and thorough answer Appropriate example chosen and features outlined At least two features for the need for the product At least two benefits of using recombinant DNA technology for the product	5
Appropriate example chosen and features outlined One or more features for the need for the product One or more benefits of using recombinant DNA technology for the product	3–4
Some relevant information about a genetically modified organism Some relevant information about the need or the benefit of the product or process	2
Some relevant information about a genetically modified organism *OR* Some relevant information about the need or the benefit of the product or process	1

23

Vision disorder	Shape of eyeball and lens	Position of image formation	Shape of corrective lens
Hyperopia	Eyeball too rounded Lens too flat	Behind retina	Convex
Myopia	Eyeball too long Lens too curved	In front of retina	Concave

Criteria	Marks
All details complete and correct, including reference to lens *AND* eyeball shape	3
Correct but only eyeball shape *OR* lens shape included *OR* One other error	2
Two errors	1

24 a The indirect transmission of an infectious disease occurs when there is no direct contact between the host and the organism to which the disease is transmitted. For example, the pathogen that causes the measles virus can spread via respiratory droplets produced during coughing or sneezing, which can travel through the air and land on surfaces. The pathogen does not require actual contact between pathogen carrier and pathogen recipient for transmission.

Criteria	Marks
Describes indirect transmission in relation to a named infectious disease	3
Describes indirect transmission of infectious disease without an example	2
Names an infectious disease that spreads via indirect transmission	1

b The pathogen that causes the measles virus is able to avoid drying out; it can stay active in air for several hours and on contaminated surfaces for two hours after respiratory droplets are produced during coughing or sneezing. This allows indirect transmission from one host to another, without direct contact between each host.

Criteria	Marks
Names an appropriate disease Provides a clear link between the adaptation of the pathogen and the ability to spread via indirect transmission	3
Provides a clear link between the adaptation of the pathogen and the ability to spread via indirect transmission	2
Identifies an adaptation	1

25 a

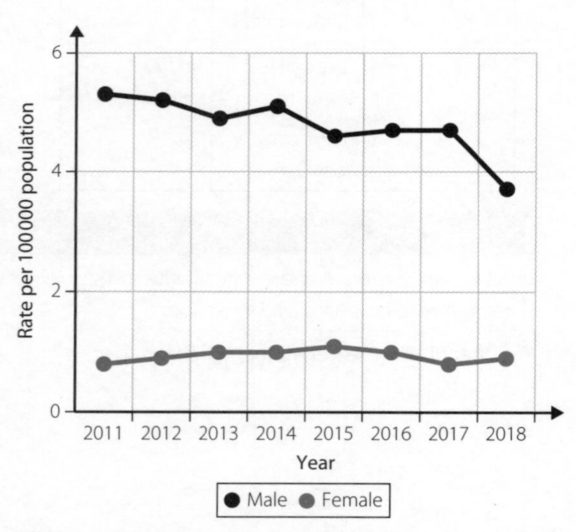

Based on Australian Institute of Health and Welfare; AMR data. Licensed under Creative Commons 3.0 https://creativecommons.org/licenses/by/3.0/au/

Criteria	Marks
Draws a line graph Axes correctly labelled and in correct position Appropriate scale Data plotted correctly	4
Draws a line graph with some minor errors *OR* Draws a column graph, with all other elements correct	3
Draws a line or a column graph with multiple errors	2
Engages with some part of the question	1

b The rate of mesothelioma between 2011 and 2018 was much higher in males than in females. For example, in 2011 the rate in males was 5.3 per 100 000 while in females it was 0.8 per 100 000. The rate of diagnosis in males reduced over time from 5.3 in 2011 to 3.7 in 2018. In females, the rate of diagnosis was relatively consistent over time.

Criteria	Marks
Describes the trend for both males and females, with reference to data	2
Describes the trend for males or for females, with reference to data *OR* Describes the trend for both males and females, without reference to data	1

26 Lifestyle disease:

Lifestyle diseases are the result of the way an individual lives their life. For example, ischaemic heart disease can be caused by a combination of factors, including insufficient physical exercise, smoking, drinking alcohol to excess and consuming an unbalanced, unhealthy diet. These factors result in hardening of the arteries (atherosclerosis), which is due to a build-up of plaque in the blood vessels. This build up restricts blood flow.

Chromosomal disease:

Chromosomal diseases are due to mutations of the chromosomes that carry the genes. Down syndrome is caused by the presence of an additional copy of chromosome number 21. This is due to abnormal cell division during gamete formation that prevents the pair of chromosome 21 correctly separating in the egg or sperm. The effects on the individual can include intellectual disability, distinct facial features, short stature, heart defects, infertility and susceptibility to infection.

Differences:

While technologies such as CRISPR are developing quickly, in general, chromosomal diseases cannot be prevented by treating the individual with the disease. Medical intervention

9780170449625

can be used to treat and manage symptoms associated with the disease but, because the mutation is present in all cells, it is very difficult to remove the causative agents.

In comparison, lifestyle diseases are generally preventable by the person suffering from the condition. A change in lifestyle, such as improving eating choices and increasing exercise, can have significant impacts on the health of the individual.

Criteria	Marks
Correctly identifies an example of a lifestyle and a chromosomal non-infectious disease Thoroughly outlines the causes and effects of each disease Thoroughly describes differences between each type of disease	5
Correctly identifies an example of a lifestyle and a chromosomal non-infectious disease Outlines the causes and effects of each disease Describes differences between each type of disease	4
Outlines the causes and effects of a named lifestyle and chromosomal non-infectious disease	3
Sketches in general terms details regarding a lifestyle or chromosomal non-infectious disease	2
Provides some relevant information	1

27 Biodiversity:

- The term *biodiversity* encompasses variation at a range of scales: genetic, species and ecosystem. In general, greater variation in each of these levels provides strength and resilience against environmental disturbances.
- For biotechnology to benefit biodiversity, it would be expected that the technology would increase biodiversity levels in the short and long term. This stability is needed for ecosystem health.

Stimuli:

- Figure 1 shows the area of Bt corn growth over time, compared with insecticide use. From 1996 to 2013, the uptake of Bt corn increased from 0% to just under 80% of all corn grown. During this time, insecticide use decreased from around 0.2 kg/ha to nearly no use in 2010. Bt corn is a transgenic species engineered to produce insecticidal proteins and hence be resistant to specific pests, which negates the use of pesticides. As the use of pesticides can have far-reaching negative environmental consequences (such as killing non-target organisms), this graph would suggest that the application of biotechnology in this case would be beneficial for biodiversity.
- Figure 2 shows that, since the 1950s, the variation in tomato chromosomes 4, 5, 6, 9, 11 and 12 has increased significantly in commercially grown tomatoes as a result of interbreeding between tomato species.
- While these data are suggestive of the ability for biotechnology to improve biodiversity, this is a simplistic viewpoint.

Biotechnologies:

- Whole-organism cloning has widespread agricultural applications. During this process, a cell is taken from a donor animal and injected into an enucleated, unfertilised egg from the same species. These cells are fused with electricity and then start dividing. The embryo is then placed into the uterus of a surrogate mother of the same species. This produces an organism genetically identical to the original donor cell. This process allows breeders to directly select for desired traits, such as the growth rate of an animal.
- The engineering of transgenic organisms can increase genetic biodiversity. For example, Bt corn was created through the insertion of the Bt gene from the bacterium *Bacillus thuringiensis*, using a bacterial vector. The addition of the Bt gene allows the corn to produce a protein that kills corn rootworm larvae, especially larvae of the western corn rootworm, a major predator of the plant. Growing this plant in agricultural settings reduces the need for spraying of pesticides.

Impact of biotechnologies on biodiversity:

- In the short term, cloning reduces genetic biodiversity, and in the long term it can also reduce species and ecosystem diversity.
- The creation of transgenic organisms can increase diversity at all scales in the short term. The long-term impacts, however, need to be carefully monitored, as altering genetic profiles across species barriers may have wide-reaching impacts on ecosystem diversity if food web balances are interrupted.

Judgement:

The statement that 'Reproductive biotechnology will benefit biodiversity' is somewhat correct, but is overly simplistic. While some biotechnologies increase biodiversity, others can decrease this variation. It is important that variation is consistent in the long term for biodiversity to benefit.

Criteria	Marks
Defines what is meant by the term *biodiversity* Thoroughly describes features of at least two named reproductive technologies, including the process, outcome and application Includes specific and thorough reference to the stimulus provided Provides a clear link between the named biotechnologies and the impact on biodiversity Provides a judgement regarding the impact of reproductive biotechnologies on biodiversity	7
Defines what is meant by the term *biodiversity* Describes features of at least two named reproductive biotechnologies, including the process, outcome and application Includes reference to the stimulus provided Provides a clear link between the named biotechnologies and the impact on biodiversity Provides a judgement regarding the impact of reproductive biotechnologies on biodiversity	6
Defines what is meant by the term *biodiversity* Describes features of two named reproductive biotechnologies Includes reference to the stimulus provided Provides a judgement regarding the impact of reproductive biotechnologies on biodiversity	5
Describes features of a named reproductive biotechnology Includes reference to the stimulus provided Sketches in general terms the link between reproductive biotechnology and biodiversity Provides a judgement regarding the impact of reproductive biotechnologies on biodiversity	4
Describes features of a named reproductive biotechnology Includes reference to the stimulus provided *OR* Describes features of a named reproductive biotechnology Sketches in general terms the link between reproductive biotechnology and biodiversity	3
Outlines in general terms features of a named reproductive biotechnology	2
Provides some relevant information	1

28 Homologous chromosomes are being pulled apart by spindle fibres to opposite ends of the cell. Sister chromatids remain joined, indicating that this is meiosis rather than mitosis. The name of this phase is anaphase I.

Criteria	Marks
Identifies that meiosis is occurring and uses expert terminology to describe the process	2
Sketches the process in general terms	1

29 Ellis-Van Creveld syndrome is more common in the Amish community compared to the global population due to the founder effect. The genetics of the initial small population that founded the community did not reflect the allele frequency found in the general population. Because the Amish are an isolated community, the frequency of the allele that causes the syndrome has been increased through inbreeding.

Criteria	Marks
Founder effect is described using expert terms and allele frequency is correctly discussed	4
The founder effect and allele frequency are identified and sketched in general terms	3
The founder effect and allele frequency are sketched in general terms but not specifically named	2
A relevant statement is made regarding population genetics	1

30 a The process occurring in the diagram is phagocytosis. This process involves a phagocyte engulfing a bacterium and using lymphocytes containing enzymes to break down the pathogen.

Criteria	Marks
Phagocytosis is described with reference to the diagram	2
Phagocytosis is sketched in general terms without naming the process	1

9780170449625

b Innate immunity is a non-specific response that targets all non-self antigens entering the body. It can involve physical, chemical and cellular mechanisms that attempt to prevent the spread of a local infection.

Adaptive immunity is targeted to a specific antigen and can be humoral or cell mediated. Adaptive immunity can fight infection that has spread throughout the body and also remembers pathogens that it has previously encountered.

Phagocytes identify non-self antigens and perform phagocytosis to destroy the foreign body. This process is indiscriminate and does not specifically target a particular antigen; therefore, it is innate. However, phagocytes can retain the antigen of a pathogen and present these antigens to memory T cells, stimulating the adaptive immune response.

Criteria	Marks
Clearly defines the innate and adaptive immune responses and describes the role of phagocytes in both responses in expert terms	4
Defines the innate and adaptive immune responses and describes the role of phagocytes in both responses in general terms	3
Sketches the link between the adaptive and innate immune system in general terms	2
Makes an attempt with some correct information	1

31 Phenotype is the expression of a genotype observable in the organism. The sequence of nucleotides in coding DNA dictates the sequence of amino acids in polypeptide chains and the resulting proteins produced.

This type of mutation is a substitution mutation, where one nucleotide is replaced by a different nucleotide.

Some amino acids are coded for by several different sequences of nucleotides and if this substitution resulted in the same amino acid being coded for, then no change in the phenotype of the individual would be seen.

This piece of DNA could be non-coding DNA and therefore any change in the sequence would not be likely to lead to a change in the phenotype of the individual.

Therefore, the statement is valid because it is possible that this type of mutation may have no effect on the phenotype of the individual.

Criteria	Marks
Describes the link between genotype and phenotype Correctly identifies the type of point mutation Discusses coding and non-coding DNA Makes a clear judgement relating to the statement	4
Correctly describes two of the above dot points with a clear judgement	3
A judgement is made with some relevant information	2
Some relevant information is given without a judgement	1

32 a The information is from a government website, meaning that it has been fact checked by experts to ensure it is correct. The information is current – it was updated in 2020.

Judgement: The information is valid.

Criteria	Marks
Makes a judgement stating that the information is valid and gives points to support the judgement	3
Makes a judgement stating that the information is valid and gives a point to support the judgement	2
Makes a judgement without supporting information	1

b Adaptive immunity provides immunity from a disease for a lifetime. This process requires an individual to be exposed to a pathogen, then specific antigen-matching B cells and T cells fight off the infection through the humoral and cell-mediated immune responses. Both produce memory B and T cells, which remain in the body post infection. These memory cells induce a heightened and rapid response to secondary infection. In this scenario it is the mother that has exposure to the pathogen. She produces antibodies via her plasma B cells to fight the infection and some of these antibodies are passed to her infant. Antibodies can remain in the body for up to 6 months and so the infant will be protected from the pathogen during this time. However, because the infant has no memory cells, it will be vulnerable to infection after these antibodies are no longer present.

Criteria	Marks
Describes adaptive immunity, production of memory cells and function of antibodies in expert terms, and relates the information to the scenario	5
Describes adaptive immunity, production of memory cells and function of antibodies in general terms and relates the information to the scenario	4
Describes some of the above features in general terms with links to the infant's immunity	3
Describes some of the above features with limited or no reference to the infant's immunity	2
Makes a relevant comment relating to adaptive immunity	1

33 In order to deduce whether the non-affected populations are at risk, a descriptive epidemiological study needs to be performed to classify the disease. This would involve collecting information about the frequency of disease, the geographic location, time period and the features of the affected population, including the sex and age of the birds.

If the disease is infectious and can be transmitted between populations through direct contact or vectors, the other populations may be at risk. If the disease is a genetic non-infectious disease, then the other populations may be at risk because they interbreed and could inherit the disease.

However, if the disease is an environmental non-infectious disease, then other populations would only be at risk if they have been exposed to the same environmental conditions.

The stimulus states that the penguin populations interbreed but have their own feeding grounds, and that four of the seven populations carry the disease. The four affected populations are close together, whereas the other three populations are further apart from the affected populations. It is highly likely that the disease is non-infectious, given that contact through interbreeding does not spread the disease. Therefore, it can be assumed that the disease may be due to environmental causes, originating from the feeding grounds, so the non-affected populations are only at risk if they meet the same environmental conditions.

To determine the exact cause of the disease, an analytical epidemiological study would have to take place. This could be a case-control study, where the environmental conditions of an affected population are compared to those of a non-infected population. The study could involve sampling:

- the drinking water near the populations for contamination
- the food sources/types in the feeding grounds

~continue in right column

- the water/pollution in the feeding grounds
- blood from dead penguins from the affected population, conducting post-mortem toxicology analysis, and comparing the result to those from a healthy blood sample.

To reach valid conclusions, the study would have to incorporate a large sample size from a range of different aged penguins of both sexes from both the infected and non-infected populations.

A detailed comparison could then be made from both populations and a likely cause could be pinpointed. If this cause was isolated or removed, the population could then be monitored to see whether there was any improvement in the health of the population. Ethical considerations to minimise disruption to the populations and minimise any harm to the endangered populations would also need to be implemented.

Given that the penguin populations are interbreeding, but not all populations carry the disease, it is possible that long-term exposure to the environmental contaminants is responsible for causing the disease. Therefore, it is unlikely that the non-infected populations are at risk of contracting the disease.

Criteria	Marks
Demonstrates an extensive knowledge of infectious and non-infectious diseases and correctly explains why the disease is likely a non-infectious environmental disease, with reference to the stimulus material Makes a clear judgement whether the non-affected bird populations are at risk Demonstrates an extensive knowledge of the key components of epidemiological studies, including large sample size, control group and comparing environmental conditions Designs a valid and reliable case-control epidemiological study Expert terms are used throughout	9
Demonstrates a thorough understanding of infectious and non-infectious diseases and explains why the disease is likely a non-infectious environmental disease, with reference to the stimulus material Makes a judgement whether the non-affected bird populations are at risk Demonstrates a thorough understanding of the key components of epidemiological studies Designs a valid and reliable case-control epidemiological study Expert terms are used	8
Demonstrates a sound understanding of infectious and non-infectious diseases and identifies that the disease is non-infectious, with reference to the stimulus material Demonstrates a sound understanding of the key components of epidemiological studies Designs a valid and reliable case-control epidemiological study	7
Demonstrates a sound understanding of infectious and non-infectious diseases Demonstrates a sound understanding of the key components of epidemiological studies Designs an appropriate epidemiological study	5–6
Refers to infectious and non-infectious diseases Sketches an epidemiological study in general terms	3–4
Attempts to classify the disease *OR* Provides an aspect of an epidemiological study	2
Provides some relevant information	1